ANIONIC SURFACTANTS

SURFACTANT SCIENCE SERIES

CONSULTING EDITOR

MARTIN J. SCHICK
Consultant
New York, New York

ADDITIONAL VOLUMES IN PREPARATION

ANIONIC SURFACTANTS

Analytical Chemistry

Second Edition, Revised and Expanded

edited by

John Cross

University of Southern Queensland
Toowoomba, Queensland, Australia

CRC Press
Taylor & Francis Group
Boca Raton London New York

CRC Press is an imprint of the
Taylor & Francis Group, an **informa** business

First published 1998 by Marcel Dekker, Inc.

Published 2018 by CRC Press
Taylor & Francis Group
6000 Broken Sound Parkway NW, Suite 300
Boca Raton, FL 33487-2742

© 1998 by Taylor & Francis Group, LLC
CRC Press is an imprint of Taylor & Francis Group, an Informa business

First issued in paperback 2019

No claim to original U.S. Government works

ISBN 13: 978-0-367-45578-1 (pbk)
ISBN 13: 978-0-8247-0166-6 (hbk)

Visit the Taylor & Francis Web site at
http://www.taylorandfrancis.com

and the CRC Press Web site at
http://www.crcpress.com

Library of Congress Cataloging-in-Publication Data

Anionic surfactants: analytical chemistry.— 2nd ed., rev. and expanded / edited by
   John Cross.
      p. cm. — (Surfactant science series; v. 73)
   Includes bibliographical references and index.
   ISBN 0-8247-0166-6 (acid-free paper)
   1. Surface active agents—Analysis. I. Cross, John II. Series.
TP994.A58 1998
   668'.1—dc21   98-16647
                                                                         CIP

Preface

Two decades have elapsed since the publication of the previous volume in the Surfactant Science Series that concentrated solely upon the analysis of anionic surfactants (Vol. 8, 1977). During that time a huge volume of water has passed beneath the proverbial bridge—much of which was more heavily laden with detergent residues than desirable

World production of surfactants since that time has continued to grow steadily: the domestic and industrial markets have been expanding at annual rates of 2–3 and 4–5%, respectively, with the major growth being observed in eastern Europe, east Asia, and China. Despite inroads made by nonionic surfactants into the market, anionic surfactants have maintained their dominant position at the head of the total production of synthetic surfactants. The global capacity for the most popular class, the alkylbenzenesulfonates, is expected to reach between 2.7 and 3.0 Mt/y by the end of the century. This market position is not surprising given their excellent technical application behavior, favorable cost/performance characteristics, and well-researched biological properties.

During the past 20 years, new surfactants derived from new raw material feedstocks, improved process technology, and constantly changing consumer and environmental demands have kept formulations continually changing. Methyl ester sulfonates, for example, are making a significant assault on the market dominance of the linear alkylbenzenesulfonates, and liquid detergent concentrates are rapidly replacing the conventional powdered detergents on supermarket shelves.

Increasingly sophisticated techniques for the study of the behavior of surfactants both in solution and on surfaces such as skin, hair, and fabrics and the intensification of the scrutiny of the ecological and environmental impact of the surfactants themselves and of their breakdown products have resulted in a corresponding increase in the degree of sophistication required in the analytical techniques applied to these materials. For example, the well-known determination of traces of anionic surfactants in wastewaters as "methylene blue active substances" has now been reduced virtually to the level of a screening test.

The enormous advances in instrumental techniques, typified by high-performance liquid chromatography, nuclear magnetic resonance, and hyphenated techniques such as gas chromatography–mass spectroscopy, have enabled the analyst to respond to these new challenges. However, many routine and nonroutine analyses still rely upon classical techniques such as titrimetry and even gravimetry: the standard methods adopted by such bodies as the International Organization for Standardisation and the American Society for Testing and Materials are heavily weighted in this direction.

Like other volumes in this series that I have compiled, this book is aimed at the chemist who has a firm grasp of the principles of analytical chemistry but is without any specialized knowledge of surfactants. Emphasis will be placed upon detailed examination of the methods used for the identification and determination of the surfactant itself, rather than attempting to present comprehensive schemes for the full analysis (i.e., every component) of formulated products such as shampoos or heavy-duty washing liquids. Sources that offer such a detailed examination are listed as recommended reading at the end of Chapter 1. Neither is it intended that this volume should function as a laboratory manual, although appropriate experimental procedures have been included in some instances. Again this function has been covered by some of the recommended readings. Rather, this volume is a critical appraisal of the literature and methodology currently available—a tour through the techniques, from which the analyst can select those necessary to tackle the problem at hand.

Following a brief introduction, the book falls naturally into three segments. Early chapters deal with colorimetry, titrimetry, and potentiometry—techniques that permit quantification without defining the structure of the surfactant to any precise degree. A chapter on tensammetry could have been included at this point, but the chapter written by M. Bos for Vol. 53 (*Cationic Surfactants: Analytical and Biological Evaluation*) of this series was sufficiently broad in its approach and content to serve as a reference for this book without further amendment. Subsequent chapters concentrate on the major instrumental techniques used for the separation and more specific identification of the separated components: quantification is also possible in most cases. The final chapter deals with fluorinated surfactants: in these compounds, the fully fluorinated alkyl chain is both hydrophobic and oleophobic, giving rise to some unique properties beyond those of their hydrocarbon-based counterparts.

The authors in this volume comprise an international collection of experienced practitioners, all of whom have industrial backgrounds. They have neither been required to adhere to any preconceived format or guidelines in producing their chapters nor been asked to cover a predetermined set of surfactants: rather, they have been encouraged to present their material as they saw fit.

My sincere thanks go to these authors for providing their contributions on schedule (well, almost!) and to their employers for supporting their efforts: to the Series Editor, Dr. Martin Schick, for his ever-readiness to help; to the team at Marcel Dekker, Inc., for steering me through the production process in such a stress-free manner; to Dr. Norbett Buschmann for his help and interest; and to my dear wife for patiently excusing me from "other duties" while compiling this volume. The final dedication, however, must go to the many researchers, both past and present, whose published works have provided the raw material for this project.

John Cross

Contents

Contributors

John Cross Department of Physical Sciences, University of Southern Queensland, Toowoomba, Queensland, Australia

Thomas M. Hancewicz Department of Advanced Imaging and Measurement, Unilever Research US, Edgewater, New Jersey

Arnold Jensen Department of Analytical Chemistry, Unilever Research US, Edgewater, New Jersey

Henry T. Kalinoski Unichema North America, Chicago, Illinois

Erik Kissa Consultant, Wilmington, Delaware

Eddy Matthijs Procter and Gamble European Technical Center, Strombeek-Bever, Belgium

Bruce P. McPherson Department of Analytical and Microbiological Sciences, Colgate-Palmolive Company, Piscataway, New Jersey

Henrik T. Rasmussen Department of Analytical and Microbiological Sciences, Colgate-Palmolive Company, Piscataway, New Jersey

Leon C. M. Van Gorkom Department of Molecular Structure Analysis, Unilever Research US, Edgewater, New Jersey

1

Anionic Surfactants—An Introduction

JOHN CROSS Department of Physical Sciences, University of Southern Queensland, Toowoomba, Queensland, Australia

I. AN INTRODUCTION

This chapter is intended to introduce the class of chemical compounds collec-
tively known as anionic surfactants, the members of that class that have
achieved major importance, and, as far as it affects the analyst, the raw materials, processes, and by-products that may be found. Some of the principal routes
from the raw materials to surfactant are shown in Fig. 1. At this very early stage
of this book, the author recommends to all readers interested in more detail that
they consult the companion volume in the Surfactant Science Series, which
concentrates on the preparation and properties of a wide range of anionic surfactants [1].

Surfactants are by necessity large molecules with molar masses usually in excess of 300. In the simplest case they consist of a nonpolar (usually hydrocarbon) chain attached to a highly polar or ionic group. These sections are known as
the hydrophobe/oleophile/lipophile and the hydrophile/oleophobe/lipophobe, respectively, depending on the attraction—or lack thereof—to water or hydrocarbon oil.

$$|\longleftarrow\quad C_{16}\longrightarrow|$$
$$CH_3CH_2CH_2 \text{-------} SO_3^- \qquad\qquad Na^+$$

| hydrophobe | hydrophile | counter |
| lypophile | lypophobe | ion |

It is essential that the ends of the hydrophobe and the hydrophile are sufficiently
remote from each other to react with surfaces and solvent molecules independently. The relative weightings of these groups was quantified by Griffin [2] in
the form of the hydrophile-lipophile balance (HLB), a parameter that indicates
the field of application to which the surfactant is most suited [3]. For example,
cetyl alcohol (HLB = 1.0) relies upon a single -OH group to provide any hydrophilic character; it is virtually insoluble in water but will spread over the surface of water in a dam to provide a surface film that significantly reduces
evaporation. Glyceryl monostearate possesses two -OH groups plus a less polar
ester group and consequently has a higher HLB (3.8): such compounds are excellent water-in-oil emulsifiers. As the hydrophilic factor grows, the surfactant
becomes more suited for use as a wetting agent (e.g., sorbitan monolaurate, HLB
= 8.6) and an oil-in-water emulsifier (e.g., polyoxyethylene monostearate, HLB
= 11–15). Anionic surfactants (anionics) have high HLBs due to the presence of
ionic hydrophiles such as -COO$^-$ and -SO$_3^-$. HLBs range upwards from about 13
(e.g., sodium dodecanesulfonate, 13; sodium oleate, 18.5; and sodium dodecyl
sulfate, 40): the major applications are to be found in the fields of detergency and
solubilization.

Many treatises draw a distinction between "synthetic" and "natural" surfac-

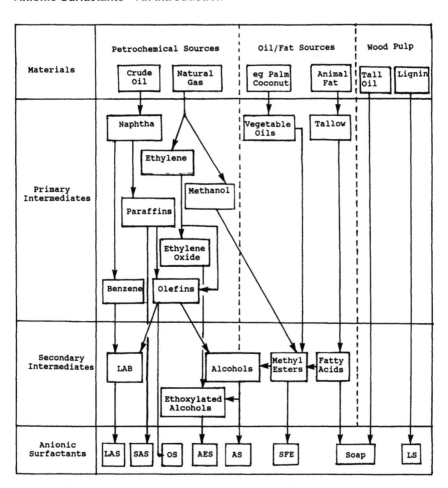

FIG. 1 Some principle routes for the production of anionic surfactants. LAB: linear alkylbenzene; LAS: linear alkylbenzenesulfonate; SAS: secondary alkanesulfonate; OS: olefinsulfonate; AES: alkyl ether sulfate; AS: alkyl sulfate; SFE: sulfonated fatty esters; LS: ligninsulfonate. (From Ref. 40.)

tants, the latter class being principally the soaps. Exactly what is natural about being boiled with sodium hydroxide is not made clear. As the public in general continues to strengthen its embrace of the concepts of "green" and "environmental capacity," surfactant products have tended to become regarded as being "of natural origin" if they contain hydrophobes derived from fats and oils of recent animal or vegetable origin, despite the intensive chemical processing that follows. On the other hand, surfactants derived from fossil fuels are regarded as "synthetic" despite the fact that their ultimate origin was, in fact, vegetable/natural [4].

It is appropriate that the next section should examine the major sources of the hydrophobes, their treatment, and the likely structures that may arise as a consequence.

II. SOURCES OF THE HYDROPHOBE

The total global consumption of soaps is estimated to be some 8.5 million tons per annum (Mtpa). The fatty acids come almost exclusively from natural sources, i.e., animal fats and vegetable oils. Those derived from the oxidation of paraffin waxes tend to contain by-products and are of inferior quality. For the production of all other hydrophobes (except for the lignosulfonates), the fossil-based hydrophobes outweigh those from fats/oils sources 10-fold [5].

A. Animal Fats and Vegetable Oils

The major fatty acid content of some of the common oils and fats are listed in Table 1. The distribution of the various homologs and ratio of saturated-to-unsaturated acids was once believed to be a fingerprint of the parent oil, but in practice there is too much variation in composition arising from the particular strain of seedstock used and environmental factors such as soil type, nutrient availability, etc. In addition, a certain degree of fractionation, intended or otherwise, is likely to occur during processing.

Whether a particular triglyceride is a solid or liquid at ambient temperatures depends largely upon the length of the hydrocarbon chains (higher intermolecular forces) and the degree of unsaturation. The enforced planar arrangement around the ethylenic linkages causes a kink in what otherwise could become a regular conformation of the chains: this, in turn, prevents the chains from packing into a neat crystalline array. The unsaturated compounds, therefore, have a lower melting range than their saturated counterparts. Soft soaps, for these reasons, may be derived from oils with a high unsaturated fatty acid content, such as cottonseed oils: alternatively, the counterion used may be be potassium instead of the more usual sodium.

Tallow-sourced acids are frequently hydrogenated (i.e., hardened) and typically contain C14, C16, and C18 saturated acids with a ratio of 5:30:65 [7]. This process is so common that many marketing companies feel that it is unnecessary to declare that this has been carried out. Consequently, the term "tallow" attached to a product could refer to either the natural or the hydrogenated version.

The production of fats and oils is growing steadily and is expected to reach 100 Mtpa by the year 2000. Some 85% of oils are consumed as food, and the remainder, usually of poorer quality, is directed towards the oleochemicals indus-

TABLE 1 Typical Percent Fatty Acid Distribution Among Some Common Oils and Fats

Fatty acid Name	Carbon no.	Unsaturated groups	Source Coconut	Palm kernel	Olive	Groundnut	Cottonseed	Sunflower	Mutton tallow	Beef tallow
Caprylic	8	—	8	3						
Caproic	10	—	7	5						
Lauric	12	—	46	47						
Myristic	14	—	15	15	0.4				2	5
Palmitic	16	—	9	7	18	6	20	6	25	28
Stearic	18	—	2	2		3	2	2	27	21
Arachidic	20	—				6				
Oleic	18	1	6	14	68	60	24	25	41	40
Linoleic	18	2	1	1	12	25	50	66	4	3

Source: Ref. 6.

try. Fatty acids are liberated from the triglycerides by saponification, by direct high pressure hydrolysis with water, or by transesterification with methanol. The crude acids may be contaminated with mono- or diglycerides and various other components, which may induce color and odor. Fractionation by vacuum distillation of the free acids or distillation of the methyl esters will produce cuts from the homologous series according to variations in volatility: such processing will have little effect upon the ratio of saturated to unsaturated components.

For conversion to alcohol sulfates or alcohol ether sulfates, the methyl esters are catalytically hydrogenated to the intermediate fatty alcohols and methanol. In contrast to fatty alcohols produced from mineral sources, those from vegetable oils and animal fats will reliably be primary alcohols with linear chains. The most useful fatty alcohols from these sources are coconut alcohol (and different cuts thereof), tallow alcohol, cetyl/stearyl alcohol, oleyl alcohol, and oleyl/cetyl alcohol [8]. Another unique feature of hydrophobes derived from natural sources is the presence of only chains with even carbon numbers, e.g., C12, C14, C16, C18, etc.

Apart from natural fats and oils, the other major natural contributor to surfactant raw materials is the wood-pulping industry. By-products include (a) tall oil, a dark, oily mixture of fatty acids and rosin acids (basically polycyclic terpene carboxylic acids), which can be treated either by acid-washing or fractional distillation to yield fatty acids containing approximately 40% and 5% rosin acids, respectively, and (b) lignin, a polymeric material as complex as wood itself, which is separated from cellulose and other compounds. During the bisulfite bleaching process a significant amount of addition occurs to produce lignosulfonic acids with molecular masses of 100,000 or more [9].

B. Petrochemical Sources

The petrochemical industry provides the largest contribution to the hydrophobes of the nonsoapy anionic surfactants. The principal primary intermediates are benzene and linear paraffins, olefins, and alcohols. Hydrocarbon chains of suitable length must, in general, be generated either by polymerization of low molecular mass alkenes or by cracking of much larger molecules. Ethylene oxide and maleic anhydride (for sulfosuccinate synthesis) also arise from petroleum sources. In contrast to the products of fats/oils processing, petrochemically derived hydrocarbons contain both odd- and even-numbered carbon chains (unless produced by polymerization of ethylene).

1. Linear Paraffins

Linear paraffins in the molecular range C10–C18 are important as hydrophobes and are found in the kerosene and gas oil fractions. They are separated from other components of similar volatility by preferential adsorption onto synthetic zeolites, which function as molecular sieves: the small cross-sectional area of *n*-alkanes allows them to enter the cavities of the adsorbent, whereas the much bulkier branched-chain, cyclic and aromatic species are too large to do so. Paraffins in this range are used to produce alkanesulfates or are dehydrogenated to alkenes.

Higher molecular mass alkanes in the C20–C30 range are components of the lubricating oils fraction. They are not amenable to isolation by molecular sieving, but can be separated from other components of the fraction by a process known as urea dewaxing (via formation of a clathrate type compound between urea and the paraffins). The principal use of these materials to the surfactant industry is as a feedstock for the production of lower molecular mass alkenes by steam cracking.

These processes, and most others outlined in this section, are discussed in some detail by Fell [5] and Hons [10].

2. Linear Olefins

Linear olefins in the range C10–C18 are important intermediates in the synthesis of alkylbenzenes, olefinsulfonates, and fatty alcohols. The ethylenic link can be at the end of the chain (α-olefins) or within the chain (ψ-olefins), depending upon the source of the hydrocarbon feedstock and the process to which it is subjected.

(a) Wax Cracking. The C20–C30 wax fraction of paraffins can be cracked to produce smaller alkenes and hydrogen. The fission occurs randomly along the length of the chain, but typically yields 40–45% of *n*-α-olefins in the desired C10–C18 range. However, also present will be significant amounts of ψ-olefins (5–10%), alkanes (1–5%), and dienes (2–4%), which serve to reduce the quality of the product. In addition, the supply of the required quantities of wax is often not reliable. For these reasons, this process is seldom used in heavily industrialised countries.

(b) Dehydration of Primary Alcohols. Primary alcohols dehydrated via a carbenium ion intermediate yield a mixture of ψ-olefins with only about 1% α-olefin. Pure α-olefin can be produced by the pyrolysis of simple esters of the alcohols, but this route involves an extra step. Recently, an aluminum oxide catalyst has been developed that results in the formation of about 93% *n*-α-olefins and 6% *n*-β-olefins.

(c) Dehydrogenation of Paraffins. Paraffins may be dehydrogenated to a mixture of ψ-alkenes by numerous processes, which usually involve a platinum-based catalyst. Unfortunately, dienes, aromatics, and products from cracking and isomerization reactions are also formed. By careful selection of catalyst and conditions and accepting a low degree of conversion, the side reactions can be largely avoided and the alkane/alkene mixture can be used for the alkylation of

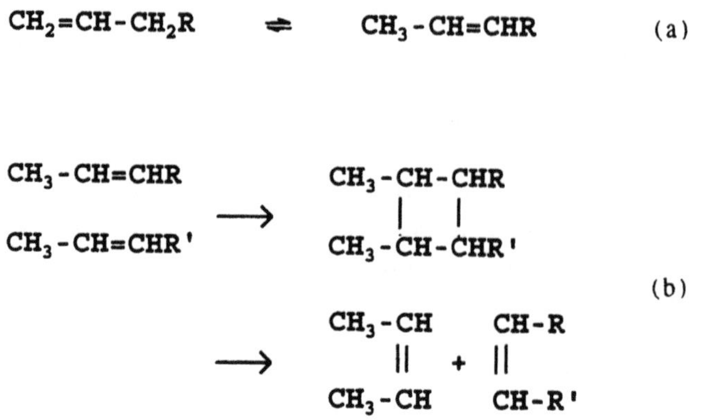

$$CH_2=CH-CH_2R \quad \rightleftharpoons \quad CH_3-CH=CHR \qquad (a)$$

$$
\begin{array}{c}
CH_3-CH=CHR \\
CH_3-CH=CHR'
\end{array}
\longrightarrow
\begin{array}{c}
CH_3-CH-CHR \\
\quad | \quad | \\
CH_3-CH-CHR'
\end{array}
$$

(b)

$$
\longrightarrow
\begin{array}{cc}
CH_3-CH & CH-R \\
\| & + \quad \| \\
CH_3-CH & CH-R'
\end{array}
$$

FIG. 2 Oligomerization of olefins: (a) internal isomerization; (b) disproportionation.

benzene: the unreacted alkane can be separated by fractionation and recycled through the dehydrogenation process.

(d) Oligomerization of Ethylene. The polymerization of ethylene produces *n*-α-olefins of high purity, especially with regard to the absence of dienes, and with even numbers of carbon atoms in the chains. The two major processes, the Ziegler using triethyl aluminium or catalysis with a transition metal complex, both result in a wide range of molecular sizes for the *n*-α-olefin plus small amounts of β-alkyl-1-olefin and linear ψ-olefins. After fractionation to isolate the desired cut (C10–C18), the remaining olefins in the C4–C8 and C18+ range may be reprocessed by isomerization to internal olefins (Fig. 2a) followed by disproportionation (Fig. 2b).

A substantial portion of the ψ-olefins produced falls within the desired molecular range [10].

3.   Branched Olefins

Propene, usually in admixture with propane, can be polymerized using a phosphoric acid catalyst at 200° and at high pressure to produce alkenes suitable for conversion to alkylbenzene and subsequently to alkylbenzenesulfonate. Fractional distillation of the product yields first unconverted propane, followed by tripropylene, which can be returned to the reactor or removed for use as motor fuel. The next fraction is an extremely complex mixture known as tetrapropylene and consists of many structural and double bond isomers of dodecene that are so closely related in physical properties that even capillary gas chromatography cannot fully resolve them [11].

Isobutene polymerization by the above process results in a simpler fraction of C12 alkenes, but they tend to depolymerize under the conditions needed for the subsequent Friedel-Craft conversion to alkylbenzene. An alternative reaction route results in a product with much less chain branching that is suitable. Since surfactants manufactured from alkylbenzenes with substantial branching in the alkyl chain are resistant to biodegradation, they are not to be found extensively in most industrialised countries. By way of comparison, the global capacity for production of linear alkylbenzenesulfonates in the early 1990s was about 2 Mtpa, whereas for the branched-chain equivalents it was less than 500 kt. Principal producers are France, Latin America, and Japan [12].

III. COMMON TYPES OF ANIONIC SURFACTANTS

The majority of surfactants to be discussed here are either sulfonates or sulfate esters. In the latter case, the hydrophobe is attached to the hydrophile by a labile C-O-S linkage; this is relatively easily hydrolyzed to the corresponding alcohol and (bi)sulfate ion by dilute aqueous acids. Sulfonates, on the other hand, contain a robust C-S linkage that is much more stable; it is broken only by drastic treatment such as refluxing with concentrated phosphoric acid. The sulfonates, therefore, find application in a variety of pH conditions that are too drastic for sulfate esters. The sulfonates will be considered first.

A. Alkylbenzenesulfonates

The behavior of alkylbenzenesulfonates, both in the washing machine and in the environment, must surely be ranked as the most intensively studied of all the surfactants. Second only to soap in its production volume, it was first reported in 1923, but large-scale production in the Western world did not commence on a significant basis until after the end of World War II. Propene was being produced in large quantities as a by-product of the oil-processing industry; it could be simply and inexpensively polymerized using a phosphoric acid catalyst. The C12 distillation cut of the mixture, tetrapropylene, could be used to produce first tetrapropylbenzene and then tetrapropylbenzenesulfonates (TPBS). These surfactants could be formulated into a detergent with an excellent price : performance ratio and in the late 1950s represented two thirds of the Western world's synthetic anionic SA production [13]. However, at this time, huge quantities of foam, albeit usually stabilized by wastes from other sources, were all too frequently to be seen on inland waters of Western Europe and the United States. This precipitated a detailed study (beyond the scope of this book, but admirably covered by Swisher [14]), from which it was ascertained that the root of the problem lay in the resistance of the highly branched alkyl chain to biodegradation. Regulations were introduced into most regions requiring that the surfactants used in detergents be degraded to an extent of at least 80% after a specified period in a wastewater treatment plant. Alkylbenzenesulfonates made from linear olefins (LAS) met this requirement easily.

The ecological suitability of LAS was the subject of a major conference in 1989, reported in detail in an issue of *Tenside, Surfactants, Detergents* [16]. A quality objective of 12–25 µg/L had been deduced from NOEC (no observable effect concentration). Occasional concentrations of 20–30 µg/L were reported in the Ruhr River at that time, and the need for continued vigilance and search for even more ecologically acceptable surfactants was stressed [17].

In addition to being the mainstay of laundry and general household products, LAS finds application in cosmetics, plastics, petroleum production, agriculture, the food and textile/dyeing industries, and metal treatment. With markets for LAS increasing in Latin America, the Near East, the Far East and Southeast Asia, the global capacity for production could reach 3 Mtpa going into the next century [15]. Alkylbenzenesulfonates with branched chains, however, have not entirely disappeared from the global scene: Fell listed the annual production as approximately 0.5 Mtpa [18,19].

1. Production of Alkylbenzenes

The alkylation of benzene with an *n*-olefin is a Friedel-Craft reaction, catalyzed by hydrogen fluoride, aluminum chloride, or aluminosilicates: a fourth process using chloroparaffins as intermediates is no longer of commercial importance. Since the ease of carbonium ion production is in the order of tertiary carbon > secondary carbon > primary carbon, there is virtually no 1-phenylalkane in the product. The result of such a reaction is a complex mixture: a C9–C14 cut of *n*-olefin, for example, can produce 30 structural isomers and homologs (Fig. 3a) in addition to dialkyltetralins (Fig. 3b). The spread of these isomers is very much dictated by the catalyst. The aluminosilicate and aluminum chloride routes result in some 25% 2-phenylalkanes, whereas hydrogen fluoride causes the proportion of these isomers to fall to around 15%. As another example, a nominal C12 olefin yields approximately 28 and 19% of 5-phenylalkanes with hydrogen fluoride and aluminum chloride respectively. (Further details are to be found in Ref. 1.)

For alkylation with tetrapropylene (Fig. 3c) (already a complex mixture), both the hydrogen fluoride and aluminum chloride catalysts promote side reactions such as polymerization, isomerization, and also fragmentation. In the latter process a C12 isomer might fragment into, say, C4 and C8 fragments, which result in the formation of *t*-butyl- and *t*-octylbenzene, respectively. Such low molecular mass alkylbenzenes are largely removed by fractionation prior to sulfation.

In summary, the alkylbenzenes used for commercial surfactant production are complex mixtures.

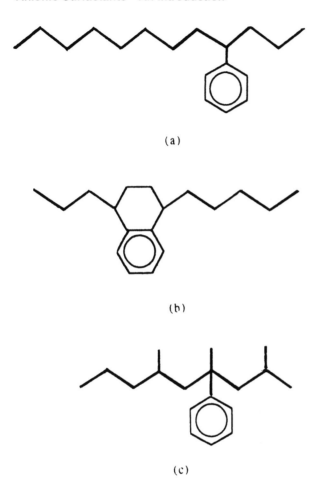

(a)

(b)

(c)

FIG. 3 Alkylbenzene structures: (a) a linear alkylbenzene; (b) a dialkyltetralin; (c) a highly branched alkylbenzene.

2. Sulfonation of Alkylbenzenes

The sulfonation can be achieved with concentrated sulfuric acid, oleum, or sulfur trioxide, the latter being preferred. Substitution occurs almost exclusively in the *para* position (Fig. 4a), since the large hydrophobic chain effectively hinders approach to the *ortho* positions. Conversion efficiency varies between 92 and 98%: attempts to increase the yield result in the formation of highly sulfonated by-products.

The actual route is quite complex. Pyrosulfonic acid (Fig. 4b) and the sulfonic anhydride (Fig. 4c) are formed as intermediates: these later react with more

FIG. 4 Alkylbenzenesulfonate production: (a) an alkylbenzene sulfonate; (b) a pyrosulfonic acid (intermediate); (c) a sulfonic acid anhydride (intermediate); (d) a dialkyldiarylsulfone (by-product).

alkylbenzene or added water to yield the desired sulfonic acid. A good-quality product typically contains:

Alkylbenzenesulfonic acid: 97–98%
Sulfuric acid: ~0.5%
Neutral oil: 1.5–2.5%

Any dialkyktetralins present are also sulfonated and in the final product will function as hydrotropes. The neutral oil consists mainly of the by-product dialkylarylsulfones (Fig. 4d) and unreacted alkylbenzenes.

B. Secondary Alkanesulfonates

Secondary akanesulfonates are produced in the order of 160,000 tons a year, mainly in Europe and Japan [20]. They occupy a position in a class between the "workhorse" alkylbenzenesulfonates and the speciality surfactants to be found in later pages [21]. Although excellent performers in the laundry, they cannot compete economically with LAS in powdered products but come into their own in concentrated liquid formulations (due to their high solubility in water), in which they are frequently to be found together with alkylether sulfates: these two types of surfactants form a particularly effective synergistic system. Major domestic applications are in (liquid) laundry detergents, dishwashing liquids, shampoos, and other personal care products. Industrial applications include cleaners, emulsifiers for PVC polymerization and products for the textile industries (washing of fibers, desizing, mercerizing, carbonization, and fatting of leather).

There are two major routes for synthesis, both photochemically induced free radical reactions. In the sulfoxidation process, the paraffin is reacted with a mixture of sulfur dioxide and oxygen. In the sulfochlorination process a mixture of chlorine and sulfur dioxide is used to produce an alkanesulfonyl chloride intermediate, which is saponified with sodium hydroxide to yield the sodium alkanesulfonate. Traces of organic chlorine compounds (approximately 0.15%) in the final product indicate this route for preparation.

Being the product of free radical attack upon the alkane, the position of substitution depends largely upon the homolytic fission energy of the C-H bonds. In practice this results in the sulfonate group group being distributed fairly evenly between the internal carbon atoms of the chain and only about 1% of the 1-isomer. Consequently, this class of surfactant is frequently termed secondary sulfonate (Fig. 5a). A C13–C18 paraffin cut, for example, has over 40 such isomeric/homologous positions to offer. In addition, about 10% of poly (mainly di-)sulfonates are formed. The biodegradability of this class of surfactants is excellent. Primary degradation occurs to the extent of 90% in 2 days and more than 99% in 4 days [21].

Another class of surfactant, the alkylethoxysulfonates (Fig. 5b), is strictly included in this group. These are produced from alkyl or alkylphenol ethoxylates

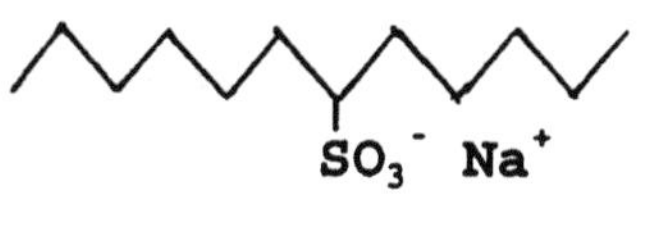

$$\overset{|}{SO_3^-} \ Na^+$$

(a)

$$R(OCH_2CH_2)_nOCH_2CH_2SO_3^- Na^+$$

(b)

FIG. 5   Sulfonates: (a) an alkanesulfonate; (b) an alkylethoxysulfonate.

by conversion of the terminal -OH group initially to a chloro derivative (with thionyl chloride) and thence to a sulfonate (with sodium or potassium sulfite). The products are too expensive for widescale use but have a high tolerance to salinity and find application in enhanced oil-recovery operations [22].

C. Olefinsulfonates

As has already been pointed out, olefins for surfactant manufacture are conveniently divided into α- and ψ-olefins according to the position of the ethylenic link (terminal or internal, respectively). Both types may be sulfonated to yield the products known as α-olefinsulfonates (AOS) and internal olefinsulfonates (IOS), respectively.

The production and application of AOS is well established, whereas that of IOS is embryonic. The composition of both types is quite complex and varies not only with the choice of olefin feedstock but also with the sulfonation method used and the reaction conditions employed [23].

1.   Sulfonation of α-Olefins

(a)   Sulfonation.   The α-olefin/sulfur trioxide reaction mixture (Fig. 6a) results initially in the formation of a reactive intermediate 1,2-sultone (Fig. 6b). During a follow-up aging process, this either (1) rearranges to a 1,3-sultone (rapid) or a 1,4-sultone (slow) (Figs. 6c and 6d) or (2) reacts to yield alkenesulfonic acids (Fig. 6e). The ratio of these products varies according to the initial ratio of sulfur trioxide to olefin. If this ratio is less than 1.0, the 1,3-sultone predominates; if it is in the region of 1.0–1.2, a larger proportion of the sulfonic acid will be formed. Excessive amounts of sulfur encourage the formation of disulfonic- and sultonesulfonic acids. As the aging time increases, the 1,3-sultone concentration decreases as it rearranges to the 1,4-sultone [24]. Prolonged aging leads to the formation of dimers and higher polymers.

(b)   Neutralization and Hydrolysis.   Heating with sodium hydroxide not only neutralizes the alkenesulfonic acid, but also causes the 1,3- and 1,4-sultones to become hydrolyzed into 3- and 4-hydroxyalkanesulfonates (Figs. 6f and 6g). The sultones remain in trace quantities only after this treatment. A typical commercial sample might be composed as follows:

Sodium alkenesulfonates: 65%
Sodium 3- and 4-hydroxyalkane sulfonates: 28%
Disulfonates*: 5%
Sodium 2-hydroxyalkanesulfonates: <1%

---

*The term "disulfonate" is strictly a misnomer. The correct description is sulfatosulfonate (Fig. 6h) [25].

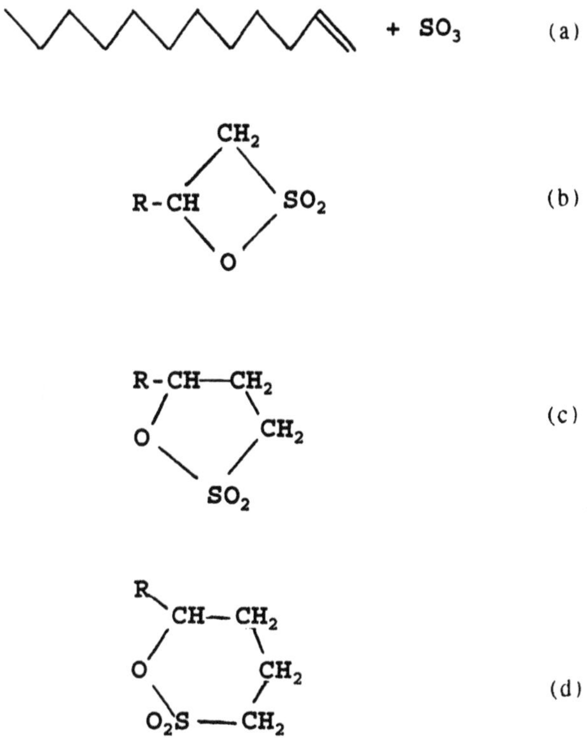

FIG. 6 Sulfonation of α-olefins: (a) α-olefin/sulfur trioxide reaction mixture; (b) a 1,2 sultone (intermediate); (c) a 1,3 sultone (intermediate); (d) a 1,4 sultone (intermediate); (e) alkenesulfonic acids; (f) a sodium 3-hydroxyalkane sulfonate; (g) a sodium 4-hydroxyalkane sulfonate; (h) a disodium sulfatoalkanesulfonate.

2. Sulfonation of Internal Olefins

One of the main purposes of aging of the α-olefin sulfonation product is to minimize the 1,2-sultone content, because this intermediate will otherwise hydrolyze to the 2-hydroxyalkanesulfonate, a compound of poor solubility. The β-hydroxysulfonates formed from internal olefins, however, have a higher solubility and consequently the aging process can be omitted. The product of such an approach is typically:

Sodium β-hydroxyalkanesulfonates: 83%
Sodium alkanesulfonates: 10%
Sodium α-hydroxyalkanesulfonates: 5%
Unsulfonated oils: 2%

$$R-CH=CH-SO_3H$$

$$R-CH=CH-CH_2-SO_3H \qquad (e)$$

$$R-CH=CH-CH_2-CH_2-SO_3H$$

$$\underset{\underset{OH}{|}}{R-CH}-CH_2-CH_2-SO_3^-Na^+ \qquad (f)$$

$$\underset{\underset{OH}{|}}{R-CH}-CH_2-CH_2-CH_2-SO_3^-Na^+ \qquad (g)$$

$$\underset{\underset{OSO_3^-Na^+}{|}}{R-CH}-CH_2-SO_3^-Na^+ \qquad (h)$$

FIG. 6 Continued

If an aging step is introduced, the β-hydroxyalkanesulfonate content falls and that of the other two surfactants rises accordingly. A comprehensive account of the factors affecting the composition of the mixture was recently presented by Radici et al. [26].

The olefin sulfonates possess both good wetting and detergency properties plus high tolerence to calcium and magnesium ions, i.e., hard water. They are used mainly in light duty liquids (C12–C16) and heavy duty powders (C16–C18). Japan is the major consumer, followed by North America, Europe, South Korea, and India; the latter country represents a major future market for AOS.

D. Ester Sulfonates

This class of surfactants goes under a variety of names, namely, α-sulfomono-carboxylic esters, α-sulfo fatty acid esters (SFE), fatty acid α-sulfonates, or alkyl (usually methyl) ester sulfonates (MES) (Fig. 7a). They are of increasing interest, particularly in Japan [4], by virtue of being derived from a renewable (natural) source coupled with good wetting and washing performance, rapid

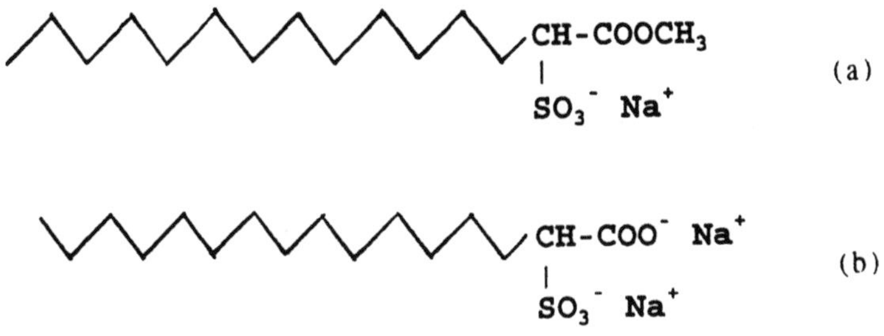

FIG. 7   Ester sulfonates: (a) a sodium alkyl (methyl ester) sulfonate; (b) disodium salt of
α-sulfofatty acid (by-product).

biodegradability and excellent performance in hard water, even to the extent of
improving the behavior of soaps in such a solvent (lime soap dispersion)
[27,28]. This latter property makes the ester sulfonates logical inclusions in
phosphate-free formulations. A mixture of linear alkylbenzenesulfonate (which
performs well in soft water but not in hard) and MES (which performs better in
hard water than in soft) provides a combination that works well as a heavy duty
detergent in both hard and soft waters, especially if an alcohol ethoxylate is present [27].

   In the initial step of the manufacturing process, fatty acid triglycerides are
trans-esterified with a low molecular mass alcohol (usually methanol, but could
be up to a C4 alcohol). The monoesters can be fractionated at this stage (low C
numbers for liquid detergents, high C numbers for powder detergents) and hydrogenated (high unsaturation results in a bad color upon sulfonation). Since the
surfactants perform better when the hydrophilic group is at the end of the chain
rather than in the middle, the use of the methyl ester ensures that the second
alkyl chain is as short as possible.

   The mechanism of the sulfonation reaction is complex [27] and involves formation of a mixed carboxylic-sulfonic acid anhydride as an intermediate. Substitution into the α-position on the fatty chain is favored not only by the activation
of that position by the adjacent carbonyl group but also by its proximity to the
cyclic structure of the intermediate.

   This reaction provides an example of the perpetual conflict between the speed
of the transformation and product quality. The reaction mixture is one that requires aging for optimum results. If the mixture is neutralized too soon, an unacceptable quantity of the di-salt of the α-sulfo fatty acid is formed (Fig. 7b),
which reduces the performance of the final surfactant mixture. This by-product is
also enhanced if the concentration of sulfur trioxide is raised to speed up the reaction.

Cullum [25] cites soaps, unsulfonated ester, and sodium methosulfate as additional by-products.

E. Other Sulfonated Surfactants

A large number of additional sulfonated surfactants has found application: the more commercially important ones are listed here. Largely they are regarded as being in the low-volume/high-cost bracket. One property that they all have in common is a good tolerance to hard water.

1. Fatty Acid Isethionates

Acyl isethionates have been known for over 60 years. Sodium isethionate (Fig. 8a) is first produced by the reaction of ethylene oxide and sodium hydrogen sulfite; this, in turn, is reacted with a fatty acid using a mineral acid catalyst to give the surfactant ester (Fig. 8b). Due to the susceptibility of the ester group to hydrolysis, especially under alkaline conditions, these surfactants are generally to be found in products of moderate pH such as cosmetics and soap/syndet combinations [28]. The only likely impurities are sodium isethionate and free fatty acid or soap [25] and sodium chloride if the preparation involved a fatty acid chloride as an intermediate.

2. Sulfosuccinate Esters

Since succinic acid is a diprotic acid, it can give rise to either mono- or diesters. Both types can be sulfonated to give two distinct series of surfactants.

The first step is the reaction between maleic anhydride and the hydrophobe, which is commonly a C12–C18 fatty alcohol or a fatty acid alkanolamide (either of which could be ethoxylated). When equimolar amounts of reactants have been consumed, the synthesis of the maleic monoester is deemed to be complete and a slight excess of sodium sulfite is added. Since neither of these reactions goes to completion, the overall product is a complex mixture comprising about 80% sulfonated monoester (Fig. 8c) plus some diester (Fig. 8d), the trisodium (nonester) salt, and unreacted raw materials [29].

To produce the diester, a higher temperature and a catalyst are employed and the water produced is continuously removed to shift the reaction equilibrium towards completion. The sulfonation step is also carried out more efficiently, so that the final product contains only minor amounts of impurities. The alcohols used for the initial diester formation are shorter, typically in the C6–C8 range. The di(2-ethylhexyl) derivative, for example, is a popular wetting agent for the textile industry.

Cullum [25] points out that ammonium bisulfite is sometimes used as the sulfonating agent, in which case small amounts of mono- and dialkyl aspartates are also formed. Typical fields of application as wetting agents include personal care products (these surfactants are particularly mild to the skin), emulsion polymerization, textile industry, paints, and agricultural sprays.

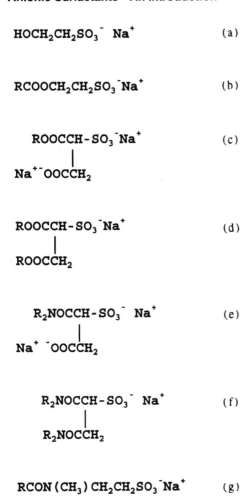

$$HOCH_2CH_2SO_3^- \ Na^+ \qquad (a)$$

$$RCOOCH_2CH_2SO_3^- Na^+ \qquad (b)$$

$$\underset{\underset{Na^{+-}OOCCH_2}{|}}{ROOCCH-SO_3^-Na^+} \qquad (c)$$

$$\underset{\underset{ROOCCH_2}{|}}{ROOCCH-SO_3^-Na^+} \qquad (d)$$

$$\underset{\underset{Na^+ \ ^-OOCCH_2}{|}}{R_2NOCCH-SO_3^- \ Na^+} \qquad (e)$$

$$\underset{\underset{R_2NOCCH_2}{|}}{R_2NOCCH-SO_3^- \ Na^+} \qquad (f)$$

$$RCON(CH_3) CH_2CH_2SO_3^- Na^+ \qquad (g)$$

FIG. 8 Production of miscellaneous sulfonated surfactants: (a) sodium isethionate; (b) a sodium acyl isethionate; (c) disodium salt of a sulfonated monoalkyl succinate; (d) sodium salt of a sulfonated dialkyl succinate; (e) disodium salt of a sulfosuccinamate; (f) sodium salt of a sulfosuccindiamide; (g) sodium of an acyl methyl taurate.

3. Sulfonated Amides

If fatty amines are used instead of alcohols to make the malate intermediates, the final product will be a sulfosuccinamate (Fig. 8e) or a sulfosuccinamide (Fig. 8f). These compounds do not have the wetting capabilities of the sulfosuccinate esters, but are used as detergents, solubilizers, and dispersants.

Acyl taurates and acyl methyl taurates (Fig. 8g), frequently oleic acid derivatives, are probably the only other members of this large group of surfactants of any general interest [25]. They are used in shampoos and syndet bars, etc.

F. Sulfate Esters

As mentioned at the beginning of this section, sulfates differ from the various sulfonate species listed above in that they are esters (mono esters of the diprotic sulfuric acid) in which the hydrophobe is joined by a C-O-S linkage rather than a direct C-S bond. The hydrophobe containing the reactive -OH group needed to form the ester is normally a fatty alcohol, but could also be a fatty alcohol ethoxylate, an alkanolamide (ethoxylate), an alkylphenol ethoxylate or a mono/diglyceride. The surfactants resulting from sulfation of these species are shown in Fig. 9.

The fatty alcohols may be derived from natural fats and oils or may be synthesised via the Ziegler or OXO processes [30]. The alcohol sulfates are the classic members of this class, having been used extensively since the 1930s. It is estimated that about 40% of the global production of fatty alcohols is converted to the sulfate ester. The product is commonly marketed as the sodium salt, but

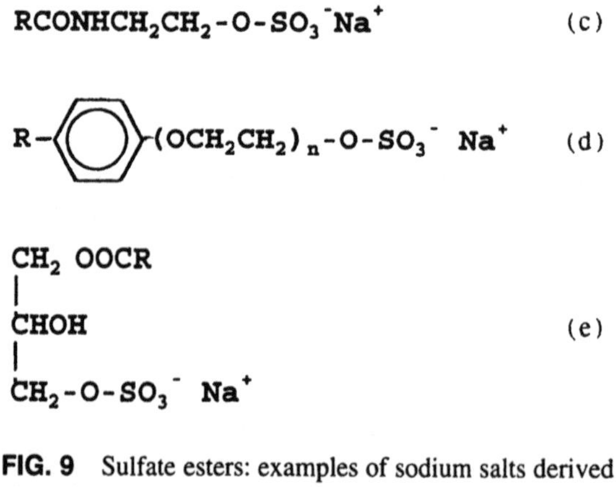

$$R-O-SO_3^- \ Na^+ \qquad\qquad (a)$$

$$R(OCH_2CH_2)_n-O-SO_3^- Na^+ \qquad\qquad (b)$$

$$RCONHCH_2CH_2-O-SO_3^- Na^+ \qquad\qquad (c)$$

$$R-\langle\bigcirc\rangle-(OCH_2CH_2)_n-O-SO_3^- \ Na^+ \qquad (d)$$

$$
\begin{array}{l}
CH_2 \ OOCR \\
| \\
CHOH \qquad\qquad\qquad (e) \\
| \\
CH_2-O-SO_3^- \ Na^+
\end{array}
$$

FIG. 9 Sulfate esters: examples of sodium salts derived from (a) fatty alcohol; (b) fatty alcohol ethoxylate; (c) fatty acid alkanolamide; (d) alkylphenol ethoxylate; (e) fatty acid monoglyceride.

the triethanolamine salt is often used in preparations such as shampoo on account of its combination of higher solubility, good foaming properties, and lower irritance to the human eye.

The application varies with the alkyl chain length. The shorter chains, C8–C10, provide excellent wetting agents and hydrotropes (viscosity modifiers), sodium 2-ethylhexyl sulfate being a particularly common example. The longer chains give rise to products with better detersive, emulsifying, and dispersing properties: the C12–C15 members perform well at low temperatures and the C16–C18 members at higher temperatures. Tallow alcohol sulfates are popular ingredients of heavy duty detergents.

Sulfate esters prepared from fully-saturated alcohols tend to contain little by way of impurities other than unsulfated matter and sodium sulfate. Sodium dodecyl sulfate, prepared from carefully fractionated dodecanol, can be recrystallized and dried to produce a surfactant approaching 100% purity, which is used as *the* primary standard anionic surfactant in analytical and basic studies alike.

The fatty alcohols may be lightly ethoxylated prior to sulfation. The average ethylene oxide chain length is usually about two, seldom more than four. The inclusion of this additional hydrophile has the surprising effect of reducing the critical micelle concentration. These derivatives are more skin compatible than their unethoxylated counterparts and are popular as ingredients in high foam preparations. The use of ethylene oxide normally results in the inclusion of some 1,4-dioxane in the final product, but this is normally only of the order of a few parts per million and well below the concentration regarded as a health hazard.

Although the sulfate ester link is regarded as labile, it is only under acidic conditions (pH < 4) that this becomes a problem. Domingo [30] sums up the alkyl sulfates thus:

It is difficult to find an industrial sector that does not use alcohol- or alcohol ether sulfates. These surfactants are rendered so versatile in their chemical structure through variations in alkyl chain distribution, the number of moles of ethylene oxide or the cation, that it is possible to find the adequate sulfate achieving the highest mark in every surfactant property. This and the relative low cost are the two main reasons for their vast industrial use.

G. Carboxylates

The carboxylate anion has already been encountered in this introduction as it occurs conjointly with a sulfonate group (e.g., in sulfosuccinates), but it does, of course, stand as a hydrophile in its own right. The sodium salts of long-chain fatty acids, better known as soaps, are, of course, *the* classic surfactants, having been used for many centuries (Fig. 10a). Many treatises on surfactants treat soaps as a class of their own, not to be included under the heading of anionic

$$RCOO^- Na^+ \tag{a}$$

$$R\ (OCH_2CH_2)_n OCH_2 COO^- Na^+ \tag{b}$$

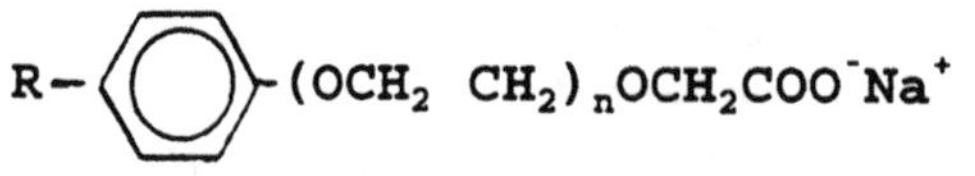

$$R-\text{⬡}-(OCH_2\ CH_2)_n OCH_2 COO^- Na^+ \tag{c}$$

$$RCONHCH_2CH_2 (OCH_2\ CH_2)_n OCH_2 COO^- Na^+ \tag{d}$$

$$RCON(CH_3)CH_2 COO^- Na^+ \tag{e}$$

FIG. 10 Carboxylate surfactants: (a) a carboxylate soap; (b) a polyether carboxylate derived from a fatty alcohol; (c) a polyether carboxylate derived from an alkylphenol; (d) a polyether carboxylate derived from a fatty acid monoethanolamide; (e) an acyl sarcosinate.

surfactants: the latter term is reserved for the synthetic members, the so-called *syndets*.

As indicated earlier, soaps may be derived from:

1. Animal fats, e.g., tallow soaps. Major constituents are palmitic acid (C16 saturated, 25–30%), stearic acid (C18 saturated, 15–20%) and oleic acid (C18 monounsaturated, or C18:1, 40–45%).
2. Vegetable oils, e.g., coconut oil. Major constituents are C12, 47%; C14, 17–20%; C16, 8–10%; C16:1, 5–6%.
3. Tall oil, a mixture of fatty acids, mainly unsaturated (70–75%), and rosin acids (25–30%) from the wood-processing industry.

The associated cation is normally sodium, but others are common. The alkanolamine (mono-, di-, and tri-) salts are more soluble than the corresponding sodium soaps, have better foaming characteristics, and are milder to the skin. Potassium salts are traditionally used as liquid soaps, but most modern "liquid soaps" are formulations of mild synthetic anionic surfactants and lather boosters, etc., and contain no soap at all [9].

Alkyl polyether carboxylates (Fig. 10b) are substantially different from soaps. Although known since the 1930s, their general application, apart from minor use in toilet preparations and the textile trade, did not flourish until until the 1980s. At this time, increased attention to environmental issues and to skin compatibility pushed them to the fore [31]. They show excellent biodegradation properties, produce the same rich foam as soap, but are much milder to the skin and foam well in hard water. They are stable to oxidizing agents such as chlorine and peroxides and have a high tolerance to electrolytes. Whereas soaps perform well only under neutral or alkaline conditions and are rendered ineffective in acid media, the polyether carboxylates, particularly those with a high degree of ethoxylation, still function as (nonionic) surfactants even when fully protonated at low pH values.

For high foaming applications coupled with mildness, such as shampoos and foam baths, C12–C14 alkyl chains and an average of 10 moles of ethylene oxide are typical. For use in cleaning preparations, wetting is improved by reducing the ethylene oxide content to 4–5 moles. Lower foaming can be achieved by reducing the alkyl chain length to C10 or less.

Other industrial applications include the textile industry (wetting agents for mercerizing, bleaching, dyeing, etc. processes), metal-working fluids (less corrosive than soap), dyes, inks, etc. (dispersing agent), and enhanced oil-recovery operations (stable to heat, electrolytes, hard water).

Variations in the hydrophobe can come from replacement of the fatty alcohol with alkylphenol or fatty acid alkanolamide (Figs. 10c,d).

The conversion of an ethoxylated fatty alcohol into a carboxylate typically involves reaction with sodium chloroacetate. In this case, the product may well contain small amounts of glycolic and diglycolic acids and a few parts per million of chloroacetate.

Another type of surfactant conveniently classified under the heading of carboxylates are the *N*-acylated amino acids, formed by the condensation of fatty acids with amino acids. The most frequently encountered examples are the acylated sarcosines (Fig. 10e). These are specialty surfactants found in toiletry preparations such as shampoos, toothpastes, and hand cleansers, in which they function as lather boosters. In contrast to the vast majority of anionic surfactants, they are compatible with cationic surfactants [9].

H. Phosphate Esters

Numerous anionic surfactants based upon phosphorus have been reported in the literature, but in general have achieved only the status of an occasional specialty surfactant. They include alkyl (ethoxylated) phosphates and polyphosphates, phosphonates (containing a direct C-P bond), alkyl phosphites, and phosphinites. A full list has been assembled and comprehensively discussed by Wasow [32]. The total global production of phosphorus-based surfactants is only about 1% of the overall surfactant output.

The most commercially important class is the alkyl phosphate esters. Phosphoric acid, being a triprotic acid, can form mono-, di-, and triesters. Although only the first two types can form salts and therefore act as anionic surfactants, in practice none of the methods of preparation results in one exclusive product. A nominal mono- or diester will contain varying amounts of the other two esters, free alcohol (or ethoxylated alcohol) and phosphate ion.

The esters are stable in neutral and alkaline conditions but can be hydrolyzed by heating with mineral acids. Among their attributes are good solubility in saline water, a resistance to hard water, an ability to disperse lime soaps, and corrosion-inhibition properties in acid media. They possess excellent skin compatability and are used as emulsifiers in cosmetics. Other applications include acting as lubricants and emulsifiers in the textile and agricultural chemical industries.

I. Fluorinated Surfactants

Surfactants in which the traditional hydrocarbon hydrophobe has been replaced by an equivalent fluorocarbon group possess some unique properties. They have been dubbed by Kissa [33] as the "supersurfactants." They can decrease the surface tension of water to a level far below that reached by their hydrocarbon counterparts. The fully fluorinated, or perfluorinated, hydrophobe is extremely resistant to chemical attack and can survive in environments in which conventional surfactants would fail. One of the most interesting points is that the perfluoroalkyl chain is not only hydrophobic but oleophobic as well, so that fluorinated surfactants can function as oil and fat repellents. A perfluoroalkyl chain joined to a hydrocarbon chain can function as a surfactant in hydrocarbon media.

If only a partial replacement of hydrogen by fluorine has occurred, then the properties of the surfactant vary considerably according to the position along the chain at which the substitution has occurred.

In general, fluorinated anionic surfactants are divided into the four main categories already discussed: carboxylates, sulfonates, sulfates, and phosphates.

Their high price confines their application to situations in which conventional hydrocarbon-based surfactants would fail. These are frequently related to their superior ability to reduce the surface tension of aqueous solutions, thus enabling a variety of surfaces normally regarded as difficult to wet to be effectively wetted. These include polythene and polypropylene: any product required to wet the surface of these polymers, e.g., paints, inks, and other coatings, will benefit from the inclusion of fluorosurfactants. Herbicide and pesticide formulations also perform the task of wetting leaves and insects more readily if they contain such compounds. Other applications include foaming agents for fighting solvent fires by spreading across burning hydrocarbons to cut off the air supply; acting as an emulsifier during the emulsion polymerization of fluoroethylene; and acting as an antifoaming agent in the strong acid (e.g., chromic) baths used in the electroplating industry [34].

The collection of surfactant types presented in the preceding pages represents the major classes likely to be encountered in a commercial product, but it is by no means a complete list. Essential details of commercially available surfactants are to be found in various compendia, some of which are listed at the end of this chapter. Reference to the new patent listings in *Chemical Abstracts* and various surfactant journals will reveal a bewildering collection of complicated molecules and processes for their production, some of which may eventually reach the market place. Undoubtedly the instrumental techniques described in the following pages will figure prominently in the elucidation of their structures.

IV. ENTER THE ANALYST

A. The Level of Analysis

The encounter between the analytical chemist and a sample believed to contain one or more anionic surfactants necessitates a logistic approach from the very beginning.

Because there so many variables involved and because the analysis can be carried out to so many different levels, it is vital that the person/organization requesting the analysis outline as fully as possible the degree of sophistication required in the answer in order to minimize time and cost. From one extreme to the other, it may suffice, for example,

1. To merely ascertain that the sample does indeed contain a significant amount of (unidentified) anionic surfactant.
2. To estimate the amount of anionic surfactant (still unidentified) in terms of an arbitrary surfactant, e.g., to state that the sample contains the molar equivalent of 14.3% sodium lauryl sulfate.
3. To estimate and identify the surfactant(s) present, possibly including a separate value for each isomer and homolog present.
4. To provide a full analysis of all active ingredients present.

The required precision of any quantitative result is another parameter that needs to be settled before work is commenced: it is pointless to languish over whether the anionic surfactant content is 21.1 or 21.2% when the customer may only require a ballpark figure, say to the nearest 5%.

B. Quality Control

Quality control of an incoming raw material or an outgoing product is an exercise that can involve many separate analyses for common by-products and impurities. A simple soap, for example, may be tested for unsaponified matter, unsaponifiable matter, total fatty acid, free alkali, chloride, volatile matter, alcohol-insoluble matter, and glycerol content. Standard procedures for all such analyses are to be found in the publications of the various standards associations such as the International Organization for Standardization (ISO), the American Society for Testing and Materials (ASTM) and the British Standards Institution (BSI), as are the accepted procedures for the all-important process of representative sampling. Many of the pertinent quality-control procedures are outlined in a companion volume in the Surfactant Science Series [9] and another excellent volume by Cullum [25], and it is inappropriate to reproduce them here.

C. Qualitative Testing for Anionic Surfactants

Phenomena such as foaming or lowering of surface tension or ability to emulsify an oil are indicative of the presence of surfactants in general, not necessarily anionic surfactants. Unless the analyst is sufficiently familiar with the type of sample to be tested, it makes good sense to apply a specific test for the presence of anionic surfactants before embarking upon the involved determination of something that might not be there.

Gabriel and Mulley [35] detailed a range of such tests involving antagonistic reactions (see Chapter 2), color restorations, chelometry, and polarography. The simplest test involves the antagonistic reaction between a cationic dye and the anionic surfactant to form a colored complex that can be extracted from water into a water-immiscible solvent. Traditionally the latter has been chloroform, but 1,2-dichloroethane and other solvents of similar density and polarity will suffice.

The first of two variations to be mentioned here uses methylene blue, which is the indicator used in the classic Epton procedure (Chapter 2) and in the standard screening test used for waste water testing (Chapter 3) [36]. To a few milliliters of aqueous test solution add an equal volume of chloroform and a few drops of 0.25% methylene blue solution containing hydrochloric or sulfuric acid. Shake vigorously and allow the two layers to (partially) settle. A deep blue color in the lower layer indicates the presence of an anionic surfactant. The mineral acid is present to convert any soap or other carboxylate into the unreactive fatty acid. If the test is carried out under alkaline conditions, soaps, etc., will react as anionic surfactants.

The second variation uses a mixture of a cationic indicator (Dimidium bromide) and an anionic indicator (Disulfine Blue VN), which was proposed by Holness and Stone [37] in 1957. This indicator system is used in the current ISO method for the titration of anionic surfactants (see Chapter 2 for details of preparation). The procedure is analogous to the one above. Normally used at pH 2, the organic layer turns pink if anionic surfactants are present and blue if cationic surfactants are present. At pH 9, carboxylates also show a positive reaction.

D. Isolation from the Sample Matrix

Surfactants occur in such a wide variety of samples, e.g., detergent powders, dishwashing liquids, hard surface cleaners, shampoos, foam baths, toothpastes, emulsions, foods, cosmetics, polishes, paints, oils, inks, effluents, wastewaters, etc., that it is impossible to lay down one simple pretreatment that should be applied prior to estimation or identification. For most purposes it will be sufficient to separate the surfactants from other components of the sample that interfere with the subsequent analytical procedures. Sometimes, however, a relatively pure specimen will be preferred, e.g., for some spectroscopic examinations or so that a weighed sample can be titrated to enable an average molecular mass to be assigned.

1. Extraction from Solids

A variety of polar solvents have been used for the purpose of extracting the surfactant(s) from a sample but leaving behind the inorganic components. They include 95% ethanol (the most frequent choice), absolute ethanol, 95% isopropanol, *n*-butanol, *t*-butanol, chloroform, acetone, and ethyl acetate [34], or mixtures such as chloroform/methanol [9]. The extraction is conveniently carried out in a Soxhlet extractor. If the separation is intended to be quantitative, Cullum [25] recommends that the first batch of solvent be cycled through 20 times and then replaced by a fresh batch; if both extracts are evaporated to dryness separately and weighed, this will give the analyst a measure of the completeness of the extraction. The bead structure of spray-dried formulations inhibits the extraction process because surfactant-rich solvent becomes trapped within the hollow beads. Cullum and Platt [38] recommend that such extractions be halted after 20 cycles and that the remaining solid be removed from the thimble, dried, dissolved in water, evaporated to dryness again, and returned to the thimble for further extraction: this procedure destroys the original structure of the solid.

The current standard ASTM method for the separation uses a hot 95% alcohol extraction for 1.5–2 hours followed by evaporation of the filtrate and re-extraction with a 1:1 v/v acetone:ether mixture [39].

2. Extraction from Emulsions and Solution

(a)   Breaking an Emulsion.   Many types of "creamy" preparations can be broken down by partition between light petroleum (40/60 or 60/80) and 50% aqueous ethanol. Essentially the petroleum ether removes fatty alcohols, etc., and the anionic surfactant(s) and inorganic components are to be found in the alcohol layer. A triple extraction with petroleum ether, followed by washing of the combined extracts with aqueous ethanol (adding the latter to the original alcohol solution) will generally quantitatively remove the neutral fatty matter [25]. If the

aqueous layer is evaporated to dryness and then re-extracted with one of the solvents listed above, then the surfactant can be obtained free of inorganic matter and usually in a fairly pure state. A preliminary infrared scan may be useful at this stage if a spectral library is available for comparision. The presence of a neutral ethoxylated species in the sample can complicate this separation process due to its high solubility in the aqueous layer.

(b)   Solvent Extraction from Neutral Solution.   If the sample dissolves readily in water, extraction with *n*-butanol (the lightest of the homologous series of alcohols to be regarded as immiscible with water) will quantitatively remove most surfactants. The butanol layer, however, will contain a considerable amount of water and as a consequence will also contain a significant quantity of inorganic salts. A pure sample of surfactant can be obtained by evaporating the butanol solution to dryness and re-extracting with ethanol or isopropanol. In some cases, the original extraction with butanol may be hampered by the formation of an emulsion: this can often be broken by the addition of a saturated solution of sodium sulfate. *n*-Butanol is by no means an exclusive extraction solvent. Holness and Stone [37] preferred *tert*-butanol, and Schmitt [9] cites the use of methylene chloride and of chloroform/methanol mixtures for the process.

(c)   Solvent Extraction from Acid Solution.   Many sulfonated anionic surfactants may be extracted from aqueous solution into diethyl ether if the former is acidified with hydrochloric acid to a concentration of at least 3 M. Fatty acids released from soaps will also be extracted. If the sulfonate is regarded as a substituted version of the weak sulfurous acid rather than a derivative of the strong sulfuric acid, it is easier to visualize the formation and subsequent extraction of the unionized sulfonic acid. This process is sufficiently efficient that a triple extraction can be regarded as quantitative and used as a basis for the estimation of a sulfonated surfactant. Cullum [25] states that the extraction is applicable to alkylbenzenesulfonates and α-olefinsulfonates (plus sultones arising from the dehydration and cyclization of any accompanying hydroxyalkanesulfonates). Disulfonates, low molecular mass sulfonates (hydrotropes), and, apparently, alkanesulfonates are not extracted.

This process may be regarded as a special case of the more general technique of ion-pair extraction in which the cation is the hydrogen ion. If a hydrophobic cation is used, then virtually all types of anionic surfactants become extractable. If an aqueous solution of the sample is extracted twice with carbon tetrachloride (to remove any component of the mixture that is naturally soluble in that solvent) followed by a third extraction after the addition of a suitable cationic surfactant, then the anionic surfactant may be quantitatively taken up by the organic layer. The latter can be evaporated to dryness and examined by infrared spectroscopy. The spectrum, however, will be the sum of the spectra of both the cationic and anionic components, and therefore the simpler the structure of the former, the easier will be the identification of the anion. Hence, although any of the quaternary titrants discussed in Chapter 2 will permit extraction, a fatty amine, e.g., dodecylamine, in its hydrochloride form is a preferred extractant.

(d)   Ion-Exchange Processes.   Ion exchange is arguably the most effective single technique for isolation of ionic surfactants from other components of formulations, including nonionic surfactants. It has been used since the early 1950s.

Those readers unfamiliar with the basics of ion exchange would benefit from a most informative and practical treatment by Cullum [25]. Another informative review of the earlier work on the separation of surfactant mixtures up to the mid-1970s, much of which is still relevent today, is to be found in an article by Gabriel and Mulley [35]. Most laboratory ion exchangers are based on polymerized styrene (phenylethylene) containing a small percentage of divinylbenzene to ensure cross-chain linking into a three-dimensional matrix. Onto the benzene rings of this polymer are substituted various acidic or basic groups. Of particular pertinence to this book is the trimethylammonium group, $(CH_3)_3N^+$-, which provides a permanent positive site onto which surfactant anions can strongly adsorb, and dimethylamino, $(CH_3)_2N$-, which only acquires a strong charge when activated with a strong mineral acid.

Two points specific to the reaction between ion-exchange resins and surfactants should be appreciated:

1.   The unsubstituted regions of the polystyrene are adsorbents in their own right and tend to retain relatively hydrophobic organic solutes from aqueous solution by hydrophobe:hydrophobe attraction. Indeed, unsubstituted, cross-linked polystyrene resins (e.g., Aberlite XAD series) are used in solid phase extraction systems to strip traces of organic materials of all types, including surfactants, from water samples. Confusing losses of surfactants onto the wrong type of ion exchanger have arisen as a result of this process.

2.   Above a certain concentration, the critical micelle concentration, aqueous solutions of surfactants cluster together into large spherical clusters (micelles) which cannot be broken up by the ion exchanger and are far too large to enter the open pores of the resin. Any modification of circumstances that discourages the formation of micelles is therefore likely to be advantageous.

For these reasons water is by no means the ideal solvent for surfactant ion exchange. Common solvents are methanol or ethanol or mixtures thereof with water.

Another facet of the resin-surfactant combination concerns the degree of cross-linking in the former. Low degrees of cross-linking theoretically result in a lower ion exchange capacity, but at the same time allows a higher degree of permeability to the large surfactant ion. Associated with minimal cross-linking is the tendency for the resin to swell in water and shrink in a nonaqueous solvent such as methanol. According to Cullum [25], a switch from organic solvent to water can create sufficient pressure during swelling to crack a glass column. As a rule of thumb, 80:20 v/v methanol:water is a good solvent for both adsorption and desorption processes.

A certain amount of selectivity over the adsorption of surfactant ions onto anion exchange resins may be achieved by variations in (1) the strength of the basic group on the resin and (2) the anion to be exchanged (e.g., hydroxide, chloride, acetate). The strongest adsorption arises from a strong basic anion exchanger (e.g., Bio-Rad AG-1, Dowex 1, Amberlite IRA 401) in its hydroxide form: such a resin will retain not only sulfates and sulfonates but also carboxylates and sarcosinates. Since the attraction between the strong basic group on the resin and the anion of the strong substituted alkyl sulfuric acid is high, recovery requires a substantial amount of eluent (conveniently 1 M hydrochloric acid in aqueous methanol). The use of a weakly basic (nonquaternary group) resin (e.g., Bio-Rad AG3, Dowex 3) results in a weaker attraction that can be broken by elution with a base (2 M ammonium hydroxide in aqueous methanol) which deprotonates the adsorption site: such a resin will retain sulfates and sulfonates but not anions of weaker acids such as carboxylates and sarcosinates.

(e) Extraction from Environmental Samples.   This topic will be dealt with in a later chapter.

V.   SOME IMPORTANT READINGS

There are numerous books on surfactants available. The analytical chemist who is a newcomer to the field of surfactants will find the following of particular interest.

A.   General Reading

1. H. W. Stache (ed.), *Anionic Surfactants—Organic Chemistry*, Marcel Dekker, Inc., New York, 1996.
Detailed accounts of the raw materials, intermediates and processes used to manufacture the major classes of anionic surfactants. Includes sections on their properties, applications, environmental behaviour, chemical analysis and toxicology of the surfactants and their by-products.

2. M. R. Porter, *Handbook of Surfactants*, Blackie, Glasgow, 1991.
Clear, straightforward accounts of basic surfactant theory and of common commercial and specialty surfactants, their structure, general and surface-active properties, and major contaminants and by-products likely to be found therein. Highlights some of the confusion over biodegradability requirements.

3. G. Hollis (ed.), *Surfactants Europa*, 3rd ed., Royal Society of Chemistry, Cambridge, UK, 1995.
A directory of surfactants available throughout Europe: some 900 products are listed.
4. *Focus on Surfactants*, Royal Society of Chemistry, Cambridge, UK.
A monthly newsletter, edited by G. Hollis, providing concise reports on developments in all aspects of surfactants.

B. Analytical Chemistry

1. T. M. Schmitt, *Analysis of Surfactants*, Marcel Dekker, Inc., New York, 1992.
A handbook for the practicing chemist with over 1200 references up to 1990. Covers characterization of surfactants and their determination in both formulated products and environmental samples.
2. D. C. Cullum (ed.), *Introduction to Surfactant Analysis*, Blackie, Glasgow, 1994.
An easy-to-read, informed tour through the analysis of common types of surfactants, including their by-products, found in formulated products. Packed with practical hints.
3. M. R. Porter (ed.), *Recent Developments in the Analysis of Surfactants*, Elsevier, London, 1991.
Concentrates on the analysis of formulated products, trace components in commodity surfactants, amphoteric surfactants, analysis in flowing streams, and measurement of low levels in environmental samples.
4. M. J. Rosen and H. A. Goldsmith, *Systematic Analysis of Surfactants*, Wiley-Interscience, New York, 1972.
The classic text on surfactant analysis. Although dated and obviously light on instrumental methodology, this book contains a wealth of information on the characteristic reactions of different classes of surfactants.
5. Two companion texts from the Surfactant Science Series devoted to the other major classes of surfactants:
J. Cross (ed.), *Nonionic Surfactants—Chemical Analysis*, Marcel Dekker, Inc., New York, 1987.
J. Cross and E. Singer (eds.), *Cationic Surfactants—Analytical and Biological Evaluation*, Marcel Dekker, Inc., New York, 1994.

REFERENCES

1. H. W. Stache (ed.), *Anionic Surfactants—Organic Chemistry*, Marcel Dekker, New York, 1996.
2. W. C. Griffin, J. Soc. Cosmet. Chem. *1*:311 (1949).

3. W. C. Griffin, J. Soc. Cosmet. Chem. *5*:249 (1954).

4. J. E. Zweig, Tenside Surf. Det. *30*:306 (1993).

5. B. Fell, in *Anionic Surfactants—Organic Chemistry* (H. W. Stache, ed.), Marcel Dekker, New York, 1996, p. 3.

6. T. P. Hilditch and P. N. Williams, *The Chemical Constitution of Natural Fats*, Chapman and Hall, London, 1964.

7. E. Jungermann (ed.), *Cationic Surfactants*, Marcel Dekker, New York, 1970.

8. X. Domingo, in *Anionic Surfactants—Organic Chemistry* (H. W. Stache, ed.), Marcel Dekker, New York, 1996, p. 225.

9. T. M. Schmitt, *The Analysis of Surfactants*, Marcel Dekker, New York, 1992.

10. G. Hons, in *Anionic Surfactants—Organic Chemistry* (H. W. Stache, ed.), Marcel Dekker, New York, 1996, p. 52.

11. B. Fell, in *Anionic Surfactants—Organic Chemistry* (H. W. Stache, ed.), Marcel Dekker, New York, 1996, p. 19.

12. G. Hons, in *Anionic Surfactants—Organic Chemistry* (H. W. Stache, ed.), Marcel Dekker, New York, 1996, p. 99.

13. G. Hons, in *Anionic Surfactants—Organic Chemistry* (H. W. Stache, ed.), Marcel Dekker, New York, 1996, p. 40.

14. R. D. Swisher, *Surfactant Biodegradation*, Marcel Dekker, New York, 1987.

15. G. Hons, in *Anionic Surfactants—Organic Chemistry* (H. W. Stache, ed.), Marcel Dekker, New York, 1996, p. 46.

16. Tenside Surf. Det., *26*(2):(1989).

17. P.-D. Hanser, Tenside Surf. Det. *26*:80 (1989).

18. B. Fell, Tenside Surf. Det. *28*:385 (1991).

19. B. Fell, in *Anionic Surfactants—Organic Chemistry* (H. W. Stache, ed.), Marcel Dekker, New York, 1996, p 2.

20. H. Hauthal, in *Anionic Surfactants—Organic Chemistry* (H. W. Stache, ed.), Marcel Dekker, New York, 1996, p. 143.

21. M. Trautmann and P. Jurges, Tenside Det. *21*:57 (1984).

22. P. K. G. Hodgson, N. J. Stewart, C. E. Grant, and A. M. Nicolls, *JAOCS, 67*:730 (1990).

23. N. M. van Os et al., in *Anionic Surfactants—Organic Chemistry* (H. W. Stache, ed.), Marcel Dekker, New York, 1996, p. 364.

24. D. W. Roberts and D. L. Williams, JAOCS *67*:1020 (1990).

25. D. C. Cullum (ed.), *Introduction to Surfactant Analysis*, Blackie, Glasgow, 1994.

26. P. Radici, L. Cavalli, and C. Maraschin, Tenside Surf. *31*:299 (1994).

27. M. J. Schwuger and H. Lewandowski, in *Anionic Surfactants—Organic Chemistry* (H. W. Stache, ed.), Marcel Dekker, New York, 1996, p. 461.

28. R. G. Bistline, *Anionic Surfactants—Organic Chemistry* (H. W. Stache, ed.), Marcel Dekker, New York, 1996, p. 634.

29. A. Domsch and B. Irrgang, in *Anionic Surfactants—Organic Chemistry* (H. W. Stache, ed.), Marcel Dekker, New York, 1996, p. 501.

30. X. Domingo, in *Anionic Surfactants—Organic Chemistry* (H. W. Stache, ed.), Marcel Dekker, New York, 1996, p. 224.

31. H. Meijer and J. K. Smid, in *Anionic Surfactants—Organic Chemistry* (H. W. Stache, ed.), Marcel Dekker, New York, 1996, p. 314.

32. G. W. Wasow, in *Anionic Surfactants—Organic Chemistry* (H. W. Stache, ed.), Marcel Dekker, New York, 1996, p. 551.

33. E. Kissa, *Fluorinated Surfactants*, Marcel Dekker, New York, 1994.

34. M. R. Porter, *Handbook of Surfactants*, Blackie, Glasgow, 1991.

35. D. M. Gabriel and V. J. Mulley, in *Anionic Surfactants—Chemical Analysis*, Marcel Dekker, New York, 1977.

36. J. H. Jones, JAOAC, 28:389 (1945).

37. H. Holness and W. R. Stone, Analyst 82:166 (1957).

38. D. C. Cullum and P. Platt, in *Recent Developments in the Analysis of Surfactants* (M. R. Porter, ed.), Elsevier, London, 1991.

39. D-23458-89 (Reapproved 1995). *Standard Test Method for the Separation of Active Ingredient from Surfactant and Syndet Compositions*, American Society for Testing and Materials, Philadelphia.

40. M. Stalman et al., Tenside Surf. Det. 32:86 (1995).

2

The Volumetric Analysis of Anionic Surfactants

JOHN CROSS   Department of Physical Sciences, University of Southern Queensland, Toowoomba, Queensland, Australia

I. INTRODUCTION

Despite the vast array of expensive computerized equipment that is seemingly obligatory in the modern analytical laboratory, the humble burette has in many instances managed to maintain a level of importance not too far below that which it enjoyed a few decades ago, albeit perhaps now as an integral part of an automated system. This is particularly true in the areas of quality control and other routine analyses. Indeed, Cullum and Platt [1] summed up the situation very succinctly with the following words:

> It is sometimes suggested that this type of analytical chemistry (titrimetry) will soon be completely superceded by the many very powerful high-tech instruments now available. But high-tech can be prohibitively expensive, especially for the smaller laboratory; and in any case, the great merit of high-tech methods is not that they can do what classical techniques can do, only better, but that they can do with ease many things that classical techniques can only do with great difficulty, or not at all.

In every type of titration, the analyte is determined by a rapid, stoichiometric reaction with a reagent possessing antagonistic properties. Obvious examples are the determination of a base by titration with an acid and of an oxidizing agent with a reducing agent. In the case of anionic surfactants the classic antagonists are the quaternary cationic surfactants, often called *quats* for short. The particular quats that have figured predominently in the literature are shown in Fig. 1. The two oppositely charged surfactant ions combine [Eq. (1)] to produce a hydrophobic salt that is insoluble in water (just how insoluble will be seen to be of

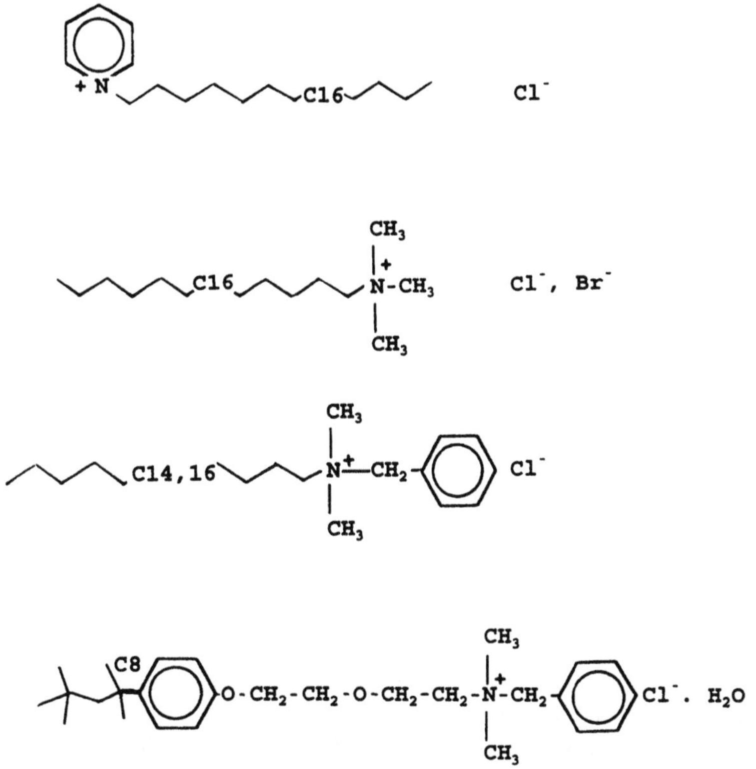

FIG. 1 Quaternary ammonium salts used for the titration of anionic surfactants: (a) cetylpyridinium chloride; (b) cetyltrimethylammonium chloride (bromide); (c) alkyldimethylbenzylammonium chloride, also known as benzalkonium chloride: alkyl usually C14 or C16; (d) *p-tert* octylphenoxyethoxyethyldimethylbenzylammonium chloride monohydrate, also known as benzethonium chloride and Hyamine 1622; (e) 1-(ethoxycarbonyl)pentadecyltrimethylammonium chloride, also known as Septonex; (f) 1,3-didecyl-2-methylimidazolinium chloride, also known as TEGOtrant.

vital importance) but soluble in an organic solvent of moderate polarity, typified by chloroform:

$$\text{Quat}^+_{(aq)} + \text{RSO}^-_{3(aq)} \rightarrow \text{QuatRSO}_{3(s)} \text{ or QuatRSO}_{3(org)} \tag{1}$$

The titrimetric procedures based upon this reaction have evolved along the following lines.

The first recorded quantitive use of the reaction was by Hartley and Runnicles in 1938 [2], who estimated cetylpyridinium chloride by titration with 0.001 M

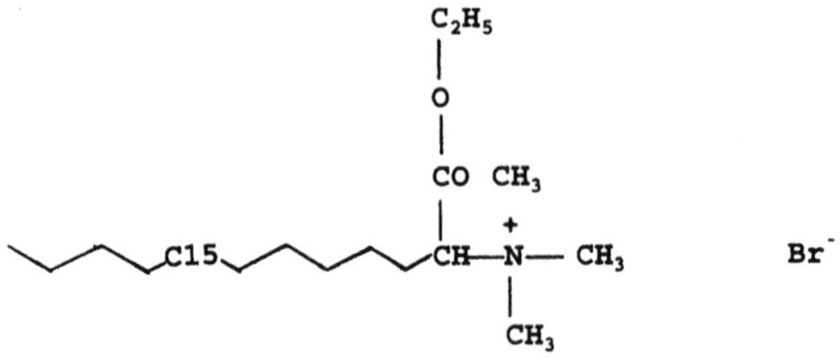

C₂H₅ group structure (e):

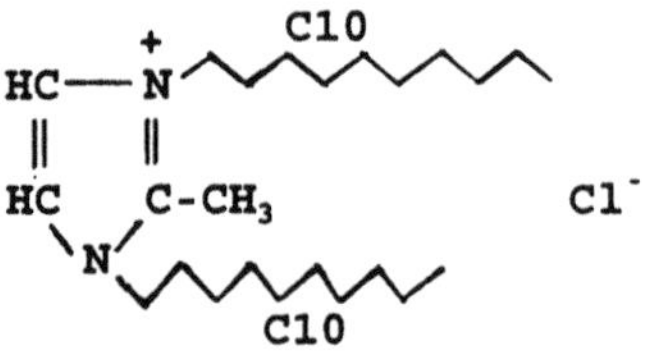

$$C_2H_5$$
$$|$$
$$O$$
$$|$$
$$CO \quad CH_3$$
$$|$$
$$CH-N^+-CH_3$$
$$|$$
$$CH_3$$

C15... chain with structure (e), Br⁻

structure (f):

HC—N⁺ with C10 chain, HC C-CH₃, N, C10 chain, Cl⁻

FIG. 1 Continued

solutions of sodium cetyl sulfate or sodium cetylsulfonate in a weakly alkaline solution until the purple (basic) form of the bromophenol blue indicator was liberated from the blue cetylpyridinium-indicator complex. This brief statement raises two points which will surprise any analyst unfamiliar with surfactants. First, surfactant titrations are performed with solutions very much more dilute (typically 0.001–0.004 M) than other more familiar reaction systems such as acid-base or redox. Indeed, attempts to utilize 0.1 or even 0.01 M solutions would be thwarted by problems with foam and limited solubility of the reagents themselves. Second, when surfactant cations react with the anionic forms of some acid-base indicators to form compounds (whose composition has been shown to conform to stoichiometric 1:1 or 2:1 salts [3–5]), there is a significant color change. This indicates a change in the energy requirements for the π-electron transitions in the indicator ion. In this chapter, compounds formed between surfactants and indicators will be referred to as complexes, without any further attempt to describe the nature of the bonding between the two entities.

II. DEVELOPMENT OF THE METHYLENE BLUE METHOD

In the 1940s, Epton [6] and Barr et al. [7] almost simultaneously reported the determination of anionic surfactants with eight or more carbon atoms in the alkyl chain by titration with a cationic surfactant using methylene blue or bromophenol blue as the indicator and using a second phase, chloroform. The transfer of the color from one phase to the other as the transition occurred between the free, water-soluble indicator ion and its chloroform-soluble complex with a surfactant ion of the opposite charge was taken to indicate the endpoint. Methylene blue (MB) will be familiar to many chemists as a redox indicator, but in surfactant work it is the cationic nature of the large organic ion that is important. The colored cation, MB^+, forms a chloroform-soluble complex with surfactant anions. This is the basis of the long-standing colorimetric method for determining traces of anionic surfactants in waters (discussed in a later chapter). Ideally, when the reaction in Eq. (1) is complete, the last trace of RSO^-_3 is removed from the complex, liberating the indicator cation, which returns to the aqueous phase:

$$MBRSO_{3(org)} + Quat^+_{(aq)} \rightarrow QuatRSO_{3(org)} + MB^+_{(aq)} \tag{2}$$

In the Epton procedure (which became the accepted standard for many years), however, the endpoint was taken as the point at which the intensity of color in both phases was judged to be equal. While this may be regarded as akin to stopping an acid-base titration using methyl red when the color reaches orange, it is pertinent to point out that the ratio of indicator to titrand in this (and other) surfactant titration is much greater than in the case of acid-base titrations. The indicator solution is prepared as follows: 0.3 g MB (mol mass 320) is dissolved in water and made up to 100 mL (8.9×10^{-3} M). Ten mL of this solution plus sulfuric acid are diluted to 1 L (8.9×10^{-5} M). Twenty-five mL of this diluted solution (2.2×10^{-3} mmol MB) are used for each titration. Bearing in mind that the maximum concentration of titrant used contains only 4×10^{-3} mmol per mL, it can be seen that approximately 0.6 mL of titrant would be required to effect the complete transfer of that amount of indicator from one phase to the other. If the titrant was less concentrated (1 and 2×10^{-3} M solutions have been used) and/or if the surfactant ion-pair reaction was not 100% complete before the indicator change became evident, then the volume of titrant required to achieve complete transfer of the indicator could be of the order of 2 mL. The choice of endpoint as the point at which the intensity of color in both cases is equal seems empirical. Although justified experimentally by Epton, the reliability of this justification has been challenged by Smith [8]. It is recognized that at this point a certain amount of titrand–indicator complex still remains in the organic layer and a correction factor can be applied [9a]. More usually, the sample size is selected such the titer falls between two relatively close limits, which overlap the titer expected during the standardization process. Should a sample titration fall outside these limits, then another titration is performed using a different aliquot of solution. Furthermore, it is recommended that the standardization be carried out against a pure specimen of the surfactant being determined [9a,b]: this effectively cancels out much of the titration error, but also enhances the degree of empiricism in the method. The optimum location of the endpoint was the subject of consideration by Smith [8,10].

TABLE 1 Apparent Purities of Cationic Surfactants by the Barr, Oliver, and Stubbings Procedure (Bromophenol Blue Indicator)

| Quaternary ammonium salt | Sodium 2-ethylhexylsulfate | Anionic surfactant | | |
		Sodium dodecylsulfate	Sodium tetradecane-2-sulfate	Sodium dioctylsulfosuccinate
Decyltrimethylammonium bromide	22.8	96.7	97.2	96.8
Dodecyltrimethylammonium bromide	22.1	99.5	99.5	99.6
Tetradecyltrimethylammonium bromide	23.2	101.2	100.2	101.2
Hexadecyltrimethylammonium bromide	23.8	100.2	99.8	100.4
Decylpyridinium bromide	26.0	97.0	96.4	96.4
Dodecylpyridinium chloride	25.6	100.1	99.7	100.8
Tetradecylpyridinium chloride	27.8	102.3	101.7	102.0
Hexadecylpyridinium chloride	26.6	102.1	101.5	101.8
(p-tert-Octylphenoxyethoxyethyl) dimethylbenzylammonium bromide		101.3	100.4	100.7
Didecyldimethylammonium bromide		99.7	99.2	100.9

Source: Abstracted from Refs. 12 and 13.

TABLE 2 Apparent Purities of Cationic Surfactants by the Epton Procedure (Methylene Blue Indicator)

Quaternary ammonium salt	Anionic surfactant			
	Sodium 2-ethylhexylsulfate	Sodium dodecylsulfate	Sodium tetradecane-2-sulfate	Sodium dioctylsulfosuccinate
Decyltrimethylammonium bromide	a	a	a	a
Dodecyltrimethylammonium bromide	93.7	90.4	86.9	94.6
Tetradecyltrimethylammonium bromide	98.4	94.8	91.6	100.5
Hexadecyltrimethylammonium bromide	101.2	97.4	93.8	105.2
Decylpyridinium bromide	a	a	a	a
Dodecylpyridinium chloride	100.1	96.2	93.1	100.1
Tetradecylpyridinium chloride	104.7	100.0	97.2	107.3
Hexadecylpyridinium chloride	108.7	106.4	102.6	114.3
(p-tert-Octylphenoxyethoxyethyl) dimethylbenzylammonium chloride		100.8	98.3	105.6
Didecyldimethylammonium bromide		97.8	95.7	105.7

[a] No endpoint was observed even when two equivalents of quaternary ammonium salt had been added.
Source: Abstracted from Refs. 12 and 13.

TABLE 3 Apparent Purities of Cationic Surfactants by Hartley and Runnicles' Procedure (Dichlorofluorescein Indicator)[a]

Quaternary ammonium salt	Anionic surfactant		
	Sodium 2-ethylhexylsulfate	Sodium dodecylsulfate	Sodium dodecane sulfonate
Decyltrimethylammonium bromide	b	b	b
Dodecyltrimethylammonium bromide	b	46.5	48.5
Tetradecyltrimethylammonium bromide	b	97.4	97.7
Hexadecyltrimethylammonium bromide	b	98.1	98.3
Decylpyridinium bromide	b	b	b
Dodecylpyridinium chloride	b	96.2	96.8
Tetradecylpyridinium chloride	b	99.7	99.1
Hexadecylpyridinium chloride	b	99.8	99.1

[a]Dichlorofluorescein was substituted for bromophenol blue.
[b]No endpoint was observed even when two equivalents of quaternary ammonium salt had been added.
Source: Abstracted from Ref. 13.

Other criticisms of the method include:

1. The endpoint is not easily recognized since the exact shade of the color in the two phases is slightly different. Careful standardization of lighting conditions is required, and even then there is considerable risk of operator-to-operator variation.
2. The exact technique of addition of titrant and degree of agitation used to ensure good mass transfer between the two phases can be critical and can be responsible for up to 3% variations in the result.
3. Deviations of up to 5% were obtained by different operators using the same solutions.
4. Use of equimolar solutions of various titrants, e.g., alkylpyridinium and alkyltrimethylammonium salts of various alkyl chain lengths, resulted in different endpoints for the same titrand [11].

The latter comment can be no better illustrated than by reference to the work of Lincoln and Chinnick [12, 13], summarized in Tables 1–3. Particular notice should be taken of these studies, since the purity of all surfactants involved was carefully controlled and checked. In one study, the bromophenol blue used by Hartley and Runnicles [2] was replaced by dichlorofluorescein. This is just one of a dozen or so alternate indicators reported upon by various researchers in their quest for an ideal indicator (listed in Ref. 14). Only one team has tackled the problem of low results from surfactants with short (C10 or less) alkyl chains by making radical changes to the organic solvent. Li and Rosen used a 40:60 mixture of chloroform:1-nitropropane to considerable effect [15].

With so much dissatisfaction with the standard methylene blue method being recorded, the reader might well question why so much space has been allotted to discussion of it and its appearance in Appendix C. The answer is simply that, in even comparatively recent publications, researchers have compared the results of their own developments with this method. This author will leave it to the reader to judge whether successful comparison is necessarily an endorsement.

III. THE MIXED-INDICATOR METHOD

With the principal task of finding a more appropriate indicator, members of the Comité Internationale des Dérivés Tensio Actifs, representing 10 of Europe's most prominent surfactant producers, combined under the chairmanship of Reid to make a thorough study of the total titration process. The two publications arising from this work are much longer and more detailed than usual: even today they could well serve as admirable models for those interested in method development and evaluation [16,17]. The outcome was a method based around the use of a mixed indicator originally proposed by Holness and Stone [18] and by Herring [19], which has become the current international standard.

The two indicators involved are the anionic diulfine blue VN (represented by HB in the following discussion) and the cationic dimidium bromide (DBr). When this mixture is added to an anionic surfactant, a pink complex is formed with the D^+ ion, which is preferentially soluble in the chloroform phase [Eq. (3)] in a manner analogous to the interaction with methylene blue in the previous method.

$$D^+_{(aq)} + RSO^-_{3(aq)} \rightarrow DRSO_{3(org)} \tag{3}$$

The HB remains in the aqueous layer and imparts to it a greenblue color. As the quat titrant is added, the main reaction [Eq. (1)] between the two antagonistic surfactants occurs, but as the supply of free anionic surfactant ions becomes exhausted, the pink complex yields its surfactant ion to the titrant and the free dimidium ion returns to the aqueous layer:

$$Quat^+_{(aq)} + DRSO_{3(org)} \rightarrow QuatRSO_{3(org)} + D^+_{(aq)} \tag{4}$$

At the same time, a slight excess of cationic titrant reacts with the disulfine blue anion to produce a blue chloroform-soluble complex:

$$Quat^+_{(aq)} + B^-_{(aq)} \rightarrow QuatB_{(org)} \tag{5}$$

The color in the chloroform layer at the endpoint, therefore, changes from pink to blue. Provided that the ratio of the two indicators is appropriate, a neutral gray color is evident at the endpoint. Although the amount of indicator(s) added is again much higher than would be encountered in a conventional acid-base titration, the endpoint is normally readily located to within a drop or two of titrant.

The team selected *p-tert*.octylphenoxyethoxyethyl)dimethylbenzylammonium chloride, more conveniently known as benzethonium chloride or Hyamine 1622 (Rohm and Haas), in preference to cetyltrimethylammonium bromide and alkyldimethylbenzylammonium (benzalkonium) chloride. The hydrophobe of Hyamine 1622 is prepared from a synthetic source, and there is no chance of the presence of homologous alkyl groups. It is marketed as its monohydrate: although there is some doubt about the exact water content as supplied, careful drying yields the anhydrous material, which many regard as a primary standard.

In addition to a vastly improved endpoint, other advantages over the methylene blue method included (1) an independence of the need to obtain a titer within two limits and (2) a lack of interference from hydrotropes (low molecular mass sulfonates, typified by toluene- and xylenesulfonates and included in some detergent formulations as viscosity modifiers). However, the endpoint, albeit much sharper than that of the methylene blue method, did not occur at the stoichimetric point. When carefully purified samples of sodium dodecyl sulfate, *n*-hexadecylsulfonate, hydroxy-hexadecylsulfonate, and alkylbenzenesulfonate were titrated, apparent recoveries were low at 96.8, 96.7, 96.3, and 96.0%, respectively. Hence the standard titrant should not be prepared by direct weighing of the Hyamine 1622, but rather by standardization against a suitable primary standard anionic surfactant. A degree of empiriscism, therefore, is still seen to exist.

Another point in which the mixed-indicator (MI) method is found lacking concerns the minimum alkyl chain length needed to show an acceptable response. In this respect it is found to be somewhat inferior to the methylene blue (MB) method [9].

Lowest Homologs to Be Determined (Recoveries > 95%)

	MB	MI
n-alkyl sulfates	C8	C10
n-alkanesulfonates	C10	C12
n-alkylbenzenesulfonates	C6	C8

Experimental details of both of the methods discussed above are given in the Appendixes to this chapter.

IV. ALTERNATE TWO-PHASE TITRATION METHODS

The two methods discussed above are the only ones to have achieved the status of international acceptance as standard methods for anionic surfactants. Both are carried out in strongly acidic conditions under which soaps and other carboxy-lates are converted to their unionized acid forms and do not contribute to the re-action. There are, however, numerous alternate two-phase methods to be found in the literature, all with their own particular degrees of merit. Many are de-signed to *include* soaps or to enhance/inhibit the contribution of minor compo-nents. Some of the earlier ones have been reviewed by Smith [20].

Cullum [21,22] preferred the use of an indicator carrying the same charge as the surfactant ion being titrated (e.g., bromophenol blue). He showed that when the first trace of quat-indicator complex appeared in the organic layer, a small amount of anionic surfactant still remained in the aqueous layer. The titration was continued until all of the color had been transferred and a blank titration us-ing the same amount of indicator was carried out to enable the gross titer to be compensated for the amount of titrant needed to transfer the indicator. Further-more, he observed a marked change in the mode of separation of the two layers at or just beyond the endpoint: the collapse of a network of bubbles gave way to the settling out of small globules. This change, attributed to a reversal of the in-terfacial charge, was probably of greater fundamental significance than any indi-cator endpoint and was sufficiently clear cut to enable an experienced analyst to perform the titration with no indicator of any type present. Using benzethonium chloride as the preferred titrant, Cullum applied the titration under both acidic and alkaline conditions, the latter providing a method for titration of soaps that is superior to the current ISO method.

Bares used bromocresol green as the indicator for the titration of soaps with Septonex (α-carboxyethyl pentadecyltrimethylammonium bromide) in alkaline solution [23]. The endpoint was taken as the point at which the intensity of the blue color in both phases was identical and an indicator blank was applied. Be-tween 1.04 and 1.05 mol of pure Septonex were required per mol of soap with a carbon chain length of C12 or greater: similar results were obtained with pure alkyl sulfates and alkylbenzenesulfonates. Once this departure from stoichiome-try had been allowed for by standardization of the titrant against a pure soap, the procedure was reliable, with a standard deviation of less than 1.5% relative. Mixtures of soaps and anionic surfactants were analyzed by performing the titra-tion under acidic conditions using the mixed indicator/Hyamine combination (anionic surfactants only) and the Septonex/bromocresol green process just de-scribed (soaps and anionic surfactants). Diederich adopted an almost identical approach, but used dichlorofluorescein instead of bromocresol green [24].

The ISO-adopted disulfine blue/dimidium bromide combination is not the only mixed-indicator system available, although no other has been subjected to so much rigorous interlaboratory testing. Wang and Panzardi [25] used a combination of the anionic indicator methyl orange (already used successfully by Uno et al. [26] and Cross [27]) and the cationic indicator Azure A (proposed by Steveninck and Riemersma [28] as an alternative to methylene blue). In the presence of anionic surfactants at pH 3, the Azure A is found as a pale blue complex in the organic layer. The blue color migrates to the aqueous layer as the last trace of indicator is displaced from the complex by the titrant (cetyldimethylbenzylammonium chloride). At the same time, a slight excess of titrant is indicated by the appearance of the yellow quat–methyl orange complex in the organic layer. This method has several features worthy of note.

1. The method was aimed at the estimation of linear alkylbenzenesulfonate in wastewaters and covers the concentration range of 10^{-5} to 10^{-4} M (i.e., approximately 3–30 mg/L), thereby overlapping the concentration range normally covered by colorimetric techniques.
2. At these low concentration levels, the rate of separation of the two phases after shaking (a significant contributor to the overall time required for a titration) is much higher than in the Epton procedure.
3. The volume of the benzalkonium chloride titrant (50 mg/L, c. 1.3×10^{-4} M) used to titrate the alkylbenzenesulfonate samples in the range of 80 to 500 mg was not exactly directly proportional to the amount of titrand present. A calibration graph, which obviously incorporates any indicator blank, was required, which helped to reduce the relative error to below 2%, a significant achievement at these very low levels of concentration.
4. Carbon tetrachloride could be used to replace chloroform as the organic phase. A separate calibration curve was required, and the relative error was somewhat higher.

V. DIPHASIC TITRATIONS WITHOUT PHASE TRANSFER

Although first the Epton procedure and currently the ISO-adopted mixed-indicator method depend upon the transfer of a colored species across a phase boundary to detect the endpoint, some workers have noted that observation of the disappearance of color from the organic layer is often made difficult by reflected color from the aqueous layer above, particularly if the sample has its own inherent color [29–32]. They found it preferable to have the color change take place in the organic layer only. Indicators suitable for this approach require an ionic form, which will combine with the surfactant of opposite charge to form a chloroform-soluble complex, and a second form, which is devoid of strongly hydrophilic groups and therefore inherently nonpolar and also preferentially soluble in the chloroform. Such an indicator is dimethyl yellow, also known as methyl yellow (4-*N,N*-dimethylaminobenzene). The yellow, basic azoid form (I)

$(CH_3)_2N$—⟨benzene⟩—N=N—⟨benzene⟩

(I)

$(CH_3)_2\overset{+}{N}$=⟨ring⟩=N—NH—⟨benzene⟩

(II)

and the pink complex formed between the anionic surfactant and the acid, quinonoid form (II) are both chloroform soluble, so that the endpoint is marked by a color change in the chloroform layer only. Jansson et al. [33] carried out an in-depth study to determine the equilibrium constants for the reactions between various organic cations (including some low molecular mass quats), dodecyl sulfate, and (di)methyl yellow. An optimum pH was calculated for each cation: nonyltrimethylammonium, the closest of the cations tested to any of the popular titrants, required a pH of approximately 1.1. Eppert and Liebscher [29,30] used 0.01 M solutions of Septonex for the estimation of alkanesulfonates. Results compared admirably with those obtained by the Hyamine 1622–MI method, but the endpoint with dimethyl yellow was deemed to be more easily and reliably discerned. They also (1) reported about the interference caused by the presence of appreciable concentrations of bromide and iodide ions (positive), chloride (neutral), and sulfate (negative) [29] and (2) demonstrated that, whereas Septonex/dimethyl yellow under strongly acidic conditions would determine only monosulfonate, Hyamine 1622–phenolphthalein under alkaline conditions would determine both mono- and disulfonate. The content of the latter could obviously be estimated from the difference between the two results [30].

Markó-Monostory and Börzsönyi compared the use of the Septonex–methyl yellow method with the Hyamine 1622–MI method for the analysis of high molecular mass (oil-soluble) sulfonates with alkyl chains of more than C20 and equivalent weights in the region of 550–850 [32]. The endpoint using dimethyl yellow was found to be more acceptable. They titrated Septonex against pure samples of potassium dodecylbenzenesulfonate and tetracosylbenzenesulfonate and found a 3–4% difference in the stoichiometry of the two reactions, in line with the observations of Reid et al. [16] that the lower molecular mass surfactants react only to the extent of 96–97%. This amounts to a reiteration of the principle of using a standard with a structure as close as possible to that of the analyte material. Chloroform was replaced as the organic phase by a 1:1 mixture of chloroform and 1,2-dibromoethane.

Another indicator in the same category as dimethyl yellow is the ethyl ester of teterabromophenolphthalein. The unionized (acidic) form is yellow, and the complex between the basic form and the quat is blue: the endpoint is characterized by a green tint. Tsubouchi et al. [31] used spectroscopic methods to determine the equilibrium constants for the reaction between lauryl sulfate (LS⁻) and Zephiramine (tetradecyldimethylbenzylammonium) cation (Z⁺):

$$Z^+_{(aq)} + LS^-_{(aq)} \rightleftharpoons ZLS_{(org)} \tag{6}$$

and for the formation of the titrant-indicator (HIn) complex:

$$Z^+_{(aq)} + HIn_{(org)} \rightleftharpoons ZIn_{(org)} + H^+_{(aq)} \tag{7}$$

The latter equilibrium is very much pH dependent. They were able to calculate the optimum pH for quantitative behavior to lie in the region 5.0–7.0, and consequently used a pH 6.0 buffer for the titration to provide a sharp endpoint and a higher level of reproducibility than the Epton procedure. They also tested a number of alternate organic solvents (nitrobenzene, butanone, butyl acetate, isopentanol, 1:2-dichloroethane, carbon tetrachloride, toluene, benzene, chlorobenzene, and n-hexane), but concluded that none of them was superior to chloroform.

Throughout this chapter thus far, the various titrants discussed have all been centred around a quaternized nitrogen atom. At least one alternate antagonist has been proposed. Tsubouchi and Mallory used tetraphenylphosphonium chloride to determine anionic surfactants in sewage at the 10^{-5} M level of concentration [34]. Five mL of sample, 5 mL of a citrate buffer (pH 4.5), one drop each of 0.03% tetrabromophenolphthalein ethyl ester (TBPPE) and 0.005% neutral red, and 1–2 mL 1:2-dichloroethane were placed in a 25-mL glass-stoppered measuring cylinder and titrated with the phosphonium quat of a suitable concentration (2×10^{-5} to 5×10^{-4} M) until the color in the organic phase changed from red to faint yellow. One drop of excess was sufficient to result in a green color due to the presence of some of the quat-TBPPE complex.

A variety of water-immiscible solvents were tested (benzene, carbon tetrachloride, chloroform, 1:2-dichloroethane, nitrobenzene, nitroethane, and methylpentanone), of which 1:2-dichloroethane was adjudged to be the most suitable. Note that the quantity of organic solvent is very much less than used in most other diphasic methods. One of the problems associated with the use of very dilute solutions of nitrogeneous quats is that significant losses from solution can be incurred by adsorption onto the anionic sites of the surfaces of glass vessels. Indeed, the fact that a clean sheet of soda glass immersed into a solution of a cationic surfactant acquired a greasy monolayer has been used as a qualitative test for the presence of such compounds [35]. The authors report that no such tendency was observed with the tetraphenylphosphonium salt.

VI. ANIONIC AND CATIONIC SURFACTANTS IN ADMIXTURE

Institutions such as hospitals use large quantities of both anionic surfactants, e.g., alkylbenzenesulfonates, in detergents and quats, e.g., benzethonium and benzalkonium chlorides, as germicides and disinfectants. Consequently their effluents contain a rich mixture of the two types, either of which could be in excess. The insoluble salt formed by the interaction of these two types would be present as a fine suspension or perhaps solubilized by the surfactant in excess. Tsubouchi et al. [36,37] developed procedures by which both components could be estimated. Two separate titrations are required, but the approach varies according to which type of surfactant is in excess: a rapid qualitative test involving the mixed indicator and chloroform will ascertain this (chloroform pink or blue for anionic or cationic, respectively; see Chapter 1). The standard titrant for quats is sodium tetraphenylborate (STPB), which binds more strongly to the quat ion than do the surfactant anions. Hence:

$$QuatLS_{(org)} + TPB^-_{(aq)} \rightarrow QuatTPB_{(org)} + LS^-_{(aq)} \tag{8}$$

A. Anionic Surfactant in Excess

(1) The excess anionic surfactant is estimated by titration of a suitable aliquot, buffered to pH 9, with a standard solution of Hyamine 1622 (1×10^{-4} or 4×10^{-5} M) using the hydrophobic indicator Victoria Blue B as the indicator and 1:2-dichloroethane as the organic phase. The endpoint is marked by the color in the organic layer changing from blue (indicator-surfactant anion complex) to red (free indicator).

(2) The cationic surfactant present, which is equimolar to the bonded anionic surfactant, is determined by titration with STPB to the same indicator, but in the presence of 4 M sodium hydroxide. The TPB ion displaces the surfactant anion from the quat-anion salt, but under the strongly alkaline conditions the indicator does not combine with the liberated surfactant ion and remains in the free (red) form. It does, however, combine with the first trace of excess TPB to yield the blue indicator complex.

The quat content can be calculated from the result of (1) alone: the total anionic surfactant content is given by the sum of the results from (1) and (2).

B. Cationic Surfactant in Excess

For this type of sample, the indicator was changed to TBPP ethyl ester without comment from the authors [37].

(1) To estimate the total (free and combined) quat present a suitable aliquot (e.g., 5 mL of a 10^{-5}–10^{-4} M sample) is mixed with 5 mL of pH 12.5 buffer, 1 drop of 0.03% indicator, and 3 mL of dichloroethane, and titrated in a stoppered vessel with 5×10^{-5} M STPB with intermittent shaking until the color of the organic phase changes from blue to colorless.

(2) To estimate the free (excess) quat, a similar aliquot is mixed with 5 mL of pH 6.0 buffer and indicator and organic solvent as above. It is then titrated with 5

$\times 10^{-5}$ M sodium lauryl sulfate until the indicator changes from blue to pale yellow.

The quantity of anionic surfactant present is given by the difference between the titers for (1) and (2). The methods are inapplicable if the quats are alkylpyridinium, alkylquinolinium, and related types since this class of surfactant decomposes under the strong alkaline conditions [14].

VII. PHOTOMETRIC DETERMINATION OF THE ENDPOINT

One of the major objections to the classic two-phase titrations is the subjective nature of the visualization of the endpoint. Various photometric adaptions to overcome the problem have been proposed, based upon changes in either absorbance or turbidity.

The work of Jansson et al. [33] has already been cited. They used an EEL titrator during the determination of various quaternary ammonium salts with sodium lauryl sulfate, selecting a number 605 filter with a nominal transmission at 545 nm (which closely corresponds to the absorbance maximum of the pink complex formed between the indicator, methyl yellow, and the lauryl sulfate ion). The detection of the first perceptible red color against the yellow background of the free indicator was signaled by a rise in the absorbance to 0.12. The precision of the endpoint location by this technique was about five times better than by visual detection.

Cantwell and Mohammed developed an apparatus that permitted continuous monitoring of either of the two phases in a vigorously agitated mixture by passing it through a spectrophotometer flow cell [38] and used it to titrate several cationic drugs by following the concentration of the titrant (picrate) in the aqueous phase [39]. They extended their work to include the titration of anionic and cationic surfactants using methylene blue and picrate, respectively, as the titrants [40]. For these analyses, it was the chloroform layer that was monitored, using wavelengths of 658 and 400 nm, respectively. The concentration of each of these species in the organic layer increased linearly as the titration proceeded due to the build-up of the titrant-surfactant salt. At and beyond the endpoint, the absorbance remained steady since additional titrant remained in the aqueous layer as the free ion. The endpoint was located by the intersection of the two straight lines of the graph of absorbance vs. titrant volume. As for the previous study, this technique removed the subjectivity of visual interpretation of the color changes at the endpoint but unfortunately required the use of large volumes of chloroform (70 mL) for each estimation.

The two techniques mentioned above still require the use of chloroform and have merely substituted an instrumental method for detection of the endpoint without reducing the complexity of the process or the time taken for an analysis. The salt formed between the antagonistic surfactant cation and anion is quite hydrophobic and, in the absence of an organic solvent in which to dissolve, will form a fine suspension which is stabilized during the early part of the titration by the excess analyte present. In other words, it creates a turbidity. Following some initial studies by Seguran [41], Hendry and Hockings [42] investigated the variation in turbidity in the single (aqueous) phase during the course of titrating several anionic surfactants with Hyamine 1622 (4×10^{-3} M). In the absence of high concentrations of salts or nonionic surfactants, the turbidity increased slowly during the initial stages of the titration, and then increasingly rapidly as the titrand concentration became reduced to a low level, before finally leveling off to a steady value. The endpoint was taken as the point of inflection between the two latter sections.

An interesting point about this titration results from the limitation of the spectrophotometer used to measure high degrees of turbidity. To keep the absorbance readings below the maximum of 1.99 permitted by their particular instrument, the working range of the method was restricted to 3×10^{-6} to 4×10^{-5} M. This corresponds roughly to the 1 to 10 ppm range, where colorimetric procedures would normally be expected to predominate.

The presence of nonionic surfactants caused complications: the magnitude of the sharp rise in turbidity near the endpoint was reduced as the concentration of added alkylphenol ethoxylate increased, disappearing altogether when the latter was present in a 10-fold molar excess. This was tentatively assigned to solubilization of the precipitate by the nonionic surfactant, but is in direct contrast to the observations of Volkmann [43]. Using 4×10^{-3} M solutions and monitoring the progress of the titration with the spectrophotometer in the transmittance mode, the latter author found that the addition of five drops of the nonionic surfactant Tween 80 caused the transmittance to rise steadily during the course of the titration and then to increase sharply at the endpoint.

Hendry and Hockings claimed that the repeatability of their technique was equal to that obtained by an experienced operator using the ISO method [44], but that the result could be obtained in less than half the time. The discrepency between the two results was, however, ±5%. It is worth noting in passing that in adapting the technique to the determination of five different cationic surfactants, the appropriate concentration range was found to be 2×10^{-6} to 8×10^{-5} M with a 4×10^{-3} M solution of sodium lauryl sulfate [45]: this extremely low range was once again dictated by the limitations of the spectrophotometer employed. As before, the repeatability was comparable to that of the standard two-phase method, but there were serious discrepencies between the results of the two methods. The results for alkylbenzylammonium chloride, benzalkonium chloride, and laurylpyridinium chloride deviated by –6, –4, and –2%, respectively. Of the two other analyte materials, *N*-hydroxyethyloleylimidazoline gave no response to the standard method and diethylheptadecylimidazolinium ethyl sulfate gave a result 22% higher. Both of these compounds were shown by HPTLC to contain substantial amounts of nonionic materials, which the authors suggest would not have been enough to upset the turbidimetric technique but could have invalidated the results of the standard method. Clearly, further study is required to tie up the loose ends before this very sensitive technique can be recommended without reserve [46].

VIII. POTENTIOMETRIC TITRATIONS

A. Introduction

At this stage it is pertinent to summarize the main points of the chapter thus far. The two-phase titration, albeit seemingly a relatively simple process and requiring only a modest assembly of apparatus, has a number of inherent problems, some of which have already been listed.

1. The endpoint location is frequently difficult to judge, i.e., is a subjective phenomenon, and may require standardized lighting conditions. Colored samples can create further problems.
2. The exact technique of addition of titrant and agitation is critical, since the titration depends upon reaction systems reaching equilibrium across phase boundaries. Operator fatigue can contribute to errors.
3. The visual endpoint is not necessarily coincidental with the stoichiometric point, and the degree of divergence between the two can vary with the nature and alkyl chain lengths of both the titrant and titrand.
4. Deviations between different operators using the same standard solutions as high as 5% have been reported (for the Epton method).
5. The methods are demanding on operator time and are not ideally suited to automation. Both of these are factors of increasing importance in modern times.
6. The increased awareness of the hazard to health and environment of chlorinated solvents has diminished the appeal of this style of titration, although automatic dispensing and removal systems are available.
7. The presence of nonionic surfactants in admixture with the anionic surfactants makes endpoint location more difficult.

Potentiometric titration has been shown to overcome most of these problems and to possess a few additional advantages of its own. Electrodes designed to respond to surfactants have been known since about 1970. Details of design, mechanism, etc. are the subject of a separate section, and only a few brief points will be made here.

It is not possible to design an electrode specific to one surfactant only. Surfactant-sensitive electrodes respond to the entire class of surfactants and perhaps to their antagonists, and also to various completely unrelated ions. The reverse is also true: in fact, some of the commercially produced electrodes that have been used successfully for surfactant titrations were designed for response to ions such as nitrate, fluoroborate, and calcium.

For use as an indicator electrode during titration, it matters not whether an electrode shows a Nernstian response to the analyte material or to the titrant. The only ions undergoing radical changes in concentration in the vicinity of the end-point are those of the titrand and titrant, and the electrode need only make a significant response to either or both of those to be useful.

Consider for a moment an entirely different potentiometric titration, which most graduate chemists would have met in their college days—the titration of halide ions with silver nitrate. The largest potential jump (Fig. 2) comes from the Ag^+ vs. I^- system, and this arises because the solubility product for AgI is very much less than that for AgBr or AgCl. A mixture of all three halides can be resolved because the precipitation of the iodide is essentially complete before that of the bromide salt commences: likewise no significant precipitation of AgCl occurs until all of the bromide has been removed. A first-derivative plot enables the

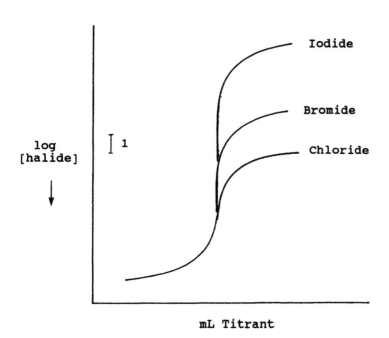

FIG. 2 The well-known titration curves for silver nitrate against halides. The iodide titration shows the largest potential jump due to the fact that silver iodide has the lowest solubility product of the three silver halides. The potentiometric titration of surfactants has much in common with this sequence.

three components to be quantified. In addition, if the endpoint were to be visualized by use of an indicator that formed a colored complex with the silver ion, the greater the potential jump at the endpoint, the more flexibility there will be in finding a suitable indicator for the process. Assuming the titrant-indicator complex was a 1:1 compound, a typical indicator would require an operating range in the order of 100 mv (corresponding to an acid-base indicator operating over a range of 1.8 pH units). It is well accepted that use of potentiometric titration curves and their derivatives can reliably locate an endpoint with much lower potential breaks than this.

B. Potentiometric Titration of Anionic Surfactants

Returning to surfactants, among the first to report on potentiometric titrations were Ciocan and Anghel [47–50], who used a sensor based upon the complex formed between ferroin and either dodecylsulfate or dodecylbenzenesulfonate. (Ferroin has been proposed by Taylor and Fryer [51] as a preferred cationic reagent to methylene blue for the colorimetric analysis of trace quantities of anionic surfactants in wastewater.) From these comparatively early studies, a number of points clearly emerge:

1. Two anionic surfactants, of substantially different character, present in the same mixture can be sequentially titrated in a manner analogous to two anions whose silver salts have substantially different solubility products (Fig. 3).
2. Using any one particular quat as the titrant, the magnitude of the potential jump varies enormously with the length of the alkyl chain of the titrand (Fig. 4), i.e., better definition of the endpoint coincides with increases in the hydrophobic nature of the anionic surfactant. A similar trend was observed using one particular anionic surfactant, sodium dodecyl sulfate and a variety of quats as titrants (Fig. 5). This latter trend is also exemplified by the results of Selig [52], using a commercial fluoroborate-sensitive electrode to follow the titration of sodium dodecyl sulfate with cetylpyridinium chloride, cetyltrimethylammonium chloride, and Hyamine 1622 (Fig. 6). These results clearly cast doubt on the adoption of Hyamine 1622 as the most appropriate choice for the universal titrant.
3. The magnitude of the potential break is severely depressed by a high ionic strength [49].
4. The magnitude of the potential break is severely depressed by the presence of water-miscible solvents such as ethanol and propanol (Fig. 7). These substances are frequently added to keep relatively hydrophobic analyte materials, such as long-chain soaps, in solution. Unfortunately, their presence also increases the solubility of the quat-surfactant anion salt. A similar decrease in the magnitude of the break and its midpoint potential occurs if the solution is heated to enhance solubility of the reactants. [49].

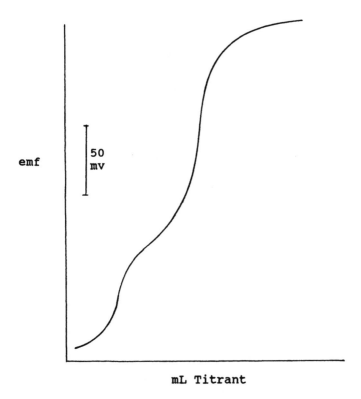

FIG. 3   The resolution of mixtures of suitably different surfactants was realized more than 20 years ago. Sodium di-2-ethylhexylsulfosuccinate (first endpoint) and sodium octyl sulfate are shown here titrated with 0.004 M cetyldimethylbenzylammonium chloride. (Adapted from Ref. 47.)

The versatility of the potentiometric method in comparison to visual methods is aptly indicated by the accompanying figures. The search for the most appropriate indicator for each titrand/titrant combination is made obselete.

In addition to overcoming the disadvantages listed at the beginning of this section, it is often suggested that the technique permits the use of much more concentrated titrant solutions, since a piston burette has no air/water interface to be upset with foam. However, not all candidates for the role of titrand are suitable for use at higher concentrations. Selig [53] reported that the titrant of choice with regard to both cost and sensitivity, cetyltrimethylammonium bromide, could not be used at concentrations of greater than 10^{-2} M as such solutions would slowly form a precipitate with time. Cetylpyridinium chloride,

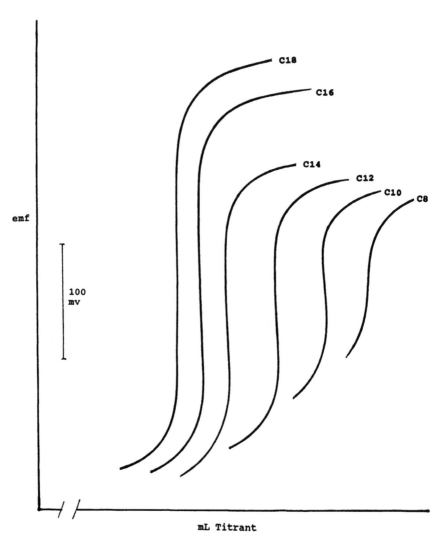

FIG. 4 Potentiometric titration of sodium alkyl sulfates with cetyltrimethylammonium chloride (0.004 M). Curves are displaced for clarity. (Adapted from Ref. 48.)

cetyltrimethylammonium chloride, tetradecylbenzylammonium chloride, and Hyamine 1622 could all be used at concentrations above 0.1 M, but the latter two were not favored due to the relatively small inflections that they produced. Cetylpyridinium bromide, cetyltrimethylammonium bromide, and cetyl-dimethylbenzylammonium chloride did not give stable 0.05 M solutions. Poten-

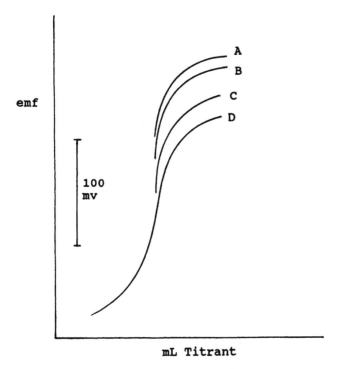

FIG. 5 Potentiometric titration curves of sodium dodecyl sulfate against 0.004 M solutions of (A) cetyldimethylbenzylammonium chloride, (B) cetyltrimethylammonium bromide, (C) C12-C14-alkyldimethylbenzylammonium chloride, and (D) cetylpyridinium chloride. Hyamine 1622 fits between curves C and D. (Adapted from Ref. 50.)

tial highly suitable titrants by account of their high hydrophobic nature, tetraoctyl- and tetradodecylammonium bromides, did not even provide stable 0.01 M solutions.

A number of different electrodes were designed and tested during the 1980s. These have been reviewed by Cullum and Platt [54] and Vytras [55]. In recent times, Buschmann, Schultz, and various colleagues have reported substantial advancements in electrode design, titrant selection, and the concept of the overall reaction [56–66]. In an authoritative summary [64], Buschmann included the following critical factors:

1. The surfactant cation-anion salt is not completely insoluble. The residual solubility is quantified by the salt's solubility product, which is a function of the hydrophobic nature of both component ions.
2. Especially in cases where the solubility product is relatively high, a signifi-

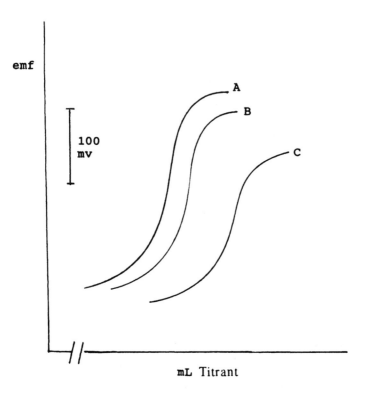

emf

100
mv

mL Titrant

FIG. 6 Potentiometric titration curves of sodium dodecyl sulfate with 0.01 M solutions of (A) cetylpyridinium chloride, (B) cetyltrimethylammonium chloride, and (C) Hyamine 1622. Curves displaced for clarity. (Adapted from Ref. 52.)

cant excess of titrant is required for quantitative precipitation if the concentrations are low.

3. Incorporation of the surfactant into micelles reduces its activity, to which the electrode responds, and hence the height and steepness of the potential break are also reduced.
4. The presence of nonionic surfactants, salts, etc. influence the formation of (mixed) micelles.
5. The shape of the potentiometric curve is determined not only by the solubility product of the surfactant ion pair, but also by the response of the electrode to both titrand and titrant ions. Although the superior titration afforded by the more hydrophobic surfactants can be attributed to the lower solubility products of the reaction products, the enhanced selectivity of such reactants towards the membrane of the ion-selective electrode (ISE) is also an important factor.

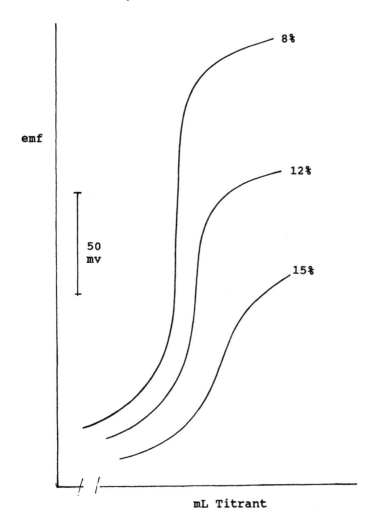

FIG. 7 Depression of the potential break by addition of alcohols. The effect of propanol on the titration of sodium stearate with 0.004 M cetyltrimethylammonium bromide at pH 12. Curves displaced for clarity. (Adapted from Ref. 49.)

6. The type of mediator used for the manufacture of the membrane determines the behavior of the ISE to a greater extent than does the ionophor.
7. Emf/concentration relationships can be "super-Nernstian."
8. There is no well-defined standard potential for any particular surfactant-ISE combination. The apparent value of E depends upon the pretreatment or immediate past history of the ISE.

C. Selection of the Ion-Selective Electrode

Buschmann and Schultz [56] constructed a membrane comprising 64% 2-nitro-phenyl octyl ether, 35% PVC, and 1% ionophore (sodium tetrakis-3,5-bis (triflu-oromethyl)phenyl borate, a derivative of sodium tetraphenylborate which is a well known titrant for cationic surfactants). The internal solution was 0.01 M with respect to both sodium tetraphenylborate and sodium chloride: a silver wire completed the internal reference system. This electrode performed very well for the reaction between Hyamine 1622 and a wide variety of anionic surfactants over a wide range of pH values (1–11).

The performance of the above ISE was compared with that of five commer-cially available electrodes (two marketed as selectively responsive to surfac-tants, two to fluoroborate and one to nitrate) for the titration of 10 mL of 0.004 M Hyamine 1622 (plus 90 mL of water, 10 mL of pH 10 buffer, and 5mL of methanol) with 0.004 M sodium lauryl sulfate. The ISEs were judged over a large number of titrations for consistency of titration volume, emf at endpoint, appearance of spurious secondary endpoints, the need for electrode recondi-tioning, and the height of the potential break. Emerging as a clear-cut winner was the ASTEC surfactant-sensitive electrode model TSE 01/91: this is equiv-alent to the Metrohm High Sense Tensid Electrode and will be referred to as the High Sense Electrode in this chapter. In the course of 100 titrations, the curves remained smooth, the 67 mv potential break* remained constant and the slight variations in titer were normally distributed about the mean value. The standard deviation was just less than 0.5% relative compared to 1.35–2.36% for the other electrodes. There was not one secondary endpoint, whereas the other electrodes demonstrated such problems in 17–40% of their runs. Extend-ing the test basis for the High Sense Electrode, more than 1000 titrations were carried out with a variety of surfactants and conditions without the need for re-conditioning.

D. Selection of the Titrant

Having settled upon the most effective ISE, Buschmann et al. [58] turned their attention to optimizing the titrant. They trialed the "standard" Hyamine 1622 against a selection of cationic surfactants in which phenoxy, ethoxy, or any other groups that added to the overall hydrophilic character of the molecule were ab-sent. The titrants selected were:

---

*The potential break is defined as the difference in potential readings between additions of 90 and 110% of the equivalent titration volume.

Hyamine 1622
Hexadecylpyridinium chloride (CPC)
Hexadecyltrimethylammonium bromide (CETAB)
Trioctylmethylammonium chloride (TOMACl)
1:3 Didecyl-2-methylimidazolinium chloride (DDMICl).

Cationic surfactants of an even less hydrophilic nature could not be used owing to a lack of solubility in water. Even TOMACl was necessarily dissolved in a propan-2-ol–water mixture containing 1% of the nonionic ethoxylated surfactant Triton X-100.

Titrands were 0.001 M of the following, buffered at pH 3:

Dodecyl sulfate
Dodecylbenzenesulfonate
Alkylether sulfate
sec-Alkanesulfonate
α-Olefinsulfonate
Estersulfonate
Di-octylsulfosuccinate

The results for the titration of sodium dodecyl sulfate are shown in Fig. 8. DDMICl proved to be the preferred titrant, producing a potential break of 185 mv and a standard deviation for the titer of only 0.19% relative (cf 60 mv and 0.76% for Hyamine 1622). Similar trends were evident for the other anionic surfactants tested. As a comment on the side, in the search for a quat titrant with a low hydrophilic character, one of the major problems is the poor solubility of such compounds in water. However, tetra(n-octyl)ammonium fluoride and hydroxide are both soluble in water to the extent of 0.05 M, whereas the bromide and even the nitrate salts are quite insoluble [78]. If the cost of such a substance was competitive with that of the more common quats, then such a reagent would surely be worth trialing.

Schultz and Gerhards [59,63] proceeded to further investigate additional advantages of the new combination of the High Sense Electrode and DDMICl (now marketed by Metrohm Ltd under the trade name of TEGOtrant). Whereas Hyamine 1622 is insufficiently responsive to octyl and decyl sulfates for analytical purposes, DDMICl produces quantifiable curves (Fig. 9). For an alkyl sulfate of mixed chain lengths, the longer-chain, more hydrophobic species are titrated first. With DDMICl it was shown possible to select titration conditions such that at least partial resolution of the different chain lengths may be achieved. The first derivative curve of the titration of a product derived from coconut oil showed three distinct endpoints, whereas with Hyamine 1622 a single, overall value was obtained. The authors, however, did not correlate the titers at these endpoints with the alkyl chain distribution of the surfactant.

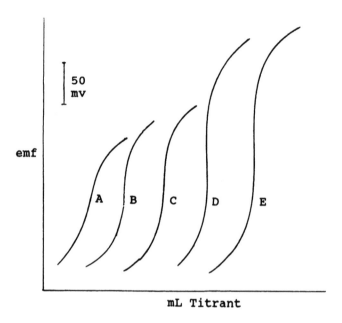

FIG. 8 Potentiometric curves in the vicinity of the endpoint for the titration of sodium dodecyl sulfate with 0.01 M solutions of (A) Hyamine 1622, (B) cetylpyridinium chloride, (C) cetyl trimethylammonium bromide, (D) trioctylmethylammonium chloride, and (E) 1:3-didecyl-2-methylimidazolinium chloride. Curves displaced for clarity. (Adapted from Ref. 58.)

Commercial alkyl ether sulfates contain a complex mixture of different alkyl and ethoxy chain lengths. A product nominally described as sodium lauryl 3EO sulfate, for example, can be shown by mass spectroscopy or HPLC to possess EO contents between 0 and 20 EO units. The higher the EO content, the more hydrophilic the surfactant and consequently the poorer the definition of the endpoint. Again, DDMIC1 provided superior curves to those of Hyamine 1622 for such substances (Fig. 10).

Alegret et al. [67] recently reported on the merits of the potentiometric titration technique in comparison to the diphasic titration. They used an electrode based upon a membrane comprising PVC (25%), o-nitrophenyl octyl ether (62.5%), and a quat dodecylbenzenesulfonate salt (12.5%) as an ionophor. In common with most other surfactant-sensitive electrodes, the response was non-Nernstian to anionic surfactants (but Nernstian to cationic surfactants in the 10^{-5}–10^{-2} M region). The potential break for sodium dodecyl sulfate (2.5×10^{-3} M) titrated with Hyamine 1622 (4×10^{-3} M) was 145 mv, which the authors point out is far superior to that produced by any commercially available ISE.

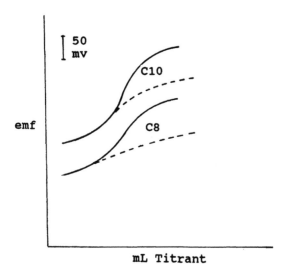

FIG. 9 Even the relatively short-chain octyl and decyl sulfates, which are almost impossible to titrate with Hyamine 1622 (dashed lines), produce resolvable curves with 1:3-didecyl-2-methylimidazolinium chloride (full lines). Curves displaced for clarity. (Adapted from Ref. 63.)

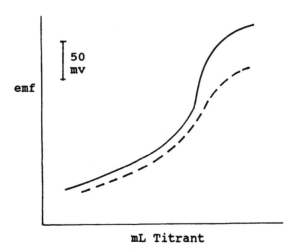

FIG. 10 The hydrophilic nature of oxyethylene groups in an anionic surfactant makes the titration more difficult. An alkyl ether sulfate with an average oxyethylene content of 3.0 moles is shown to be more readily titrated with 1:3-didecyl-2-methylimidazolinium chloride (full line) than with Hyamine 1622. Curves displaced for clarity. (Adapted from Ref. 63.)

When the starting concentration was reduced to 1×10^{-5} M (corresponding to approximately 3 mg L, the break reduced to a (still resolvable) 30 mv. A variety of the common types of anionic surfactants, including alkyl ethoxy sulfates, were successfully titrated with an overall standard deviation of 0.17% compared to 0.49% for the same suite of samples titrated by the mixed indicator method.

As with other PVC membrane electrodes, a long functional lifetime was observed. No changes were observed in the shape of the titration curves on a day-to-day basis, and during 2 years of discontinuous use no reconditioning was needed.

In 1991, Cullum and Platt [1] wrote, "over the next few years any major advance in the titrimetry of surfactants seems more likely to lie in the field of potentiometry and potentiometric titration." They were not wrong!

E.   Optimizing the ISE Performance

The following remarks apply principally to the High Sense Electrode, but are pertinent to most electrodes involving a PVC membrane. The presence of organic solvents in the (mainly) aqueous phase containing the titrand would seem to be undesirable because it would raise the solubility of the reaction product and hence reduce the sharpness of the potential break at the endpoint. However, the presence of methanol at a concentration of 5% (up to 10%) seems almost mandatory [59,68]. Methanol, which is used to wipe the membrane surface when necessary, enhances the overall titration by:

1. Reducing interfacial tension between the membrane and the solution.
2. Keeping the electrodes clean of reaction product.
3. Reducing foaming during dilution.
4. Reducing the amount of air bubbles in the system in general and adhering to the electrode in particular (a source of noise to the potentiometer).
5. Suppressing the formation of micelles.

Lower alcohols to the extent of 15, 10, and 5% for methanol, ethanol, and 2-propanol can be tolerated without noticeable effect upon the quality of the titration curves. Higher concentrations, e.g., 30% methanol and 20% ethanol, shorten the electrode life. Higher alcohols (butanol and above) are powerful membrane poisons and should be avoided, as should the presence of any oily substances in general [60]. After titrating any mixture containing significant amounts of lower alcohols or any amounts of polyols, etc., it is recommended that the electrode be soaked in water for a few minutes before reuse. Acetone, chloroform, tetrahydrofuran, etc. will become imbibed into the PVC and leach out the mediator and ionophore: such substances should obviously be avoided.

It is not uncommon to find a recommendation to add a nonionic surfactant such as an alkylphenol ethoxylate to the solution in order to solubilize the rather sticky precipitate and prevent its deposition on the electrode surfaces. However, the quality of the titration curve deteriorates upon addition of such compounds and, provided that methanol is present, Schultz recommends that no such addition be made [68].

Thorough mixing is an obvious necessity in any type of titration and in this case must be performed mechanically. To avoid the introduction of air bubbles while maintaining rapid mixing, it is recommended that a propeller stirrer be used in preference to the more common magnetic bar [68].

Prior to use, an ISE benefits from vigorous shaking downwards, followed by two or three conditioning runs. It is always advisable to allow the electrode to stand in the titrand solution for half a minute or so prior to commencing the titration to allow the system to equilibrate. The ISE should not be allowed to stand in the presence of excess titrant at the end of the titration for too long, as its response may diminish: if this occurs, it should be rinsed with methanol and allowed to recondition by immersion in a 4×10^{-3} M solution of sodium lauryl sulfate for a few minutes [69].

F. Reliability of the Endpoint

Most of the reported work on potentiometric titration, especially in recent years, has involved the use of an autotitrator, which has associated processing equipment to enable the endpoint to be located via the first derivative process. Cullum and Platt [70] emphasize that the analyst should always scan the original primary curve "because autotitrators will nearly always find a spurious endpoint when there is no real one, and inspecting the curve is the only way to be sure that the instrument is telling the truth." Schultz [68] reinforces this attitude, recommending that both the titration curve and the first derivative be recorded. The peak in the latter should be symmetrical with respect to the inflection point. If it is not so and the software has identified more than one endpoint, repeat the titration. Spikes in the first derivative curve are almost always due to interferences during measurement, and results of such curves should not be used. If a titration curve reveals more than one endpoint where only one is expected, it is virtually certain that none of the endpoints have been detected accurately. It is further recommended that the titer should always be evaluated from the maximum rate of change in potential, i.e., as indicated by the first derivative curve: titrating to a fixed, predetermined potential is not a reliable alternative [57].

G. The Reference Electrode

All potential measurements are made between two electrodes. One of these, the indicator electrode, ISE, responds to changes in the concentration of titrant and/or titrand and the other, the reference electrode, produces a fixed potential. Common examples of the latter are the calomel and silver/silver chloride electrodes. Just as it is important to keep the membrane of the indicator electrode free from contamination by the "sticky" reaction product, the same is equally true for the junction by which the reference electrode connects with the titration solution. It is advisable to use a system that can be dismantled for cleaning if and when necessary [66].

Wang et al. [71,72], titrating long-chain quats with sodium tetraphenylborate, designed an ingenious way of introducing a platinum electrode into the system without it being in direct contact with the solution. They inserted it via a plastic connector into the titrant stream between the burette stopcock and the tip: the latter dipped below the surface of the titrand solution to ensure conductive continuity. An analogous system could be envisaged for the titration of anionic surfactants. Incidentally, it is worth noting in passing that since the High Sense and comparable electrodes can be used for titration of both classes of ionic surfactants, traces of potassium chloride leaching from standard reference electrodes could react with tetraphenylborate and lead to an inferior endpoint. Shoukry [73] saw a need for a system that produced a clear-cut endpoint without the need to purchase sophisticated equipment to produce derivative curves. He replaced the traditional ISE-reference electrode combination with two indicator electrodes: one responding to the decrease of titrand ions and other to the increase in titrant ions. A millivolt recorder then becomes the only expensive item needed (Fig. 11). Although developed for the quat-tetraphenylborate system, the general approach could be adapted to suit other titrations.

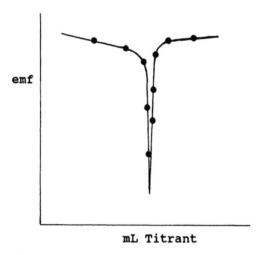

FIG. 11 A potentiometric titration using two indicator electrodes. (Adapted from Ref. 73.)

IX. ALTERNATE METHODS OF ENDPOINT DETECTION

This chapter has concentrated on the diphasic/visual indicator methods, which are still regarded as the internationally accepted standard procedures, and the potentiometric methods, which, although having been promoted for some 20 years, are now receiving increased attention as the shortcomings of the diphasic methods are becoming more widely recognized.

However, a number of alternate approaches have been reported, some of which are mentioned below. Giving them such a brief treatment is not meant in any way to demean the scientific merit of such studies, but rather reflects the fact that the surfactant industry has not, in general, embraced them to any great extent. Like potentiometry, some of them have particular appeal for use with highly colored samples.

Thermometric titrations are by no means restricted to the highly exothermic reactions such as strong acid vs. strong base. Modern thermistors can reliably measure temperature rises of the order of 0.01. Both Jordan et al. [74] and Hattori and Yoshida [75] have used this technique to good effect for surfactant determinations.

Bos [76a,b] developed a routine quality-control procedure for the assay of 0.5–2.5 µmol amounts of quats, including benzethonium chloride, by titration with sodium lauryl sulfate using a tensammetric pulse technique to locate the endpoint. With proper choice of measuring potential, differential titrations were possible for some binary mixtures, and it would appear that an analogous situation could be likely for a mixture of anionic surfactants being titrated with a suitable quat (DDMIC1?).

X. OPTIMUM pH FOR THE TITRATION

The PVC matrix electrodes operate over a very wide pH range. Alegret et al. [67], for example, stated that their electrode performed well over a pH range of 2 to 12 (but was used mainly at pH 2.2–3.0). Most anionic surfactants, apart from the carboxylates, can also be titrated over a similarly wide range. Schultz [69] recommends pH 3 for most major classes (alkylbenzenesulfonates, alkyl sulfates, alkyl ether sulfates, alkanesulfonates, isethionates, etc.) and pH 10 for carboxylates. An unexplained anomaly has been reported in the case of the sulfosuccinates. The sulfonate group (only) is titrated at pH 1–3 and both sulfonate and carboxylate groups at pH 10, as would be expected: however, at the intermediate pH of 5, apparently neither of the anionic groups are reactive [63].

XI. SPECIAL CASES

Some classes of anionic surfactants possess specific properties which can be used to permit the estimation of those particular substances in admixture with others. The classic example is a soap/syndet mixture which can be resolved by titrating under acidic and alkaline conditions. A few other such cases are discussed in this section.

A. Disulfates and Disulfonates

The behavior of an additional hydrophile in a surfactant is by no means as predictable as the first. While alkyl disulfonates alone are generally regarded as unreactive in a two-phase titration, their presence in a mixture with monosulfates causes a variable response.

Wickbold [77] depressed the interference completely by addition of sodium sulfate (2 g) to the titration mixture, whereas Liebscher and Eppert [30] found that disulfonates reacted quantitatively if titrated with Hyamine 1622 under alkaline conditions using phenolphthalein as the indicator (magenta color transfered from the aqueous layer to the organic layer at the endpoint). From the difference between these two results an estimate of the di- (and poly-) sulfonates may be made.

One of the most recent and comprehensive studies on this topic is that of Buschmann and Starp [65]. They synthesized pure samples of the 1:2, 1:5, and 1:12 isomers of disodium dodecyl disulfate and disodium dodecane-1:12-disulfonate and investigated the titrimetric behaviour of these substances both alone and in admixture with sodium dodecyl sulfate and sodium dodecanesulfonate. As would be expected, the titration of both monosulfated and monosulfonated species with Hyamine 1622 resulted in 100% recoveries by both the mixed-indicator method and by potentiometric titration using the High Sense Electrode. There was wide divergence, however, shown by the difunctional surfactants. In the mixed indicator method, the recovery worsened as the distance between the two sulfate groups increased approximately 95%, 55%, and 35% for the 1:2, 1:5, and 1:12 isomers, respectively: moreover, the endpoint was no longer sharp. The 1:12 disulfonate showed no response whatsoever.

A similar trend was observed with potentiometry. The large, vertical potential breaks of 200 and 150 mv for the monosulfate and monosulfonate, respectively, gave way to much smaller (but still resolvable) breaks of 80, 50, and 30 mv for the 1:2, 1:5, and 1:12 disulfates, respectively: no potential jump was observed in the case of the disulfonate. The recoveries of the disulfates, however, was 10–20% above theoretical when small quantities (4 μmol) of analyte were used, but reduced to the 100% level for larger amounts (40 μmol, equivalent to 10 mL of 4×10^{-3} M Hyamine 1622).

Different behavior again was observed in the case of mixtures of the mono- and difunctional species. Synthetic mixtures were prepared from pure sodium dodecanesulfonate and the 1:12 disulfonate. The consumption of Hyamine 1622 titrant in the diphasic titration increased constantly with the amount of disulfonate added, but the actual consumption of titrant corresponded to that required by the monosulfonate plus only about 5% of that theoretically possible from the disulfonate. Potentiometric titrations were much more useful, leading to curves with only one resolvable break, corresponding to complete reaction from both mono- and disulfonates. In the course of this investigation, the authors determined the precipitation phase diagrams for the various titrant/titrand pairs and were able to estimate such parameters as the concentrations of active ions at the endpoints and solubility products, etc. to explain these apparent anomalies [61,65].

DDMICl was again proven to be a more discriminating reagent than Hyamine 1622. As the titrant in the two-phase titration, it determined about 30% (compared to 5%) of the disulfonate in a mono/disulfonate mixture. In potentiometric titrations it resulted in two potential breaks, the first for monosulfonate and the second for disulfonate (both sites reacting), both of which were resolvable unless the relative amount of disulfonate was unreasonably high.

B. Acid Extraction of Sulfonates

A variety of sulfonates and their co-products can be extracted from an aqueous solution, made strongly acidic with hydrochloric acid, with diethyl ether [22], or petroleum ether [78]. Monosulfonated surfactants are in general converted to the unionized free acid and extracted, whereas the disulfonates and any short-chain sulfonates (hydrotropes) are not. Of the hydroxyalkanesulfonates, which are components of the mixture marketed as olefin sulfonates, the 3- and 4-hydroxy isomers become cyclized to the corresponding sultones and are also extracted. The 2-isomer, which would yield a sultone based on a strained four-membered ring, does not cyclize under these conditions and, by virtue of the additional hydrophilic character imposed by the hydroxyl group, remains in the aqueous layer. Other anionic surfactants to be extracted are the mono- and disulfosuccinates and, of course, fatty acids from soaps: the presence or absence of the latter will be known from the results of titration at different pH values.

This extraction may be quantified. Cullum [22] recommends the following procedure: Weigh out a sample containing about 0.01 mol of active matter and dissolve in a minimal amount of water, transfer quantitatively to a separatory funnel and dilute to about 70 mL with water. Add 30 mL of concentrated hydrochloric acid (approximately 10 M), so that the overall acid concentration is approximately 3 M (van Os et al. [78] recommend 6 M). Extract the solution with three 50-mL portions of diethyl ether and combine the extracts. Wash the combined extracts with 3 M hydrochloric acid and evaporate the ether layer to near-dryness in a beaker. Repeat the evaporation with two further 25-mL quantities of ether. This triple evaporation effectively removes the hydrochloric acid which has been transferred from the aqueous layer: it is important that these evaporations do not go to total dryness, in which case the sulfonic acids would begin to decompose. Dissolve the residue in water (50 mL) and titrate with standard sodium hydroxide (1 M) to a phenolphthalein endpoint. The amount of base used is directly equivalent to the amount of monosulfonic acids present.

If the beaker used for the evaporation/titration was tared, the neutralized solution could be evaporated to dryness and constant mass. The mass of the product is due to the sodium salt of the sulfonic acid and, in conjunction with the amount of base used for the titration, can be used to calculate a mean molecular mass. This assumes that no neutral oil (unsulfonated fatty matter) or other substances have been extracted at the same time.

Fatty acids from soaps and acyl sarcosines will be extracted along with the alkylbenzenesulfonic acid, but as acids they are to weak to contribute to the titer when the neutralization is carried out in aqueous solution. They will, however, respond quantitatively along with the sulfonic acids if the residue is dissolved in methanol/acetone and titrated potentiometrically with a stronger base such as tetramethylammonium hydroxide in isopropanol/methanol or sodium hydroxide in methanol [79].

C. Hydrolyzable Surfactants

Some classes of surfactants are hydrolyzable under acid or alkaline conditions. This is one of the factors that formulators must consider when designing a product required to perform under extreme pH conditions. To the analyst, this can have several advantages:

1. The hydrolysis is usually quantifiable (e.g., saponification value of fats), yielding useful data.
2. The products of hydrolysis (e.g., fatty acids and alcohols) can be separated for examination by GLC, IR spectroscopy, etc.
3. Determination of total anionic surfactants in a mixture before and after hydrolysis can in some circumstances permit an estimation of each active component without the need for prior separation.

Alkaline hydrolysis is frequently carried out in alcoholic solution: this not only provides for a more rapid reaction but also prevents foaming. By contrast, acid hydrolysis is almost invariably carried out in aqueous solution to avoid formation of the ethyl esters of any fatty acids present by transesterification with the ethanol in the solvent. Foaming is best avoided by heating the reaction mixture in a boiling water bath (without permitting the reaction mixture itself to boil) until the destruction of the surfactant has substantially progressed. Since the hydrogen ion is the catalyst, the rate of the hydrolysis varies considerably with the concentration of mineral acid used: as a general rule, the hydrogen ion concentration should be at least 1 M to ensure complete hydrolysis of the common alkyl sulfates within 2 hours. The response of most common types of anionic surfactants to acid and alkaline hydrolysis, as listed by Cullum [22], is summarized in Table 4.

The hydrolysis of alkyl sulfates, described in Appendix A, results in an increase in acidity, which is directly proportional to the quantity of this class of surfactant present:

$$ROSO_3^- + H_2O \xrightarrow{H^+} ROH + H^+ + SO_4^{2-} \qquad (9)$$

An alternative approach is to measure the difference between the amount of anionic surfactant present both before and after hydrolysis by one of the titration procedures described herein. If the titrations are carried out under acidic conditions (carboxylate contribution suppressed), then any difference between the pre- and posthydrolysis can be attributed to the loss of a hydrolyzable surfactant, typically alkyl sulfate, and the posthydrolysis titer itself to a nonhydrolyzable surfactant such as alkylbenzenesulfonate.

If the antagonistic titrations are also performed under alkaline conditions, additional useful information may be generated. If a fatty acid has been liberated by the hydrolysis, e.g., from a sulfosuccinate ester, then one surfactant will have been effectively replaced by an equivalent amount of another and no net change in titer will be observed. In one case, that of the α-sulfonated fatty esters, acid hydrolysis leaves the sulfonate group intact but liberates a carboxylic acid: this results in a doubling between the pre- and posthydrolysis titers if the titration is carried out under alkaline conditions but no change if carried out under acidic conditions. Battaglini et al. [80] have reported an analysis scheme along these lines for quality control in the production of this class of surfactant.

The situation becomes more complicated if the surfactant is only partially degraded under the reaction conditions selected. The international standard methods for the determination of hydrolyzable and nonhydrolzable surfactants specify boiling 25 mL of the surfactant solution (approximately 4×10^{-3} M) with 5 mL of 10 M aqueous sodium hydroxide for 30 minutes [81] and with 5 mL of 5 M sulfuric acid for 3 hours [82] for alkaline and acid reactions, respectively. Cullum comments that these conditions may not be sufficient for all types of surfactants and recommends that, if samples of unknown character are to be tested, the procedure be carried out in duplicate (at least) with one sample being heated for 1 hour longer than the other: if the two results indicate that the shorter hydrolysis time is insufficient, then the process should repeated using longer reaction times until the second titration is not significantly greater than the first.

TABLE 4 Hydrolytic Behavior of Anionic Surfactants

Surfactant type	Acid hydrolysis (approx. 1 M H$^+$)	Alkaline hydrolysis
Sulfate esters: alkyl sulfates, alkylphenol ether sulfates	Complete hydrolysis by refluxing for 2 hours with 0.5 M sulfuric acid. Increase in acidity can be used for estimation (see Appendix A).	No effect
Sulfated fatty acid alkanolamides	Sulfate group will hydrolyse completely by refluxing for 2 hours with 0.5 M sulfuric acid. Hydrolysis of the amide linkage takes much longer (8 hours with 3 M sulfuric acid or 6 M hydrochloric acid).	Monoalkanolamide are fairly resistant to attack from boiling 0.5 M potassium hydroxide in ethanol, although dialkanolamides hydrolyse more readily. Boiling for 6 hours with 1 M potassium hydroxide in ethylene glycol is required for thorough hydrolysis. Even this latter treatment has no effect upon the sulfate ester linkage.
Sulfated mono- and diglycerides	Sulfate group removed as above. Carboxylate ester groups are also hydrolyzed. Products are fatty acid, glycerol and bisulfate.	Carboxylate group is hydrolyzed to produce a soap and sodium glyceryl sulfate.
Alkane- and alkylbenzene sulfonates	No effect	No effect

Fatty ester α-sulfonates	Carboxylate ester hydrolyzed, but sulfonate unaffected	The C-S bond is slightly more labile than in other sulfonates. Several hours of boiling with 1 M potassium hydroxide will remove the sulfonate group. The carboxylate ester group is attacked readily.
Sulfonated amides, e.g., sulfosuccinamates, sulfosuccinamides, and acyl methyl taurates	Amide groups are hydrolyzed by boiling for 6 hours with 6 M hydrochloric acid under pressure (150–160°). Sulfonate groups are unaffected.	Mild conditions—little or no effect. Quantitative hydrolysis requires 6 hours of refluxing with 1 M potassium hydroxide.
Sulfonated esters, e.g., acyl isethionate or alkyl sulfosuccinate	Acyl ester group hydrolyzed in both cases to give fatty acid plus isethionate and sulfosuccinate, respectively	As for acid hydrolysis, but carboxylate soaps are formed instead of free fatty acids
Acyl sarcosinates	Amide group hydrolyzed by boiling for 6 hours with 6 M hydrochloric acid under pressure (150–160°)	Mild conditions—no effect. Quantitative hydrolysis requires 6 hours of refluxing with 1 M potassium hydroxide.
Phosphate esters	No effect under mild conditions. Concentrated acids and elevated temperatures are required for quantitative hydrolysis.	No effect

APPENDIX A: PRIMARY STANDARD ANIONIC SURFACTANT

The internationally recognized primary standard for many years been a grade of sodium dodecyl sulfate known as specially purified sodium lauryl sulfate and marketed by British Drug Houses (BDH). It is prepared by the sulfation of a carefully purified dodecanol guaranteed to contain only tiny traces of the C10 and C14 homologs. The purity is claimed to be not less than 99%, but Reid et al. [16] reported it to assay at >99.5% by the method outlined below and >99.8% by sodium content. Occasional assays in this author's laboratory have placed it at 99.7%. The purity of the same product from another source has been found to be 99.2% over a considerable period of time: the remainder consists of water, lauryl alcohol, and sodium sulfate [69]. Having already made the recommendation that standardization of the titrant should be made against a standard of close or identical structure to the analyte material, the question is raised concerning the availability of an alkylbenzenesulfonate standard, since the latter is the most widely used surfactant. In the early work of the Commission Internationale d'Analyses, it did appear that the latter gave rise to slightly different results to the specially purified sodium lauryl sulfate, but subsequent further testing [17] showed that if more attention was paid to the purification of the sulfonate, then both gave the same result (for the standardization of Hyamine 1622 by the mixed-indicator method). However, in view of the subsequent information that has since come to light from the systematic studies on potentiometric titrations presented herein, a standard of comparable purity but with a longer hydrophobe, e.g., sodium hexadecyl sulfate, would be preferred if it ever became available because of the sharper potential break it would exhibit at the endpoint.

For most purposes, the specially purified sodium lauryl sulfate may be taken as pure, with a molecular mass of 288.4. However, if a precise check is required via an independent route, a measure of the acid released on hydrolysis may be made using the procedure outlined below. Weigh accurately approximately 5 g of the material into a 250-mL round-bottom flask. Pipette in 25 mL of sulfuric acid (approximately 0.5 M) and heat under reflux. If the solution is allowed to boil with even modest vigor, then the process will be beset by problems with foam, so it is good practice to immerse the flask in a boiling water bath for 20 minutes or so until most of the reaction has occurred and then to boil under reflux for 90 minutes to ensure the reaction is complete. The liberated lauryl alcohol will steam-distill into the condenser, but this will not affect the analysis. Remove the heat source, allow the the mixture to cool, rinse down the condenser, add a few drops of phenolphthalein, and titrate with standard sodium hydroxide (1 M) until neutral (A mL). Titrate 25 mL of the sulfuric acid directly with the sodium hydroxide (B mL).

During the hydrolysis, the acidity increases due to formation of the hydrogen sulfate ion:

$$ROSO_3^- + H_2O \xrightarrow{H^+} ROH + HSO_4^- \tag{10}$$

Although 1 M titrants are not frequently used in analytical chemistry, if the sulfuric acid used is any less concentrated, then the hydrolysis time will be correspondingly lengthened.

$$\% \text{ Purity} = \frac{28.84 \ (A{-}B) \ \text{M}}{\text{sample weight}} \tag{11}$$

where M = exact molarity of the sodium hydroxide.

The specially purified sodium lauryl sulfate will be used to standardize the quaternary ammonium surfactant of choice. If the standard solution is likely to be stored for future use, it may be stabilized against bacterial attack by, for example, the addition of 0.5% methanal [69].

There are alternative methods of standardizing the quat via sodium tetraphenylborate, which, in turn, can be standardized gravimetrically as its potassium salt or by potentiometric titration with standard solutions of silver or thallium ions [14,58]. However, it is best if the standardization and application titrations are of identical nature: apart from helping to reduce the effects of anomalies in stoichiometry, etc., in this case it also compensates for minor errors caused by variations in the drainage characteristics between surfactant and non-surfactant solutions in pipettes, etc.

APPENDIX B: SURFACTANT SOLUTIONS—SOME GENERAL REMARKS

By their very nature, surfactant solutions tend to foam. The huge surface area of a rich foam, with its high concentration of surfactant adsorbed at each face of the thin film of water enclosing each bubble, can severely reduce the concentration of surfactants in very dilute solutions. This, of course, is the principle behind the solvent sublation technique used for stripping surfactants from wastewaters, etc.

For more concentrated solutions, such as those used as titrants, foam can still be a nuisance during mixing, transfer, and making up to volume in volumetric flasks. Generation of foam during dissolving can be minimized by swirling or gentle action from a magnetic stirrer, rather than by the extreme alternative of up-and-down shaking, and by gentle action during transfer. For the purpose of making up to volume, the foam in the neck of the volumetric flask can be negated by a few drops of ethanol or by blowing in air from a wash bottle containing a few mL of acetone. Cationic surfactants, in particular, become strongly adsorbed onto surfaces exhibiting negative sites, e.g., glassware. The more dilute the solution, the greater will be the relative loss of surfactant by this process. For precision work with $1{-}4 \times 10^{-3}$ M solutions, it is recommended that volumetric flasks and burettes be filled with a solution of titrant and left for 24 hours to saturate the adsorption sites; these solutions may then be discarded and replaced by another freshly prepared in the same glassware [60,69].

APPENDIX C: THE METHYLENE BLUE METHOD

This is *the* classic method proposed by Epton in 1948 [6]. Its inclusion here is not a recommendation for use, but is justified on the grounds that it is still used as a standard against which newer methods are judged (e.g., Ref. 83). This version [9a] uses Hyamine 1622, but alternate titrants such as cetyl pyridinium chloride or cetyltrimethylammonium bromide may be substituted. The latter alternative is used in the ASTM Test Method D1681-92 [84].

Reagents

Titrant: 0.004 M Hyamine 1622 (Rohm and Haas). Dissolve 1.772 g of the solid (oven-dried at 105°C) in water and dilute to 1 L in a standard volumetric flask. Standard sodium lauryl sulfate, 0.004 M: Dissolve 1.154 g of specially pure sodium lauryl sulfate (BDH) in 100 mL of water by gentle swirling, transfer to a 1-L standard volumetric flask, make up to volume, and mix thoroughly. Indicator: The indicator is made up in two stages. Initially, dissolve 0.30 g of methylene blue chloride monohydrate in water, make up to 100 mL, and mix thoroughly. Prepare the working solution by diluting 10 mL of the above solution to about 300 mL, adding 6.6 g of concentrated sulfuric acid and 50 g of sodium sulfate (anhydrous), swirling until dissolved making up to 1 L and mixing thoroughly.

Procedure

Weigh out accurately an amount of sample containing about 4 mEq of anionic active material, dissolve in water, neutralize to phenolphthalein if necessary, and make up to 1 L in a standard volumetric flask. Mix thoroughly. Pipette 20 mL of the sample solution into a suitable titration vessel (a stoppered conical flask, measuring cylinder or bottle, 100 or 250 mL capacity: the common Erlenmeyer flask will not suffice, since vigorous up-and-down shaking, rather than the conventional swirling, is necessary to mix the two phases into a finely divided emulsion to ensure efficient transfer of the solutes across the phase boundaries.). Add 25 mL of the indicator solution and 15 mL of chloroform.

Add 5 mL of titrant, stopper the vessel, shake vigorously for 30 seconds and allow the two layers to (begin to) separate. The lower layer should be blue. Continue the titration with the addition of 1-mL portions of the titrant. As the endpoint is approached, the emulsion formed between the two layers breaks and separates more readily as increasing amounts of the titrand are "neutralized." At this stage, titrate dropwise with vigorous mixing between additions. The endpoint is taken as the point at which the intensity of the blue color, when viewed in reflected light after 1 minutes standing, is judged to be the same in both phases.

Notes

1. Since the organic layer is still blue at the declared endpoint, some anionic surfactant, typically 0.5 μmol, remains untitrated and a correction of approximately 0.1 mL could be applied. However, if the amount of sample taken is adjusted so that the titer for the sample is similar to that obtained for the standard (approximately 20 mL), the error is largely self-compensating.
2. If the presence of alkanesulfonates is suspected, the error caused by accompanying di- and polysulfonates can be largely annulled by the addition of a further amount of sodium sulfate (2 g) to the titration mixture.
3. 1:2 dichloroethane may be substituted for chloroform as the organic phase.
4. The method gives low recoveries for alkyl sulfates, alkanesulfonates, and alkylbenzenesulfonates if the alkyl chain lengths are less than C8, C10, and C6, respectively.

APPENDIX D: THE MIXED-INDICATOR METHOD

This is the current international standard method, first presented by Reid et al. [16] and subsequently incorporated into ISO 2271-1989 [85] and BS 3762 [86].

Reagents

Titrant and standard sodium lauryl sulfate solutions: as for Appendix C above. Indicator: 0.5 g of the cationic indicator dimidium bromide and 0.25 g of the anionic indicator known alternatively as disulfine blue VN or acid blue 1 are dissolved separately in 30 ml hot 10% ethanol, transfered to a 250-mL volumetric flask, diluted to volume and mixed thoroughly (stock solution). The working indicator solution is prepared by diluting 20 mL of the stock solution to 500 mL with water containing 20 mL of sulfuric acid (2.5 M).

Procedure

The procedure is analogous to the previous method. The titration mixture consists of sample aliquot (20 or 25 mL), water (10 mL), indicator solution (10 mL), and chloroform (10 mL). Titrate as above. The chloroform layer in the early stages will be a magenta color (dimidium–surfactant anion complex). At the endpoint it should be a neutral gray color and will in the presence of excess titrant turn blue.

Notes

1. The endpoint is much sharper than is the case when methylene blue is used and the requirement to organize the sample aliquot to give a titration in the 15- to 20-mL range is less important. There is no inherent indicator blank.

2. As in the previous case, the contribution of di- and polysulfonates to the titer of monosulfonates may be largely depressed by the addition of 2 g of anhydrous sodium sulfate or by replacing the 10 mL of water by 10 mL of 20% sodium sulfate solution. This modification is covered by the standard methods ISO 6121:1988 [87] and BS 6829 [88].

3. The standard methods quoted above [44,85,86] offer a choice between the traditional stoppered bottle/measuring cylinder and a sealed system, first described by Hoffmann et al. [89], in which are fitted inlet ports for automatic addition of the chloroform, indicator, and titrant and a drain for the removal of the spent titration mixture to waste. The necessary agitation is supplied by a mechanical stirrer. With this apparatus, the exposure of the operator to chloroform and operator fatigue are considerably reduced. Endpoint detection, however, is still visual.

4. Buschmann [90] has pointed out the close similarity between the structures of the indicator dimidium bromide and ethidium bromide: the latter, which differs from the indicator by only one methylene group, is well known as a strong mutagenic agent. He recommends caution during handling and that the waste aqueous phase be passed through activated charcoal before discharge.

5. The method gives low recoveries (<95%) for alkyl sulfates, alkanesulfonates, and alkylbenzenesulfonates if the alkyl chains are less than C10, C12, and C8, respectively [9a].

APPENDIX E: STANDARD POTENTIOMETRIC METHOD

This procedure is covered by ASTM D4251-89, which was reapproved in 1995 [91]. A commercial automatic titrator assembly (e.g., Metrohm E536) including a 5-mL burette is recommended. The indicator and reference electrodes stipulated are a nitrate ion–selective electrode (Orion 93-07 Nitrate ISE or equivalent) and a silver/silver chloride reference electrode: a calomel electrode is also satisfactory.

Reagents

The reagents, Hyamine 1622 and sodium lauryl sulfate, are essentially the same as for the previous two methods, but are 10 times more concentrated—0.04 M. The titrant also contains 4 mL of 5% sodium hydroxide per liter.

Procedure

To a 150-mL beaker add an aliquot of standard sodium lauryl sulfate or sample sufficient to result in a titer of no more than 4 mL of titrant (approximately 0.15 mEq) and dilute to about 50 mL with water. Ensure that both electrode tips are below the surface and commence addition of titrant at a rate of 0.5 mL per minute, reducing the rate as the inflection point is approached. Autotitrators can be preset to achieve this automatically. Constant stirring is essential throughout the process.

Notes

1. New indicator electrodes require at least one hour's conditioning in 0.01 M sodium nitrate before use. Previously used electrodes are conditioned by a dummy titration. At the end of each titration the electrode should be rinsed with water, ethanol, and water again. Prolonged contact with alcohol or any other organic solvent will result in failure of the membrane.
2. As mentioned earlier in this chapter, reference electrodes with ground glass sleeve-type junctions are preferred to those using a ceramic or porcelain plug: the latter type become clogged with the precipitate.

APPENDIX F: HIGH-PRECISION POTENTIOMETRIC TITRATION

The superior potential breaks, enabling endpoints to be located with greater precision, obtained by the use of 1:3 didecyl-2-methylimidazolinium chloride (DDMIC1) and Metrohm's High Sense Electrode have already been discussed. This combination has yet to be adopted into any officially recognized form of standard method, but the advantages are obvious.

The titrant was developed by Th. Goldsmith AG and is marketed by Metrohm Ltd under the title of TEGOtrant A100.

Reagents

Sodium lauryl sulfate, 0.004 M: As for Appendixes C and D. TEGOtrant A100, 0.004 M: Dissolve 1.60 g of reagent in water and make up to 1 L. Use the same flask repeatedly for making up the reagent and allow it to stand for 24 hours before use [69].

Procedure

Select an aliquot of sample to provide a consumption of at least 10 mL of titrant and dilute to 50 mL with distilled water. Add 10 mL of the appropriate buffer solution (usually pH 3) and 5 mL of methanol. Insert the reference and indicator electrodes and the burette tip below the surface of the solution and autotitrate, preferably in dynamic mode. As in all potentiometric titrations, the addition of titrant well beyond the endpoint is required to enable the derivative technique of endpoint location to be applied.

Notes

1. As in Appendix E, a new or dry electrode is best conditioned by a dummy titration before being used in a genuine estimation. At the end of the titration series, rinse the electrode with methanol followed by distilled water and, if necessary, wipe off with a soft paper towel soaked in methanol. Store the electrode dry.
2. As outlined also in Appendix E, avoid ceramic or asbestos plug junctions.
3. Finally, inspect the direct potentiometric curve and its derivative visually before accepting any titration value calculated by an autotitrator assembly (see Sec. VIII. F).

APPENDIX G: ASPECTS OF POTENTIOMETRY AND ELECTRODES

The following section was prepared just before this volume went to press, as it became obvious that the chapter on the potentiometry of anionic surfactants, commissioned to a leading international authority on the subject, was not going to materialize. It does not purport to cover the topic with any degree of depth or breadth, nor is it written with the insight of personal experience: it is merely intended to complement the material already presented in this chapter on potentiometric titration and to expand on electrode construction. A relatively recent and well-referenced review of all electrochemical techniques for the analysis of surfactants has been prepared by Gerlache et al. [91]. It is assumed that the reader has a basic understanding of potentiometry, the Nernst equation, etc.: if this is not the case, then reference should be made to a modern text on analytical chemistry such as that of Christian [92].

Early ISEs, such as those described by Ciocan and Anghel [48–50], were of the liquid membrane type. A porous membrane of suitable composition such as graphite [50], teflon, or hydrophobic glass frit [93] was sealed into the end of a glass tube and impregnated with an electroactive substance (ionophore) dissolved in a water-immiscible solvent such as dichlorobenzene. Typical ionophores were salts formed by the reaction between surfactant anions (e.g., dodecyl sulfate and dodecylbenzenesulfonate) and cationic metal complexes (e.g., ferroin) or cationic indicators (e.g., methylene blue or crystal violet). The interior of the electrode body was filled with a standard solution of anionic surfactant and sodium chloride, into which was sealed a silver/silver chloride electrode.

The emf of such electrodes, measured against a reference such as a calomel electrode, usually showed a linear response to variations in log concentration over a 3- to 4-decade range, but the slope was sub-Nernstian, being typically 30–45 mv per decade in comparison to the theoretical value of 59 mv.

Two major problems with this type of electrode, identified by Gerlache and others, are drift in potential with time and the tendency of the liquid ion exchanger to be solubilized from the membrane by surfactants at concentrations near to and above the critical micelle concentration.

Polymer Membranes and Mediators

The liquid membranes have been substantially replaced by polymer membranes, which are far more robust and less prone to solubilization problems. The polymer base is most commonly a high molecular mass polyvinyl chloride, although alternatives such as polyvinyl butyral [95] and silicone gum rubber [94] have been reported. The membrane itself contains a high proportion (typically 60–70%) of a liquid popularly referred to as a plasticizer. However, this liquid does far more than contribute to the structural properties and durability of the membrane. As pointed out by Buschmann [64] and mentioned above, the potential developed by the electrode is due more to the selective extraction of the surfactant ions into the lipophilic membrane (a function of the plasticizer) than to any reaction between the surfactant ion and the ionophore. As a consequence, the electrode behavior depends heavily upon the type of plasticizer selected and ISEs show a greater response to the more lipophilic (long-chain) surfactants than to their shorter-chain counterparts (see below). Gerlache et al. [91] summarized the situation with the statement that while the proportion of plasticizer used in manufacturing the membrane determines the stability and longevity of the electrode, it is the nature of that substance that determines the slope and linear range of the response. The term "mediator," as used by Moody and Thomas [96], seems a much more appropriate descriptor.

These latter authors investigated a wide range of mediators. Membranes utilising low-viscosity solvents (nitrobenzene, p-nitroethylbenzene, o- and p-nitrotoluene, and nitroxylene) tend to shrink during storage (3–6 weeks), whereas those made with high-viscosity solvents (o-nitrophenyl phenyl ether, o-nitrophenyl n-octyl ether, and o-nitrophenyl butanoate) retained their flexibility even after 30 months. The most popular mediators are o-nitrophenyl phenyl ether [96], o-nitrophenyl n-octyl ether [56], o-nitrophenyl 2-ethylhexyl ether [94], tritolyl (tricresyl) phosphate [97], and dioctyl phthalate [98]. Buschmann and Schultz [56] have pointed out that the esters of diprotic acids, such as phthalic and sebacic acids, which are commonly used as plasticizers in the normal sense of the word, may not be ideal as mediators as they could be susceptible to hydrolysis when used in strongly acidic or alkaline solutions.

Ionophores

If the ISE is required only to indicate the endpoint of a titration, the inclusion of an ionophore in the electrode is optional. Dilley [99], after examining the symmetry of titration curves obtained by use of a variety of membrane electrodes, theorized that the precipitated surfactant ion pair formed during a titration could be absorbed into the membrane via the moderator. He constructed an electrode using a membrane comprised solely of 40% PVC and 60% tricresyl phosphate. After one or two dummy titrations to condition the membrane, the electrode did indeed function in the anticipated manner. The application of this concept has been confirmed by Vytras et al. [95], White [100], and Selig [98].

If the electrode is to be used for direct potential measurement or for critical micelle concentration studies, then it is essential that the ionophore be evenly distributed throughout the membrane, and to this end small quantities, typically 0.1–1.0%, are included in the mixture at the time of casting. Popular ionophores include benzethonium dodecyl sulfate [22], hexadecylammonium dodecyl sulfate [93], and hexadecylammonium pentane-1-sulfonate [97].

Schmitt [93] provides brief details of a membrane electrode utilizing PVC which has been chemically modified by substitution of either quaternary ammonium or sulfonate groups. The modified PVC is dissolved in tetrahydrofuran and reprecipitated by pouring the solution into a large volume of water containing a surfactant ion of opposite charge to that of the ion covalently bonded to the polymer. The precipitated polymer, now incorporating the stoichiometric amount of surfactant ion, is washed, dried, and used to make the membrane.

Membrane Fabrication

The membrane will be seen to be the critical component of the whole electrode. It is cast from a "cocktail" of the polymer, mediator, and ionophore (if any) all dissolved in tetrahydrofuran. Dilley [99] dissolved 6.0 g of mediator (tricresyl phosphate) in 75 mL of solvent and stirred the mixture rapidly using a magnetic stirrer while slowly introducing 4.0 g of a high molecular mass PVC powder such that the solid did not coagulate into lumps before it dissolved. The cocktail was allowed to stand until all of the air bubbles had dispersed and was then poured gently into a Petri dish in a fume cupboard. The loose-fitting lid was then put in place and the solvent was allowed to evaporate in a stream of air for 24 hours. The resulting membrane, typically 0.3–0.5 mm thick, was peeled from the dish and the last traces of solvent were removed by treatment in a vacuum oven at 30° for 1 hour. A suitable section cut from this membrane was used to make the electrode. Moody and Thomas [96], Buschmann and Schulz [56], and Cullum [22] prepared smaller circular membranes by pouring a small volume of cocktail into a glass ring (30–35 mm in diameter) standing on a glass plate. Slow evaporation was again an essential part of the procedure. Moody and Thomas achieved this by placing a wad of 20 filter papers on top of the ring and a heavy weight on top of the papers.

Both Dilley [99] and Moody and Thomas [101] stress that the tetrahydrofuran

TABLE 5 Membrane Preparation: Some Cocktail Compositions

PVC (g)	Mediator (g)	Ionophore (g)	Tetrahydrofuran (mL)	Ref.
0.17	0.36 o-Nitrophenyl phenyl ether	0.04	6	96
4.0	6.0 Tricresyl phosphate	—	75	99
0.072	0.132 o-Nitrophenyl octyl ether	0.002	2	56
0.25	0.75 Tricresyl phosphate	0.01	4	22

used should always contain a stabilizer (e.g., 0.025% tri-*t*-butyl-*p*-cresol). Buschmann [56] required the evaporation to take place in the dark, having found the cocktail solution to be unstable to light. The resultant membranes, however, are quite stable to light.

The ionophore, if included, can be prepared by precipitation from the mixing of aqueous solutions of the two appropriate reagents, filtering, washing, and drying. It can be dissolved in either the tetrahydrofuran or the mediator. Cullum [22] avoided this preparation step by dissolving equimolecular amounts of pure, dry sodium dodecyl sulfate, and benzethonium chloride (0.1063 and 0.1719 g, respectively) in tetrahydrofuran, mixing the two solutions, making up to 100 mL in a standard volumetric flask, and centrifuging out the precipitated sodium chloride. The resultant solution contained 0.25% benzethonium dodecylsulfate, 4 mL of which, containing 10 mg of ionophore, was used to dissolve the other ingredients of the cocktail.

A few cocktail recipes are given in Table 5.

Electrode Construction

The membrane is attached to the square-cut end of a PVC tube, which will function as the electrode body, by use of a cement (5% PVC in THF) or by merely dipping the end of the tube in tetrahydrofuran for a few seconds and then pressing firmly onto the membrane. After an hour or so, superfluous polymer can be trimmed off. The internal solution used is 10^{-3} M in both sodium dodecyl sulfate and sodium chloride, into which a silver wire coated with silver chloride is dipped.

Coated Wire Electrodes

An alternative, simpler method of construction avoids the use of an internal solution. Instead, the membrane is coated directly onto a conductor such as an aluminium wire [95] or a graphite rod [97]. To avoid direct contact between the test solution and the conductor, the latter must be sheathed right down to the polymer coating. The exposed end of the conductor is dipped into an appropriate cocktail (Dowle et al. [97] used 0.1% n-hexadecyltrimethylammonium pentane-1-sulfonate in a 60:40 mixture of tricresyl phosphate and PVC) and allowed to dry for 30 minutes: the process is repeated twice.

Gerlache et al. [91] comment that CWEs, as they are popularly known, have a relatively short lifetime and exhibit more potential drift than the standard type of electrode.

Electrode Applications

The ideal electrode responds to a specific analyte in a Nernstian fashion over a wide range of concentrations up to the critical micelle concentration, gives a response which does not drift during the course of the day, shows no memory effect from previous test solutions, is unaffected by ions other than those of the surfactant, and gives the same response to equimolar solutions of a wide range of surfactant types. Regretfully, the ideal electrode has yet to be invented. It has al-

TABLE 6 Selectivity Coefficients of an Ion-Selective Electrode

Surfactant	K_{DS^-/A^-}^a
Sodium octyl sulfate	1.7×10^{-3}
Sodium decyl sulfate	6.7×10^{-2}
Sodium dodecyl sulfate	1.0
Sodium tetradecyl sulfate	13
Sodium hexadecyl sulfate	82
Sodium dodecylbenzenesulfonate	21

[a]Responses of each surfactant ion, A, measured against that of dodecylsulfate, DS.
Surfactant concentrations and pH constant at 10^{-3} M and 6.0, respectively.
Membrane: bis(dimethylglyoxime)-1,10-phenanthroline cobaltate(III) in o-dichlorobenzene/n-decanol.
The actual values of the selectivity coefficients vary with the nature of the mediator and ionophore, but the general trend is consistent.
Source: Ref. 49.

ready been pointed out that lipophilic surfactants are preferentially dissolved by the membrane mediator. This is observed as a large variation in selectivity coefficient for a relatively small change in chemical structure (Table 6). As a result, any attempt to quantify a surfactant of unknown identity or a mixture of surfactants by direct potentiometry is quite futile.

Clearly, a single surfactant, in the absence of any others, can perhaps be estimated, but only if the electrode has been calibrated using that particular surfactant. Even then, many electrodes also respond in practice to the presence of a variety of nonsurfactant ions. A standardization, therefore, should not only be carried out using a surfactant identical to the analyte, but the standard solutions should be made up in a sample matrix identical to that of the test solution. The most reliable approach to this complex problem is to use the method of standard additions (see Ref. 22 for details). Any calibration can only be valid up to the critical micelle concentration, the point at which individual surfactant molecules begin to cluster together into large aggregates.

None of the above problems in any way diminish the application of surfactant-sensitive electrodes to detecting the endpoint of a potentiometric titration, or to studies leading to the determination of the critical micelle concentration of a single surfactant.

REFERENCES

1. D. C. Cullum and P. Platt, in *Recent Developments in the Analysis of Surfactants* (M. R. Porter, ed.), Elsevier, London, 1991, p. 15.
2. G. S. Hartley and D. F. Runnicles, Proc Roy. Soc. (London) *Ser A168*:420 (1938).
3. G. Zografi, P. R. Patel, and N. D. Weiner, J. Pharm Sci. *53*:544 (1964).
4. E. Kulling, *Promotiosarbeit Nr 3050*, Eidgenossichen Technische Hochschule, Zurich, 1961.
5. C. W. Ballard, J. Isaacs, and P. G. W. Scott, J. Pharm. Pharmacol. *6*:971 (1954).
6. S. R. Epton, Trans. Faraday Soc. *44*:226 (1948).
7. T. Barr, J. Oliver, and W. V. Stubbings, J. Soc. Chem. Ind. *67*:45 (1948).
8. W. B. Smith, Analyst *84*:77 (1959).
9a. E. Heinerth, in *Anionic Surfactants: Analytical Chemistry* (J. Cross, ed.), Marcel Dekker, New York, 1977, p. 221.
9b. E. Heinerth, Fette Seifen Anstrichm. *72*:385 (1970).
10. W. B. Smith, J. Soc. Cosmet. Chem. *14*:513 (1963).
11. Scientific Committee, J. Soc. Cosmet. Chem. *15*:33 (1964).
12. P. A. Lincoln and C. C. T. Chinnick, Lab. Pract. *3*:364 (1954).
13. P. A. Lincoln and C. C. T. Chinnick, *Proc. 1st Inter. Congr. Surface Activity*, Vol. 1, Paris, 1954, p. 209.
14. J. Cross, in *Cationic Surfactants:Analytical and Biological Evaluation* (J. Cross and E. Singer, eds.), Marcel Dekker, New York, 1994.
15. Z. Li and M. J. Rosen, Anal. Chem. *53*:1516 (1981).

16. V. W. Reid, G. F. Longman, and E. Heinerth, Tenside *4*:292 (1967).
17. V. W. Reid, G. F. Longman, and E. Heinerth, Tenside *5*:90 (1968).
18. H. Holness and W. R. Stone, Analyst *82*:166 (1957).
19. D. E. Herring, Lab. Pract. *11*:113 (1962).
20. W. B. Smith, Analyst *84*:77 (1959).
21. D. C. Cullum, *3rd Int. Congr. Surface Active Substances*, Cologne, 1960c, pp. 42–50.
22. D. C. Cullum, *Introduction to Surfactant Analysis*, Blackie, Glasgow, 1994.
23. M. Bareš, Tenside *6*:312 (1969).
24. R. Diederich, Seifen Öle Fette Wasche *97*:914 (1971).
25. L. K. Wang and P. J. Panzardi, Anal. Chem. *47*:1472 (1975).
26. T. Uno, K. Miyajima, and H. Tsukatni, Yakugaki Zasshi *80*:153 (1960).
27. J. T. Cross, Analyst *90*:315 (1965).
28. J. V. Steveninck and J. C. Riemersma, Anal. Chem. *38*:1250 (1966).
29. G. Eppert and G. Liebscher, Z. Chem. *18*:188 (1978).
30. G. Liebscher and G. Eppert, Z. Chem. *19*:69 (1979).
31. M. Tsubouchi, N. Yamasaki, and K. Matsuoka, J. Am. Oil Chem. Soc. *56*:921 (1979).
32. D. Markó-Monostory and S. Börzsönyi, Tenside Deterg. *22*:265 (1985).
33. S. Jansson, R. Modin, and G. Schill, Talanta *21*:905 (1974).
34. M. Tsubouchi and J. H. Mallory, Anal. Chim. Acta *114*:249 (1982).
35. D. Hummel, *Identification and Analysis of Surface-Active Agents*, Interscience, New York, 1962.
36. M. Tsubouchi and Y. Yamamoto, Anal. Chem. *55*:583 (1983).
37. M. Tsubouchi and J. H. Mallory, Analyst *108*:636 (1983)
38. F. F. Cantwell and H. Y. Mohammed, Anal. Chem. *51*:218 (1979).
39. H. Y. Mohammed and F. F. Cantwell, Anal. Chem. *51*:1006 (1979).
40. H. Y. Mohammed and F. F. Cantwell, Anal. Chem. *52*:553 (1980).
41. P. Seguran, Tenside Deterg. *22*:67 (1985).
42. J. B. M. Hendry and A. J. Hockings, Analyst *111*:1431 (1986).
43. D. Volkman, GIT Fachz. Lab. *28*:278 (1984).
44. *Determination of Anionic-Active Matter by Manual or Mechanical Direct Two-Phase Titration Procedure*, ISO 2271, International Organization for Standardization, Geneva, 1989.
45. J. B. M. Hendry and H. Read, Analyst *113*:1249 (1988).
46. D. C. Cullum and P. Platt, in *Recent Developments in the Analysis of Surfactants* (M. R. Porter, ed.), Elsevier, London, 1991, p. 13.
47. N. Ciocan and D. F. Anghel, Anal. Lett. *9*:705 (1976).
48. N. Ciocan and D. F. Anghel, Fresenius Z. Anal. Chem. *290*:237 (1976).
49. D. F. Anghel, G. Popescu, and N. Ciocan, Mikrochim. Acta (II): 639 (1977).
50. N. Ciocan and D. F. Anghel, Tenside Deterg. *13*:188 (1976).
51. C. G. Taylor and B. Fryer, Analyst *94*:1106 (1969).
52. W. Selig, Fresenius Z. Anal. Chem. *300*:183 (1980).
53. W. Selig, Fresenius Z. Anal. Chem. *312*:419 (1982).
54. D. C. Cullum and P. Platt, in *Recent Developments in the Analysis of Surfactants* (M. R. Porter, ed.), Elsevier, London, 1991, p. 16.
55. K. Vytras, Ion Select. Electrode Rev. *7*:77 (1985).

56. N. Buschmann and R. Schultz, Jorn. Com. Esp. Deterg. 23:323 (1992).
57. N. Buschmann and R. Schultz, Tenside Surf. Deterg. 30:18 (1993).
58. N. Buschmann, U. Görs, and R. Schultz, Jorn. Com. Esp. Deterg. 24:469 (1993).
59a. R. Schultz and R. Gerhards, Am. Lab. (July):40 (1994).
59b. R. Schultz and R. Gerhards, Int. Lab. 24 (10):10 (1994).
60. N. Raulf, N. Buschmann, and D. Sommer, Fresenius Z. Anal. Chem., 351:526 (1995).
61. N. Buschmann, Jorn. Com. Esp. Deterg. 26:215 (1995).
62. N. Buschmann, Tenside Surf. Deterg. 32:504 (1995).
63. R. Schultz and R. Gerhards, Tenside Surf Deterg. 32:6 (1995).
64. N. Buschmann, Riv. Ital. Sostanze Grasse 73:219 (1996).
65. N. Buschmann and H. Starp, Tenside Surf. Deterg. 34:84 (1997).
66. N. Buschmann, Anal. Eur. (March/April): 35 (1996).
67. S. Alegret et al., Analyst 119:2319 (1994).
68. R. Schultz, Determination of Ionic Surfactants in Cosmetic Products, Metrohm, Herisau, Switz., 1995.
69. Applications Bulletin No. 233/2e, Metrohm, Herisau, 1995.
70. D. C. Cullum and P. Platt in Recent Developments in the Analysis of Surfactants (M. R. Porter, ed.), Elsevier, London, 1991 p 19.
71. C. N. Wang, L. D. Metcalfe, J. J. Donkerbroek, and A. H. M. Cosijn, J. Am. Oil Chem. Soc. 66:1831 (1989).
72. J. J. Donkerbroek and C. N. Wang, presented at the CESIO 2nd World Surfactants Congress, Paris, 1985.
73. A. D. Shoukry, Analyst 113:1305 (1988).
74. J. P. Jordan, P. T. Pei, and R. A. Javick, Anal Chem. 35:1534 (1965).
75. T. Hattori and H. Yoshida, Anal. Sci. 2:209 (1986).
76a. M. Bos, Anal. Chim. Acta 138:11 (1982).
76b. M. Bos, in Cationic Surfactants: Analytical and Biological Evaluation (J. Cross and E. Singer, eds.), Marcel Dekker, New York, 1994, p. 229.
77. R. Wickbold, Tenside 8:130 (1971).
78. N. M. van Os et al., in Anionic Surfactants: Organic Chemistry (H. W. Stache, ed.), Marcel Dekker, New York, 1996, p. 433.
79. H. Meijer and J. K. Smid, in Anionic Surfactants: Organic Chemistry (H. W. Stache, ed.), Marcel Dekker, New York, 1996, p. 347.
80. G. T. Battaglini, J. L. Larsen-Zebus, and T. G. Baker, J. Am. Oil Chem. Soc. 63:1073 (1986).
81. Method for the Determination of Hydrolyzable and Non-Hydrolyzable Anion-Active Matter Content after Hydrolysis under Alkaline Conditions, ISO 2869, International Organization for Standardization, Geneva, 1973.
82. Method for the Determination of Hydrolyzable and Non-Hydrolyzable Anion-Active Matter Content after Hydrolysis under Acidic Conditions, ISO 2870, International Organization for Standardization, Geneva, 1973.
83. S. Motomizu et al., Bunseki Kagaku 42(7):T105 (1993).
84. Standard Test Method for Synthetic Anionic-Active Ingredient in Detergents by Cationic Titration Procedure, D1681-92, American Society for Testing and Materials, Philadelphia, 1992.

85. *Standard Test Method for the Determination of Anionic-Active Matter by Manual or Mechanical Direct Two-Phase Titration Procedure*, ISO 2271, International Organization for Standardization, Geneva, 1989.

86. *Analysis of Formulated Detergents: Method for Determination of Anion-Active Matter Content*, BS 3762: Section 3.1, British Standards Institution, London, 1990.

87. *Technical Alkanesulfonates—Determination of Alkane Monosulfonate Content by Direct Two-Phase Titration*, ISO 6121, International Organization for Standardization, Geneva, 1988.

88. *Method for the Determination of Alkane Monosulfonate Content*, BS 6829, Section 2.3, British Standards Institution, London, 1989.

89. A. R. Hoffmann, W. W. Boer, and G. W. G. Schwartz, Fette Seifen Anstrichm. *78*:367 (1976).

90. N. Buschmann, J. Am. Oil Chem. Soc. *72*:1243 (1995).

91. M. Gerlache, J. M. Kaufman, G. Quarin, J. C. Vire, G. A. Bryant, and J. M. Talbot, Talanta *43*:507 (1996).

92. G. D. Christian, *Analytical Chemistry*, 5th ed., Wiley, New York, 1994.

93. T. M. Schmitt, *Analysis of Surfactants*, Marcel Dekker, New York, 1992, p. 359.

94. A. G. Fogg, A. S. Pathan, and D. Thorburn Burns, Anal. Chim. Acta *69*:238 (1974).

95. K. Vytras, M. Dajkova, and V. Mach, Anal. Chim. Acta *127*:165 (1981).

96. G. J. Moody and J. D. R. Thomas, in *Nonionic Surfactants: Chemical Analysis* (J. Cross, ed.), Marcel Dekker, New York, 1987, p. 121.

97. C. J. Dowle, B. G. Cooksey and W. C. Campbell, Anal. Proc. (London) *25*:78 (1988).

98. W. S. Selig, Fresenius Z. Anal. Chem. *329*:486 (1987).

99. G. C. Dilley, Analyst *105*:713 (1980).

100. A. D. White, Am. Lab. (July):74 (1984).

101. G. Moody and J. D. R. Thomas, in *Cationic Surfactants: Analytical and Biological Evaluation* (J. Cross and E. Singer, eds.), Marcel Dekker Inc, New York, 1994, p. 186.

3

Trace Analysis of Anionic Surfactants in Laboratory Test Liquors and Environmental Samples

EDDY MATTHIJS Procter and Gamble European Technical Center, Strombeek-Bever, Belgium

I. INTRODUCTION

Surfactants are the active ingredients of laundry and cleaning detergents, hard surface cleaners, and shampoos. After use, consumer product ingredients are discharged directly into municipal sewers and thence to treatment plants, where most of the surfactants are removed by adsorption and biodegradation. Small amounts, however, may enter surface waters via the discharge of treated effluent and soils as part of the sludge when applied to agricultural land as fertilizer. In order to assess the potential risk of the surfactants in the various environmental compartments the exposure concentration (PEC, the predicted environmental concentration) must be determined and compared to the predicted no-effect concentration (PNEC). For large-volume chemicals, a tiered-based risk assessment is often used, which is driven by the projected annual consumption volume and in which subsequent steps require increasingly refined and detailed safety data representing an increased relevance to real-world systems. As part of this process, selective and sensitive analytical methods are required to assess the fate of anionic surfactants during sewage treatment and to determine the exposure concentration in the environmental compartments of interest.

For a long time, colorimetric methods such as methylene blue active substances (MBAS) have been used as the standard methods for determining the concentration of anionic surfactants in environmental samples. These methods, which are based on simple colorimetric measurements, are rapid and cost-effective. However, because of their nonspecificity, they tend to overestimate the true exposure concentration of a specific anionic surfactant. Therefore these methods are increasingly considered as screening methods and have been replaced by more specific methods based on high-performance liquid chromatography (HPLC), gas chromatography (GC), and mass spectrometry (MS).

II. ENVIRONMENTAL MONITORING

The collection of samples represents the first important step of every environmental analytical method. Samples must be representative of the environmental compartment from which they have been taken. Depending on the goal of the environmental monitoring, the chemists must decide on the sampling strategy, frequency, type of samples (single grab or flow proportional samples), distribution in time and/or space, type of container, etc. It is equally important to ensure that the concentration and the chemical identity of the environmental samples are not altered during storage. Therefore, appropriate sample-preservation techniques to eliminate or minimize degradation of the surfactants between the time of collection and analysis should be used. In most cases addition of 3% formaldehyde (37% active, v:v) has been found effective for maintaining the initial concentrations of the major types of surfactants for several months if the samples are subsequently stored in a refrigerator at 4°C. For some surfactants such as alcohol ethoxysulfate (AES), addition of up to 8% formaldehyde (37% active, v:v) is required for efficient preservation during storage of the samples, particularly those containing a high biomass such as sewage samples [1]. The preservation agent must be added to the sample container prior to the collection of the sample. Failure to do so may result in significant losses of parent surfactant.

Quality assurance samples should be prepared in the field by standard addition of known amounts of the target surfactant to selected environmental samples directly after their collection. These samples allow chemists to determine the efficiency of both the sample preservation and the storage procedures. In addition, standard amounts of the surfactants should be added to a limited number of samples in the laboratory prior to the analysis in order to check the efficiency of the analytical procedures.

In environmental samples, surfactants are typically distributed between the liquid phase and the suspended solids. For the determination of the total surfactant concentration, the environmental sample must be vigorously shaken to ensure adequate mixing and suspension of the particular matter. If the dissolved and adsorbed fraction must be determined, the particulates must be removed, preferably by centrifugation, and both phases are then analyzed separately. Solid samples such as sludges are preserved by addition of 8% formaldehyde followed by centrifugation, drying, and grinding of the solids.

III. COLORIMETRIC PROCEDURES

A. General Principle

Colorimetric methods are typically based on the formation of a chloroform-extractable ion-association complex between the anionic surfactant and a cationic dye, followed by spectrophotometric measurement of the intensity of the extracted colored complex. The excess of cationic dye that did not react with anionic surfactants remains in the water layer and is not extracted by the organic solvent. The measurement is very simple, rapid, and cost-effective. All anionic surfactants containing a sufficiently long alkyl chain, of both natural and synthetic origin, respond in the reaction, which is therefore nonspecific and can be positioned as a sumparameter for total anionic surfactants. The colorimetric methods overestimate the true concentration of a specific synthetic anionic surfactant, particularly in complex environmental samples. Negative interferences may occur due to competition of cationic surfactants already present in the environmental sample. The impact of the negative interferences is usually less important than the impact of the positive interferences [2]. The concentration of anionic surfactants is calculated from a calibration curve established with an appropriate reference material, usually linear alkylbenzene sulfonate (LAS). No information can be obtained on the homolog and isomer distributions of the surfactants studied. The colorimetric methods can be used to provide accurate determinations of known surfactants in relatively clean water samples. In addition, they can be used for determining the concentration of parent surfactant in biodegradation tests in which control samples are available to correct for the presence of interfering materials. These colorimetric procedures are still part of the European directive for determining the biodegradability of anionic surfactants. Their lack of specificity and sensitivity make the colorimetric methods less applicable to the analysis of environmental samples in which concentrations are usually low and many interferences are encountered.

B. Methylene Blue Active Substances

MBAS is the most widely accepted sumparameter for measuring the concentration of anionic surfactants in environmental samples. The most commonly used procedure is based on the Longwell and Maniece method [3] and Abbott's modification thereof [4]. The procedure includes a double extraction. In the first step the aqueous solution containing the surfactant is mixed with an alkaline solution of methylene blue using a phosphate, borate or carbonate buffer; this mixture is then extracted with chloroform. The chloroform extract is then back-extracted with an acid solution of methylene blue. This extraction step is repeated twice to ensure complete extraction of the complex. The first step eliminates negative interferences from proteinaceous matter, while the second step removes interferences of substances such as nitrate, sulfate, chloride, and carboxylates, which form complexes with methylene blue that are less efficiently extracted into chloroform. The absorbance of the chloroform extract is measured at 650 nm and compared against a calibration curve prepared from a reference surfactant, usually LAS. The method permits the determination of anionic surfactants in the range of 0.01 to 0.02 mg/L MBAS. This procedure forms the basis of the method adopted by the International Organisation for Standardisation (ISO) [5] and of the official German method [6] for the examination of water, wastewater, and sludge with respect to anionic surfactants. A detailed outline of the latter method is given in Appendix A. Because of its simplicity and good reproducibility, the MBAS procedure has been selected as the analytical method for assessing the primary biodegradability of anionic surfactants as laid down in the EEC directive [7]. Since these biodegradation tests always include a blank unit, correction can be made for any interfering materials present in the biological test liquor.

For a long time the MBAS response has been positioned as an acceptable measure for the concentration of LAS, the most commonly used anionic surfactant in detergents. Although the double extraction procedure eliminates a large part of the interferences, the MBAS reading largely overestimates the true concentration of LAS in environmental samples, as reported by Matthijs and De Henau [8] and Hennes and Rapaport [9] and summarized in Table 1. For aqueous

TABLE 1 Contribution of LAS to the Methylene Blue Active
Substances

Environmental matrix	Percentage LAS in MBAS
Raw sewage	55–93
Biologically treated sewage effluent	18–53
River water	11–50
River sediment	2–38
Aerobic-activated sludge	3–14
Digested anaerobic sludge	37–66
Sludge-amended soil	4–7

Source: Refs. 8 and 9.

environmental samples the contribution of LAS to the MBAS reading decreases
with the degree of biological treatment, and within the same type of matrix it appears to vary strongly from one wastewater treatment plant to the other. The
overestimation is mainly due to naturally occurring anionic surfactants and the
active ingredients used in detergent formulations such as sulfated and other sulfonated surfactants. Soaps do not respond in the MBAS method, since the fatty
acids are so weakly ionized that an extractable ion pair is not formed under the
acidic conditions of the method. LAS typically undergoes biodegradation via oxidation of the alkyl chain [10] by ω-oxidation of the terminal methyl group, followed by β-oxidation resulting in the formation of molecules with a shorter alkyl
chain length and an end-standing carboxylic acid defined as sulfophenylcarboxylates. Since the extraction efficiency in the MBAS procedure depends on the hy-
drophobicity of the molecule, it is unclear to what extent the biodegradation
intermediates with various alkyl chain lengths still respond in the MBAS reading. The data in Table 1 also demonstrate that the direct MBAS analysis of
methanol extracts from sludge, sediment, and soil samples largely overestimates
the LAS concentration. Several workers have attempted to develop alternative
sample preparation techniques to eliminate part of the interferences. These include solvent sublation with ethyl acetate [11] to eliminate nonsurfactant anionic
material, indirect procedures [12,13] in which the anionic surfactant is selectively
extracted as its 1-methylheptyl amine salt to isolate it from interfering proteins,
hydrolysis [14] to remove interfering organic sulfates and other hydrolyzable
compounds, and the use of macroreticular XAD-2 resin and cation-exchange
steps [15] prior to the MBAS determination. The latter procedure is referred to as
"interference-limited" MBAS (IL-MBAS). The use of a cation-exchange step
eliminated the negative interference of quaternary ammonium compounds and
long-chain amines as well as lipophilic material. The XAD-2 resin eliminated interferences due to inorganic salts. Osburn [15] and Sedlak and Booman [16] re-

ported that the results for influent and sludge samples obtained by the IL-MBAS method were in good agreement with the data obtained by the desulfonation/gas chromatographic method. For effluent and river water samples, the IL-MBAS results were significantly higher than the corresponding desulfonation/gas chromatographic data. Although all of these sample preparation procedures improve the specificity of the MBAS method, substantial interferences by undefined compounds remain. In addition, the extra clean-up steps result in lengthy procedures which are less suitable to routine applications.

C.   Other Cationic Dye and Organometallic Cationic Systems

Methylene blue is by far the most widely used cationic dye for determining the concentration of anionic surfactants in biodegradation liquors and environmental samples. Other cationic dye systems such as methyl green [17], azure A [18], toluidine blue [19], rosaline [20] and crystal violet [21] have been validated. Except for the azure A procedure, none of these cationic dye variants has found broad application. Tonkelaar and Bergshoeff [22] have applied the azure A principle to determine trace levels of anionic surfactants in surface waters. This procedure is particularly attractive because it consists of a rapid single extraction of the acidified aqueous test sample with chloroform. Reagent blanks are low, resulting in a high sensitivity. Comparison with the MBAS method showed that the azure A method always resulted in lower levels when applied to drinking and river waters [22], which may indicate that the azure A procedure is less subject to positive interferences. Tonkelaar and Bergshoeff [22] relate the lower azure A reading to the fact that the primary degradation products of LAS or other anionic surfactants interfere to a lesser extent with azure A than with methylene blue. The applicability of the azure A method for sewage samples has been questioned, mainly because of the expected high interference from proteins. Still, given that MBAS and azure A have to be considered as screening methods, both overestimating the true LAS content of the environmental sample, the author considers the azure A method as an attractive alternative for MBAS in view of its simplicity, which is mainly related to the use of a single extraction step. A detailed description of the azure A method is provided in Appendix B.

Cationic metal chelates such as copper ethylenediamine [23,24] have been used for the indirect determination of anionic surfactants in water samples. The procedure is based on chloroform extraction of the complex formed between anionic surfactants and bis(ethylenediamine) copper (II) ion. The copper is then back-extracted from the chloroform layer into dilute acid and its concentration is determined by flame atomic absorption spectrometry. The method has been reported to be applicable to natural and wastewaters that contain anionic surfactants in the concentration range of 0.02 to 0.5 mg/L. The use of other organometallic cationic systems for the determination of anionic surfactants in

aqueous samples such as tris(1,10-phenanthroline) iron (II), better known as ferroin [25] and even [^{59}Fe] ferroin [26], have been reported in the literature. These
procedures may have an advantage in terms of speed of analysis, but they will
suffer from the same positive interferences as previously described for MBAS.
In addition they have found only limited application and should therefore not be
seen as a possible alternative for the widely used MBAS method.

IV. SPECIFIC ANALYTICAL METHODS

During the last 10 years numerous analytical methods have been developed for
the selective determination of anionic surfactants in various environmental matrices including raw and treated sewage, river water, and sediment and soil. This
has been particularly the case for LAS, which represents the major anionic surfactant used in laundry and cleaning products. Most of these methods are based
on the use of gas and liquid chromatographic techniques, in many instances coupled to mass spectrometry. Contrary to the colorimetric sumparameters, the instrumental methods have been developed for a specific surfactant and can as
such be used to determine the environmental fate and exposure of the main type
of anionic surfactants in a variety of environmental compartments. This chapter
aims to provide a broad overview of the currently available methods for the main
types of surfactants including LAS, AES, alcohol sulfate (AS), and secondary
alkanesulfonate (SAS).

A. Linear Alkylbenzene Sulfonate

LAS is the major anionic surfactant used in domestic detergents. The commercial material consists of a homolog distribution from C10 to C13 and of various
phenyl positional isomers. For a long time gas chromatography has been the major technique used for the specific determination of LAS in environmental samples. The gas chromatographic procedures permit the determination of the
individual alkyl homologs and phenyl positional isomers. This has been very
useful in elucidating the mechanism of biodegradation of both the homologs and
isomers as part of laboratory-scale biodegradation studies. The technique is very
sensitive and specific and can accurately determine microgram amounts of LAS
in a broad range of environmental matrices. It requires, however, complicated
and tedious sample pretreatments to convert the analyte into volatile derivatives
and is therefore less appropriate for routine work. High-performance liquid chromatography (HPLC) has now become the most widely used method for the determination of LAS in environmental samples. HPLC does not have the superior
resolution of GC, but it permits the separation of the main LAS homologs without preliminary derivatization and offers a practical tool for environmental monitoring by combining selectivity and speed of analysis (Table 2).

TABLE 2 Main Features of the GC Methods for the Determination of LAS in Environmental Samples

Reference	Chemical	Isolation	Derivatization detection	Matrix	Detection limit
Swisher [28] Sullivan et al. [29]	LAS	Methylene blue extraction	Desulfonation FID	Influent	min. 1 mg required
Waters et al. [30]	LAS	Methylene blue extraction—SCX—acid hydrolysis—1-methylheptyl amine extr.	Desulfonation FID	Influent-effluent- river water	10 µg/L
Osburn [15]	LAS	Liquids: XAD-ethyl ether extraction Sludge: acid hydrolysis—ethyl ether extraction—SAX—ethyl ether extraction Sediment: CH₃OH extr. + as above	Desulfonation FID	Influent-effluent- river water Sludge-sediment	10µg/L 1 µg/g
Hon-Nami et al. [31]	LAS	Methylene blue extraction—SCX- derivatization—silica clean-up	Methyl sulfonate ester MS	River water	3 µg/L
McEvoy et al. [34]	LAS	Alkaline methylene blue extraction—TLC —CH₃OH extraction	Sulfonyl chloride FID and MS	Sludge	—
Trehy et al. [35]	LAS-SPC[a] DATS[b]	Liquids: C8 SPE Solids: CH₃OH extraction—C8 SPE	Trifluoroethyl derivatives MS	Influent-effluent- river water Sediment	1 µg/L
Tabor et al. [36]	LAS	Liquids: C18 SPE Solids: CH₃OH extraction	Trifluoroethyl derivatives MS	River water Sediment	0.1 µg/L
Field et al. [37]	LAS	SFE tetrabutylammonium ion pair	Butyl ester MS	Sludge	—

[a]SPC: Sulfophenylcarboxylic acid—biodegradation intermediates of LAS.
[b]DATS: dialkyltetralin sulfonates—by-products resulting from the manufacturing process.

1. Gas Chromatography

LAS is not sufficiently volatile to permit its direct analysis by GC, and therefore suitable derivatization and desulfonation techniques have been developed. Setzkorn and Carel [27] have described a microdesulfonation procedure in which LAS is heated in concentrated phosphoric acid at about 200°C. Following desulfonation, the resultant alkylbenzenes are solvent extracted prior to their gas chromatographic separation. This results in a very typical and readily identifiable chromatographic pattern for the individual homologs and isomers. Swisher [28] has modified the microdesulfonation/gas chromatographic technique for the determination of LAS in wastewater samples. The technique permits the determination of the individual alkyl homologs and phenyl positional isomers. The chromatographic pattern obtained from environmental samples was compared with that of intact LAS and allowed determination of the degree of biodegradation of the individual components. The author preferred a methylene blue extraction as the initial concentration step instead of evaporation of the entire aqueous sample because it is more rapid and the final chloroform extract can be readily evaporated directly in the desulfonation flask. The main drawback of the proposed technique is that a large amount of LAS (about 1 mg) is required for accurate determinations. Sullivan and Swisher [29] used this technique to determine LAS concentrations in the Illinois river. They used a duplicate sample of the water as a type of internal standard after addition of a known amount of LAS. Prior to the desulfonation, C8-LAS was added to the residue as a second internal standard, since this compound is not found in detergent-finished products. Their study showed that only 10–20% of the MBAS reading is related to LAS. The authors only used a very limited sample clean-up, which consisted of a direct desulfonation of the methylene blue extract prior to the gas chromatographic analysis.

Waters and Garrigan [30] found, however, that the above procedure resulted in chromatograms in which the typical homolog/isomer pattern was disturbed by interfering material, which affected identification and quantification of the peaks. The authors further improved the microdesulfonation/gas chromatographic method by the introduction of additional concentration and clean-up steps, including concentration of LAS by a large-scale solvent extraction with methylene blue, ion exchange removal of the methylene blue cation, acid hydrolysis of alcohol sulfates and natural hydrolyzable material, followed by selective extraction of LAS as its 1-methylheptyl amine salt. The isolated salt was then transferred to a desulfonation flask and boiled with phosphoric acid at 190–200°C, following which the liberated alkylbenzenes were analyzed by capillary column gas chromatography. This procedure resulted in relatively clean chromatograms for a variety of environmental samples, in which the LAS homologs and isomers were readily identifiable on the basis of their relative retention time. The authors designed a desulfonation apparatus with a lower dead volume in order to be able to handle smaller amounts of LAS. The technique, which permits the determination of microgram amounts, was applied to the analysis of LAS in samples of UK river water. The method has a limit of detection of about 10 µg/L and permits the determination of all individual isomers within the C10–C15 alkyl homolog range. The authors demonstrated that, on average, only 26% of the MBAS reading in UK river water samples was due to the presence of LAS.

Osburn [15] extended the microdesulfonation/GC methodology to the analysis of LAS in sediment and sludge samples, which allowed further elucidation of the fate of LAS during sewage treatment. An aqueous suspension of the sludge sample was hydrolyzed by boiling it in hydrochloric acid. LAS was then isolated from the hydrolyzed solution by extraction with ethyl ether. Prior to the microdesulfonation (3 hours at 215°C) and gas chromatographic analysis, the extract was purified by anion-exchange chromatography and a series of two ethyl ether extractions. Sediment samples were first Soxhlet-extracted with methanol for 16 hours. The methanol extract, containing the LAS, was then evaporated to dryness and resuspended in an aqueous hydrochloric acid solution. The sample was hydrolyzed by heating and then further treated as described for the sludge samples. C9 LAS, containing a natural isomer distribution, was used as an internal standard for quantitation purposes.

In addition to the desulfonation technique, LAS molecules may be converted to volatile derivatives. Trace analysis of the methyl sulfonate esters of LAS combined with GC/Mass spectrometry (MS) has been investigated by Hon-Nami and Hⁱ ⁿya [31]. The detection limit of the method was around 3 µg/L. LAS was first isolaᵢed from aqueous environmental samples by chloroform extraction of its methylene blue complex. The methylene blue cation was subsequently removed by cation exchange chromatography. The isolated LAS homologs were then converted to their sulfonyl chloride derivatives by reaction with phosphorus pentachloride at 110°C for 10 minutes. These were subsequently converted into their corresponding methyl sulfonate derivatives by reacting them with methanol at 70°C for 20 minutes. Prior to the gas chromatographic analysis, the derivatives were further purified over a silica column to remove interfering organic material. The assignment of the GC peaks was performed on the basis of retention times and mass spectra. The GC analysis was performed on a packed column and resulted in an acceptable separation of the phenyl positional isomers, although it has to be recognized that the use of a capillary column, as applied by Waters and Garrigan [30] and Osburn [15], would have resulted in an improved chromatographic separation. Hon-Nami and Hanya have used their method to determine the LAS concentration in river water and sediment samples [32] and in estuary and bay water samples [33].

McEvoy and Giger [34] applied GC/MS for the quantitative analysis of LAS in sludge samples. The work was based on the procedure of Hon-Nami and Hanya [31] described above. They found, however, that GC of the methyl sulfonate esters did not offer any advantage over the GC analysis of the sulfonyl chlorides. Two different methods were described: a screening method to verify whether LAS was present and a confirmatory method for the quantitative determination of LAS. In the screening method about 1 mL of wet sludge was dried. The residue was then reacted with phosphorus pentachloride for 20 minutes at 110°C in the presence of hexane. Prior to the GC/MS analysis, the excess phosphorus pentachloride and the reaction by-product, phosphorus oxychloride, were removed by evaporation of the derivatized solution. The screening method allows semiquantitative determination of the LAS concentration. In the confirmatory method, the wet sludge was mixed with an alkaline aqueous solution of methylene blue and extracted several times with chloroform after addition of pentadecyl benzenesulfonate as an internal standard. The combined chloroform extracts were evaporated to dryness and the residue was then dissolved in dichloromethane and purified by thin layer chromatography on a silica plate, using dichloromethane as mobile phase. The LAS–methylene blue ion-pair complex was scraped from the plate and extracted from the silica with methanol. The extract was evaporated to dryness, and the residue was then further treated as described in the screening method. Quantitation was achieved via high-resolution capillary column gas chromatography with flame ionization detection (FID) and using the internal standard approach. In case of co-eluting compounds, quantitation of the individual components were made by GC/MS. The concentration of LAS determined in 12 sewage sludge samples varied between 2.9 and 11.9 g/kg dry sludge.

Trehy et al. [35] described a GC/MS procedure for the determination of LAS and dialkyltetralin sulfonates (DATS) as well as their biodegradation intermediates, sulfophenylcarboxylates (SPC) and alkyltetralincarboxylates, in river water and sediment. DATS are impurities resulting from the manufacturing process. The method requires a two-step derivatization procedure. The sulfonates and carboxylates are converted to their sulfonylchlorides and acid chlorides, respectively, by treatment with phosphorus pentachloride and then to their corresponding trifluoroethyl derivatives by reaction with trifluoroethanol. The high electron affinity of the trifluoroethyl groups results in an enhanced sensitivity and selectivity when using electron capture/negative chemical ionization and permits the determination of the environmental concentrations of LAS, DATS, and biodegradation intermediates thereof at low microgram per liter concentrations. The main advantage of chemical ionization is that there is less fragmentation than in electron impact ionization, thereby facilitating the detection of the parent molecular ion. This simplifies the interpretation of the spectra and results in an increased sensitivity in selected ion monitoring mode. Derivatized LAS and DATS have characteristic mass spectrometric fragmentation patterns, which simplifies their identification in complex environmental matrices. The authors reported that the MS source temperature should not exceed 110°C to avoid extensive fragmentation of the target ions and to maximize the concentration of the more informative higher mass fragments. Aqueous environmental samples (5 mL for influent and 50 mL for effluent and river water) were concentrated and purified over an octyl solid phase extraction (SPE) column. LAS and DATS were eluted from the column with acetonitrile. Following evaporation of the eluate, phosphorus pentachloride was added to the residue and allowed to react for 10 minutes at 100°C. The reaction product was extracted with pentane, which was subsequently evaporated, and then further reacted with dried trifluoroethanol for 20 minutes at 70°C, followed by pentane extraction and capillary column GC/MS analysis. LAS and DATS were isolated from sediment samples by sonication with methanol. The extract was diluted with deionized water, purified over a C8 SPE column, and then treated as described for the liquid samples. C9 DATS (1-butyl-4-methyltetralinsulfonate), which was added to the environmental samples prior to extraction and derivatization, was used as an internal standard. The peak areas for all isomers within a specific alkyl homolog (C10–C14) range were summed both for LAS and DATS, and the response factors for each homolog series were calculated compared to the internal standard. The limit of detection for 50 mL samples was 1μg/L. Recovery of standard additions was in excess of 80% and 64% for aqueous samples and sediments, respectively.

Tabor et al. [36] used GC/MS analysis of the trifluoroethyl esters to determine the concentration of LAS and its homolog and isomer distribution in the Mississippi river. The method used is to a large extent based on the procedure reported by Trehy et al. [35]. For river water samples a C18 SPE column was used instead of a C8 column, and elution of the LAS was achieved by the combined use of acetonitrile and methylene chloride. The trifluoroethyl esters were prepared via treatment with phosphorus pentachloride and trifluoroethanol successively. LAS identification was based on retention time and mass spectra of the derivatives of a C10–C14 LAS standard. Since the authors aimed to analyze only the dissolved LAS concentration, the water samples were filtered on site through a 0.7-μm glass fiber filter. Isolation of LAS from sediment samples was achieved by three successive extractions with methanol on a rotating mixer followed by the direct derivatization of the evaporated combined extracts. The extraction procedure used is less efficient and considerably more labor intensive than the traditional Soxhlet extraction procedure. In addition, sample clean-up is less rigorous than the process reported by Trehy et al. [35]. Three quantitation standards consisting of single compounds that do not occur in commercial LAS mixtures were added during different stages of the analysis. C9 LAS was added before the C18 SPE step for aqueous samples and before the methanol extraction step for sediment samples. C8 LAS was added before the derivatization step and 2,2,2,-trifluoroethyl-*p*-toluenesulfonate was added as internal standard before the GC/MS analysis. The analytical detection limit for LAS was 0.1 μg/L for a 900-mL sample. Recovery of added LAS was generally in excess of 80%. This survey showed that the dissolved LAS concentration in the Mississippi river ranged from 0.1 to 28.2 μg/L. The concentration of LAS in the river sediments ranged from 0.01 to 2 mg/kg.

Field et al. [37] described a supercritical fluid extraction (SFE) method for the isolation of LAS from sewage sludge samples, followed by GC/MS analysis. LAS was quantitatively extracted from the sludge by SFE as its tetrabutylammonium ion pair. Upon injection of the ion pair in the GC, the corresponding butyl esters were formed. The SFE consisted of a two-step procedure; static extraction using 400 atm of supercritical CO_2 for 5 minutes was applied during which the ion-pair reagent was mixed with the sewage sludge sample under supercritical conditions; this was then followed by a dynamic extraction step for 10 minutes at 80°C to recover the analytes. The application of the ion-pair/SPE technique using tetraalkylammonium coupled with ion-pair derivatization was previously reported by Hawthorne et al. [38] and significantly reduces the sample preparation steps and time as well as the analysts' exposure to hazardous reagents. The ion-pair reagent enhanced the extraction efficiency of sulfonated surfactants into supercritical CO_2 by decreasing their polarity and allowed rapid formation of the sulfonate alkyl esters in the GC injection port. The LAS concentrations reported by Field et al. [37] ranged from 3.83 to 7.51 g/kg of dry sewage sludge and were comparable to the values previously reported by McEvoy and Giger [34] as well as the results obtained by using HPLC techniques (see later).

2. High-Performance Liquid Chromatography

HPLC has been widely applied to measure the concentration of LAS in a broad range of environmental samples. The main advantage of HPLC over GC is its simplicity, since LAS, due to the presence of an aromatic functionality, can be separated and detected without prior derivatization or desulfonation. This makes it an attractive technique for routine application. The majority of the HPLC methods have been developed for use as part of environmental monitoring exercises and aim to determine the total LAS concentration and its alkyl homolog distribution. Most of these methods provide little or no information on the isomer distribution, which results in relatively simple chromatographic pattern and peak identification as well an increased sensitivity.

Nakae et al. [39] described a reversed-phase HPLC method with fluorescence detection for the determination of dissolved LAS in river water without prior preconcentration or sample cleanup (see also Table 3). Five hundred μL of Millipore filtered samples were injected onto a C18 analytical column. Chromatographic separation was obtained using a methanol:water (80:20, v:v) mobile phase at a flow rate of 1 mL/min and containing 0.1 M sodium perchlorate as a phase modifier. The addition of sodium perchlorate is required to obtain sufficient retention of the LAS homologs on the analytical column. Sodium dodecyl sulfate was added to

TABLE 3 Main Features of the HPLC Methods for the Determination of LAS in Environmental Samples

Ref.	Chemical	Isolation	HPLC conditions: column–detection–mobile phase	Matrix	Detection limit
Nakae et al. [39]	LAS (dissolved)	None	C18—fluorescence—isocratic CH₃OH-H₂O-NaClO₄	Clean solutions	100 µg/L
Linder and Allen [41]	LAS (dissolved)	None	C18—fluorescence—isocratic THF-H₂O-Pic A	Biodegradation liquors	50 µg/L
Kikuchi et al. [42]	LAS	Liquids: CH₃OH extr. solids—C18 SPE of extract + liquid phase	C18—fluorescence—isocratic CH₃CN-H₂O-NaClO₄	Sea water	0.1 µg/L
				Sediment	0.03 µg/g
		Solids: CH₃OH extraction—C18 SPE		Fish	0.3 µg/g
Matthijs et al. [8]	LAS	Liquids: C8 SPE	C18—UV—gradient	Influent-effluent-river water	10 µg/L
De Henau et al. [43]		Solids: CH₃OH extr.—SAX—C 8 SPE	CH₃CN-H₂O-NaClO₄	Sludge-sediment-soil	0.4 µg/g
Holt et al. [44]	LAS	Solids: CH₃OH extr.—SAX—C8 SPE	C18—fluorescence—isocratic CH₃OH-H₂O-NaClO₄	Soil	0.2 µg/g
Holt et al. [50]	LAS	Liquids: C18 SPE—SAX	C18—fluorescence—isocratic CH₃OH-H₂O-NaClO₄		
Matthijs et al. [1]	LAS		C18—fluorescence—isocratic CH₃OH-H₂O-NaClO₄	Influent-effluent-river water	2 µg/L
Matthijs et al. [55]	LAS	Liquids: solvent sublation—C8 SPE—SAX—C8 SPE	C18—fluorescence—isocratic CH₃OH-H₂O-NaClO₄	Sea water	0.4 µg/L

Reference	Analyte	Sample preparation	Separation—detection—elution	Matrix	Detection limit
Marcomini et al. [56]	LAS	Solids: CH_3OH/NaOH extraction	C8—fluorescence—gradient IPA-H_2O-CH_3CN	Sludge-sediment-soil	1 µg/g
Marcomini et al. [57]	LAS	Liquids: CH_3OH extr. solids—C18 SPE of extract + liquid phase	C8—fluorescence—gradient IPA-H_2O-CH_3CN	Influent-effluent	20 µg/L
Marcomini et al. [59]	LAS-SPC[a]	Liquids: C18 SPE	C18—fluorescence—gradient CH_2CN-H_2O-CH_3OH-$NaClO_4$-trifluoroacetic acid-tetrabutylammonium phosphate	Influent-effluent-river water	—
Castles et al. [60]	LAS	Liquids: C2 SPE—SAX	C1—fluorescence—isocratic THF-H_2O-$NaClO_4$	Influent-effluent-river water	2 µg/L
Di Corcia et al. [62]	LAS	Liquids: graphitized carbon black (GCB)	C8—fluorescence—isocratic CH_3OH-H_2O-$NaClO_4$	Influent-effluent-river water	0.8 µg/L
Di Corcia et al. [51]					
Di Corcia et al. [63]	LAS-SPC[a]	Liquids: GCB	C8—fluorescence—gradient CH_2Cl_2-CH_3OH-tetramethyl-ammonium chloride	Influent-effluent	1 µg/L LAS, 0.8 µg/L SPC
Crescenzi et al. [64]	LAS-DATS[b]	Liquids: GCB	C8—fluorescence—gradient CH_3OH-H_2O-$NaClO_4$	Influent-effluent-river water	0.5 µg/L

[a]SPC: Sulfophenylcarboxylic acid—biodegradation intermediates of LAS.
[b]DATS: dialkyltetralin sulfonates—by-products resulting from the manufacturing process.

the samples and the standard solutions to reduce losses of LAS by adsorption in the sample loop and injection system. The detector used 225 nm and 295 nm wavelengths for excitation and emission, respectively. Quantification was achieved with external standardization. The method resulted in a partial isomer distribution for each alkyl homolog. A more efficient separation of the phenyl isomers was obtained when using an acetonitrile:water (45:55, v:v) mobile phase containing 0.1 M sodium perchlorate and by employing a Hitachi Gel 3053 analytical column instead of a LiChrosorb RP-18 column [40]. The latter procedure, however, was not demonstrated to be applicable to river water samples. The high injection volume used (500 μL) when analyzing river water is atypical for conventional reversed-phase HPLC and may result in peak broadening. Because of the absence of any sample pretreatment, the method could only be applied to relatively clean samples containing LAS concentrations in excess of 100 μg/L. This seriously limits its application to river waters, particularly those situated in areas with a good network of well-operating municipal sewage treatment plants.

Linder and Allen [41] used HPLC with fluorescence detection to follow the biodegradation of LAS in activated sludge and river die-away studies. In a procedure similar to that of Nakae et al. [39], Millipore filtered samples were directly injected onto a C18 analytical column. The mobile phase consisted of a mixture of tetrahydrofuran:water (51:49, v:v) containing Pic A (tetrabutylamine phosphate) and was run at a flow rate of 1 mL/min. The fluorescence detector was operated at excitation and emission wavelengths of 232 nm and 290 nm respectively. The method was successfully used to determine the disappearance of parent LAS and the appearance of the sulfophenylcarboxylated intermediates. Detection limit was around 50 μg/L for injections of 100–250 μL of nontreated samples. The authors expected that their technique, when coupled with proper concentration steps, would also be applicable to the determination of low levels of LAS and biodegradation intermediates in river waters as well as sewage treatment plant influents and effluents.

Kikuchi et al. [42] developed an analytical procedure for the determination of LAS in sea water, sediment, and fish. Sea water samples were split into liquid and solid phases by filtration. The filter cake was extracted with methanol and the extract was added to the liquid phase. The solution was then passed through a C18 SPE column. The column was subsequently rinsed with water to remove salts, and the LAS was eluted with methanol. The LAS concentration in the eluate was determined using an HPLC separation based on the original work of Nakae et al. [39]. The chromatographic separation was achieved on a C18 analytical column using a mobile phase of acetonitrile:water (60:40, v:v) containing 0.1 M sodium perchlorate. The authors demonstrated that all individual alkyl homologs had similar molar responses for fluorescence detection. Sediment and fish samples were extracted three times with cold methanol. The extract was diluted with distilled water and then further purified using a C18 SPE column as described for the liquid samples. The authors reported good recovery of spiked ^{14}C-labelled C12 LAS from sediment and fish based on liquid scintillation counting. Overall recovery of LAS was in excess of 80% for all types of matrices tested. The salts present in the water sample did not affect the analytical recovery. For fish samples the capacity of a single C18 SPE column was not sufficient and a combination of four columns was needed for efficient recovery from a 50-g wet fish sample. Detection limits were in the order of 0.1 µg/L for sea water, 0.03 µg/g for sediment, and 0.3 µg/g for fish.

Matthijs and De Henau [8] and De Henau et al. [43] developed an HPLC method for the determination of LAS in aqueous environmental samples, sediments, sludges, and soils. Aqueous samples were used without filtering and passed directly onto a C8 SPE column. The column was washed with water and a 40% aqueous methanol solution to eliminate polar substances. This resulted in an improvement in the first part of the HPLC chromatogram. Complete elution of LAS from the column was then achieved with methanol, while hydrophobic matter that would otherwise remain on the HPLC column was retained. This step also efficiently isolated the LAS associated with the suspended solids which were retained on the top of the column. As such, the method determined the concentration of total LAS. The eluate was then dried on a steambath under a stream of nitrogen and kept in a dry state until analysis by HPLC. The isolation of LAS from solid samples such as sludge, sediment, and soil was achieved by extraction with methanol under reflux. Recoveries in excess of 84% were obtained, and it was expected that a continuous extraction of the solids in a Soxhlet apparatus would enhance this result. An aliquot of the methanol extract was then passed through an anion-exchange column to eliminate nonionic surfactants, particularly alkylphenol ethoxylates, which would otherwise interfere in the HPLC separation. LAS was eluted from the column with acidic methanol (methanol:hydrochloric acid, 20:5; v:v). The eluate was diluted with water and adjusted to neutral pH prior to further purification on a C8 SPE column as described for the liquid samples. The HPLC separation is based on reversed-phase separation on a C18 column and using water:acetonitrile as mobile phase containing 0.15 M sodium perchlorate as phase modifier. This is basically the procedure as described by Nakae et al. [39,40], who employed an isocratic system for the LAS determination. The use of a gradient system, as reported by Matthijs and De Henau [8], resulted in a better separation of the individual LAS homologs. Sodium perchlorate was added to obtain sufficient retention of the alkylbenzene sulfonates on the analytical column. A UV detector (230 nm) was used. The capacity factors of the LAS homologs were directly proportional to the concentration of sodium perchlorate in the mobile phase. The concentration of the inorganic salt was limited to 0.15 M to prevent crystallization from the acetonitrile/water mobile phase. Quantitation was achieved by reference to an external calibration curve. The detection limit for total LAS was 10 µg/L for aqueous samples and 0.4 µg/g for solid samples. A comparison was made between the MBAS response and the specific LAS concentration. It was shown that the percentage LAS in the MBAS reading represented only 75, 38, and 30% for samples of raw sewage, effluent, and river water, respectively.

Holt et al. [44] used the method reported by Matthijs and De Henau [8] to determine the fate of LAS in sludge-amended soils. A 4-hour Soxhlet extraction with methanol was employed instead of extraction under reflux with the same solvent. This resulted in a more complete recovery of LAS from soil samples. The recovery of added LAS was generally in excess of 90% compared with 84% for the original reflux technique. In order to minimize contamination it was found necessary to preextract the entire equipment, including the thimbles, with methanol. Fluorimetric detection was found to be more sensitive than UV detection for LAS standards. However, the limit of detection for soil samples with fluorescence (0.2 µg/g) was comparable to the originally reported detection limit with UV detection (0.4 µg/g), due to the magnitude of the blank determinations, which had typical values of about 1 µg LAS.

The method of Matthijs and De Henau [8] was successfully used by Ramon et al. [45] to determine the concentration of LAS in the Tajo River Basin and also formed the basis for the work of Moreno et al. [46] in their study of the biodegradability of LAS in the sewer. In the latter work, the original method was modified by addition of a cation-exchange resin and an additional C18 column, which resulted in increased sensitivity and specificity. Modifications to the original method were also reported by De Henau et al. [47] and by the UK Standing Committee of Analysts [48]. Samples of raw and treated settled sewage were evaporated to dryness on a steambath instead of adding them directly to the top of an SPE column. This modification was made to ensure full recovery of adsorbed LAS in influent samples with a high solids content. The residue was extracted several times with methanol, which was then reduced to a few milliliters by evaporation and diluted with water prior to concentration/clean-up over a reversed-phase SPE column. This approach was successfully used by Feijtel et al. [49] to determine the fate of LAS during sewage treatment as part of a wide AIS (Association Internationale de la Savonnerie et de la Détergence)/CESIO (Comité Européen des Agents de Surface et Intermédiaires Organiques) exercise. In this case the methanol extract of the evaporated aqueous samples was first passed through an anion exchange resin prior to the C18 SPE clean-up. The AIS/CESIO task force on Environmental Surfactant Analysis found that this extra anion-exchange step significantly improved the first part of the HPLC chromatogram by reducing the fluorescence background. Holt et al. [50] found that the evaporation step and the subsequent methanol extraction were not generally applicable to all types of liquor samples. Specifically, for influent samples collected from the Owlwood wastewater treatment plant in the United Kingdom, (as part of the AIS/CESIO monitoring exercise) poor recoveries of added LAS were reported. This is believed to be mainly due to the presence of a high concentrations of calcium and magnesium salts in these samples. It was found that direct application of the unfiltered sample to a C18 SPE column, as described in the Matthijs and De Henau [8] paper, complemented with an anion-exchange step, resulted in good recoveries of LAS from all types of sample studied. Also in this case, the introduction of a strong anion exchange resin was found to reduce the fluorescence background in the LAS chromatograms. This procedure has also been used by other AIS/CESIO member companies as part of the industry monitoring exercise. In addition to the UK [50] and the Dutch [1,49] exercise, the AIS/CESIO method was also successfully applied to determine the fate of LAS during sewage treatment in Italy [51], Spain [52], and Germany [53]. This method, which is considered as the standard method in Europe for determining LAS in a broad range of environmental samples, is detailed in Appendix C.

Inaba and Amano [54] used the above HPLC principles to determine the concentration of LAS in Japanese lake water and sediment. They introduced a sample preparation based on extraction of the sample with 4-methyl-2-pentanone in the presence of an excess of potassium chloride to enhance the extractability of LAS. The sample preparation step is simple, but it is doubtful whether it will result in chromatograms of sufficient quality to determine the concentration of LAS in a broad range of environmental samples.

Matthijs and Stalmans [55] expanded the scope of the method reported earlier by Matthijs and De Henau [8] to the determination of LAS in sea water and estuary samples. The method is based on solvent sublation of the unfiltered samples prior to purification by solid phase extraction and anion-exchange chromatography. The sea water samples (1 L) were solvent sublated with ethyl acetate after addition of sodium chloride and sodium hydrogen carbonate to enhance the extractability. The ethyl acetate phase was then evaporated to dryness and the residue was dissolved in deionized water prior to clean-up over a C8 SPE column and an anionic-exchange column. A final treatment over a second C8 SPE column resulted in a further improvement of the quality of the chromatogram. The recovery of added LAS averaged 84%, and the detection limit was reported to be 0.4 μg LAS/L.

Marcomini and Giger [56] reported an HPLC method for the simultaneous determination of LAS, nonylphenol ethoxylates (NPEO), and nonylphenol (NP) in sewage sludge samples, sludge amended soils, and river sediments. An appropriate amount of sample was transferred into a preextracted paper thimble. C8 and C15 LAS were added as internal standards. A 4-hour Soxhlet extraction with methanol was sufficient to almost completely extract LAS from the sludge samples. Under these conditions, recovery of NP and NPEO, however, averaged only 50%. Quantitative extraction of the nonionic surfactants required addition of sodium hydroxide pellets (20%, w:w) to the dried sludge prior to extraction. The methanol extract was then partly evaporated and diluted up to a ratio 1:1:2

methanol:water:acetone. The acetone contained sodium dodecyl sulfate (SDS) to prevent adsorption of the analytes in the chromatographic system. The effectiveness of SDS as a competitive adsorbing reagent had already been demonstrated by Nakae et al. [39]. After centrifugation of this solution to remove finely suspended material, reversed-phase HPLC on an octylsilica column was performed using gradient elution with a mobile phase system consisting of isopropanol, water, and acetonitrile. Also, in this case, sodium perchlorate was used as a phase modifier. The traditionally used octadecylsilica column was not suitable for the simultaneous determination of LAS, NPEO, and NP because of the long residence times, which resulted in band broadening of the NPEO/NP peak. LAS was separated into the individual alkyl chain lengths while the NPEO/NP coeluted in a single peak. The detection limit for LAS with a fluorescence detector was about 80 ng injected, which corresponds to a concentration of 20 μg/g for sludge samples and 1 μg/g for sediment and soil samples. Information on the oligomeric distribution of NPEO was obtained by normal-phase HPLC on an aminosilica column using an isocratic elution with an hexane:isopropanol (98.5:1.5, v:v) mobile phase.

Marcomini et al. [57] expanded the scope of their HPLC method [56] to the simultaneous determination of LAS, NPEO, and NP in aqueous environmental samples including sewage influent and effluent. They employed a sample concentration and clean-up procedure, which is very similar to the one previously reported by Kikuchi et al. [42]. In principle, samples of influent (50 mL) and effluent (250 mL) were filtered, the filter cake was extracted with methanol, and this extract was added to the liquid sample followed by passage through a C18 SPE column. C8 LAS and C15 LAS were added to the environmental sample as internal standards as well as SDS (7×10^{-3} M) or sodium chloride (8%, w:w) to avoid losses of the analytes during the filtration step. Desorption of the analytes from the C18 SPE column was achieved with acetone which was subsequently diluted with water containing SDS (2×10^{-2} M). The HPLC separation of the alkyl homologs of LAS was achieved using the conditions as described by Marcomini et al. [56]. The detection limit for LAS was 20 μg/L. Brunner et al. [58] used this method to determine the occurrence and behavior of LAS, NPEO, and NP during sewage treatment. Samples of influent, effluent, and sludges were analyzed.

Marcomini et al. [59] further developed a method for the simultaneous determination of LAS, NPEO, and their biodegradation intermediates sulfophenylcarboxylic acids (SPC) and nonylphenol carboxylic acids (NPEC), respectively. They used a combined ion-suppression/ion-pair RP-HPLC system on a C18 column with an aqueous acetonitrile and aqueous methanol mobile phase containing alternatively sodium perchlorate, trifluoroacetic acid, and tetrabutylammonium dihydrogen phosphate as phase modifiers. They found that by combining mobile phases of different compositions and different phase modifiers they could alter the fractionation of the analytical column to obtain an optimal separation for the different analytes. The technique is very suitable for indepth monitoring of laboratory-scale biodegradation studies of LAS and NPEO simultaneously because the disappearance of the parent compounds as well as the formation and subsequent disappearance of the biodegradation intermediates can all be followed, elucidating the biodegradation pathway and kinetics. However, for environmental monitoring purposes, the proposed technique looks too complicated for routine applications, and it would be more appropriate to focus the HPLC separation on one type of surfactant only.

An HPLC method for the determination of LAS in aqueous environmental samples was described by Castles et al. [60]. Acidified (pH 3–4) aqueous samples were concentrated by adsorption onto a C2 SPE column, followed by elution with methanol. It was then further purified by adsorption onto a strong exchange column (SAX) followed by elution with acidified methanol. The eluate was evaporated and the residue was reconstituted in a solution of methanol:water (50:50, v:v). Sodium dodecyl sulfate was added to aid the resolubilization of LAS. C9 LAS was added as internal standard to the original environmental sample, C15 LAS was added to the SAX eluate. The authors reported that the use of a C2 SPE column in the concentration and clean-up step results in a lower fluorescence background compared to the use of a C8 or C18 SPE column. The HPLC chromatographic separation was achieved on a C1 reversed-phase column, using an isocratic elution with a mixture of tetrahydrofuran:water (45:55, v:v) containing 0.1 M sodium perchlorate as phase modifier. The fluorescence detector used excitation and emission wavelengths of 225 and 290 nm, respectively. The use of a C1 analytical column resulted in chromatograms in which all isomers of a given alkyl homolog co-eluted into a single peak, increasing the sensitivity, the interpretation and the quantitation of the method. The detection limit is about 2 µg/L for total LAS when using a 200-mL sample. Field et al. [61] used this HPLC method to determine the fate of LAS and DATS in groundwater. Fast atom bombardment mass spectrometry and tandem mass spectrometry were used to confirm the HPLC results, to differentiate between linear and branched alkylbenzene sulfonate, and to confirm the identity of DATS.

Di Corcia et al. [62] reported on an HPLC method for the determination of LAS in aqueous environmental samples with a simple and efficient sample preparation step based on the use of graphitized carbon black (GCB) (Carbopack B). In principle, this sorbent behaves as a reversed-phase material; however, the presence of positively charged chemical groups on its surface enables it to act as both a nonspecific and an anion-exchange adsorbent. Because the ion-exchange sites on the Carbopack material are limited, saturation of the active sites with subsequent losses of LAS must be avoided. Therefore the sample volume was chosen with due consideration to the nature of the sample. Predescribed volumes of 10, 50, and 200 mL were used for analyzing raw sewage, effluent and river water, respectively. Care was essential when treating highly contaminated samples, e.g., raw sewage samples containing high concentrations of organic acids from industrial origin may lead to losses of LAS. Unfiltered samples were passed through the Carbopack B column. The column was then washed with water, methanol, and a mixture of methylene chloride:methanol (80:20, v:v) acidified with formic acid. LAS was subsequently eluted with a mixture of methylene chloride:methanol (80:20, v:v) basified with tetramethylammonium hydroxide. The eluate was dried and the residue was reconstituted in a mixture of methanol:water (80:20, v:v) prior to injection into the HPLC. The chromatographic separation was achieved on a C8 column using an isocratic system with a mobile phase of methanol:water (80:20, v:v) containing 0.1 M sodium perchlorate. The fluorescence detector was again set at 225 and 290 nm for excitation and emission, respectively. In contrast to the findings of Castles et al. [60], the authors reported that the use of a C8 column resulted in co-elution of all isomers of each LAS homolog into a single peak. This again was seen as an advantage because it simplified the interpretation and quantitation of the chromatograms and increased the sensitivity. The reported limit of detection in river water samples was 0.8 µg/L based on a 200-mL sample volume. Di Corcia et al. [51] applied this methodology to estimating the LAS concentration in samples from sewage treatment plants and compared it with the proposed AISE/CESIO methodology [8,49,50]. The authors reported very comparable results for LAS concentrations and homologs by both procedures when applied to aliquots of the same influent, effluent, and river water samples. Comparison of the HPLC chromatograms reported for both procedures showed, however, that the AISE/CESIO procedure results in a cleaner chromatogram, especially in the C9–C10 alkyl homolog range, which can be better interpreted and quantified. This is most likely related to the use of a strong anion-exchange resin in the AISE/CESIO procedure, which is more efficient in removing interfering fluorescent material than the newly proposed Carbopack B column. The work reported by AISE/CESIO, e.g., by Holt et al. [50], indeed indicated that the use of a SAX column significantly improved the first part of the HPLC chromatogram by reducing the fluorescence background. In terms of speed of analysis, it is clear that the AISE/CESIO method is more time consuming than the Carbopack B procedure because of the extra cleaning step used.

Di Corcia et al. [63] used the principle of the graphitized carbon black (GCB) sample procedure for the simultaneous determination of LAS, NPEO and their carboxylated biodegradation intermediates in samples of raw and treated sewage. The analytes were fractionated on the GCB column by differential elution. The unfiltered environmental samples (10 and 100 mL for influent and effluent samples, respectively) were acidified to pH 3 with concentrated hydrochloric acid (in order to increase the retention volumes of the dicarboxylated forms of SPC) and passed through the GCB column, which was subsequently rinsed with distilled water and methanol. Neutral and very weakly acidic compounds such as NPEO and NP were then selectively eluted from the column with a solvent mixture of methylene chloride:methanol (90:10, v:v); in the next stage of the stepwise desorption, a mixture of 25 mmol/L of formic acid in methylene chloride:methanol (90:10, v:v) was used to elute weakly acidic analytes such as the carboxylated forms of NPEO, and finally a solvent mixture of 10 mmol/L of tetramethylammonium hydroxide in methylene chloride:methanol (90:10, v:v) was used to elute the more acidic LAS and its SPC biodegradation intermediates from the GCB column. Esterification of the carboxylic acids of SPC in the latter eluate was avoided by addition of formic acid in acetonitrile directly after elution from the GCB column. The extract containing LAS and SPC was then evaporated using a water bath at 30°C. The residue was reconstituted in 200 μL of a mixture of water:methanol (50:50, v:v) acidified with 0.2% of trifluoroacetic acid. Twenty μL of this solution was then injected onto a C8 analytical HPLC column. LAS and SPC were separated using a gradient elution program, which consisted initially of 90% solvent A [methylene chloride:methanol (70:30, v:v)] and 10% solvent C [10 mmol/L of tetramethylammonium chloride in methylene choride:methanol (90:10, v:v)]. The concentration of solvent C was then linearly increased to 80% within 50 minutes. The fluorescence detector was set at 225 and 290 nm. The detection limits of the method for LAS and SPC were 10 μg/L and 8 μg/L, respectively, for influent samples (10 mL sample volume) and 1 μg/L and 0.8 μg/L, respectively, for effluent samples (100 mL sample volume). Recovery of standard additions of the target analytes was generally in excess of 89%.

Crescenzi et al. [64] reported on an HPLC method for the simultaneous determination of LAS and dialkyltetralin sulfonates (DATS) in samples of raw and treated sewage, river water, and groundwater. DATS are alicyclic by-products of commercial LAS. The authors used the differential elution procedure as described by Di Corcia et al. [63] to separate LAS and DATS from other organic compounds prior to their liquid chromatographic separation and quantitation. The entire sample was concentrated on a GCB column, and the basic, neutral, and weakly acidic compounds were washed out with a methylene:methanol (90:10, v:v) solvent system acidified with 15 mmol/L of formic acid. LAS and DATS were subsequently eluted from the column with a methylene:methanol (90:10, v:v) solvent mixture made basic with 10 mmol/L tetramethylammonium hydroxide. The HPLC separation of LAS and DATS was achieved on a C8 analytical column and using a gradient elution system. The initial mobile phase consisted of 35% of solvent A (1 mol/L aqueous sodium perchlorate) and 65% of solvent B (methanol). The concentration of solvent B was then increased to 80% within 15 minutes. A fluorescent detector was again used as above. Analogously to LAS, it was assumed that C10–C13 DATS homologs exhibit the same molar response factor. It was found that the use of sodium perchlorate in the mobile phase resulted in higher fluorescence molar quantum efficiencies than the ion-suppression system using, e.g., trifluoroacetic acid. As reported previously by Di Corcia et al. [62], the C8 column packing enabled the co-elution of all isomers of each LAS homolog into a single peak. Identification of the DATS homologs in the chromatogram was made by coupling the HPLC to a mass spectrometer and using selective ion monitoring with m/z values corresponding to the protonated molecular ions relative to both C10–C13 DATS and LAS. The mass spectrometric identification work was performed using trifluoroacetic acid as phase modifier instead of sodium perchlorate because of incompatibility of the electrospray interface to handle solvents containing a relatively high concentration of nonvolatile salts. This could be done because both chromatographic systems result in a similar type of chromatographic profile. The authors used the method to compare the kinetics of biodegradation of LAS and DATS. Detection limits were 10 µg/L for influent, 1 µg/L for effluent, 0.5 µg/L for river water, and 0.1 µg/L for groundwater.

B. Alcohol Ethoxysulfate and Alcohol Sulfate

AES and AS are anionic surfactants which are used in laundry and cleaning detergents as well as in cosmetic products such as shampoos. The typical commercial AES has an alkyl chain length distribution from C12 to C18 and an ethylene oxide (EO) distribution from 0 to 12 with an average EO of 3. AES contain generally up to 20% nonethoxylated material (alkyl sulfate). Since these surfactants are used at much lower volumes than LAS, less attention has been given to the development of component-specific methods to determine their fate in the environment. Furthermore, the development of specific methods is not straightforward because of the lack of any characteristic groups such as an aromatic ring in these molecules. To date, the concentrations of AES and AS have generally been measured by the nonspecific MBAS sumparameter. This approach is acceptable in such studies as following the fate of AES and AS in laboratory-scale biodegradation, when a blank unit is run to compensate for background reading. However, MBAS is not suitable for measuring AES and AS in environmental samples because these anionic surfactants represent only a very small fraction of the total MBAS response.

Neubecker [65] developed an analytical method for the trace analysis of AES in wastewater and surface waters based on the selective cleavage of the sulfate functionality followed by gas chromatographic analysis of the alkyl bromides formed upon reaction of the desulfated material with hydrogen bromide. The unfiltered environmental sample was concentrated using an anion exchange resin. Hydrophobically adsorbed material was removed with a methanol rinse and the retained anionic surfactants were then eluted from the resin with 10% hydrochloric acid in methanol. The volume of the eluate was then reduced to about 10 mL by evaporation on a steam bath. Losses of AES occurred if the eluate was taken to dryness. During the evaporation process the AES is hydrolyzed into the corresponding nonionic ethoxylates. Sulfonated anionic surfactants such as LAS are not hydrolyzed under these conditions. The acidic concentrated eluate was then diluted with water and after addition of magnesium sulfate as salting-out agent, the resulting alkylethoxylates were extracted into chloroform. The chloroform extract was then taken to dryness at about 60°C under a stream of nitrogen. C17 bromide was selected as internal standard because it is not present in environmental samples. The residue was then reacted with hydrogen bromide in acetic acid for 16 hours at 90°C to convert the alcohol ethoxylates in the corresponding alkyl bromides. In this step an amount of 1,2-dibromoethane equivalent to the number of ethoxylate units was also formed. The alkyl bromides were then extracted in toluene and the individual alkyl homologs were quantified by gas chromatography on a column packed with 10% OV-17 on Chromosorb W using temperature programming and FID detection. Quantitation was made using a calibration curve established with a series of alkyl bromide standards in toluene. All areas were normalized to the C17 bromide internal standard response. Since the degree of ethoxylation of AES in environmental samples is not known, an average EO value of 7 was assumed when calculating the AES concentrations from the alkyl bromide responses. Recovery of added AES homologs ranged from 68 to 78%. The sensitivity of the method was estimated to be 1 µg/L for each homolog. Analysis of aqueous environmental samples indicated that the GC response represented between 6 and 13% of the MBAS reading. Contrary to the findings of Matthijs and De Henau [8] for LAS, the percentage of AES in the MBAS reading was similar for influent, effluent, and river water samples. Although this method has been reported to be specific for the determination of AES, it must be recognized that the technique will not resolve AES from AS.

A gas chromatographic method for the selective determination of alkyl sulfates in natural waters has been described by Fendinger et al. [66] (see also Table 4). The environmental samples were concentrated on a C2 SPE column after addition of C17 AS as internal standard. Following a rinse with high-purity water, AS were eluted from the column with methanol and directly passed through a strong anion exchange (SAX) resin. This step eliminates nonionic interferences as well as fatty alcohols. The elimination of the latter is important because they would react in the final derivatization procedure. Elution of AS from the SAX column was achieved with a 20% hydrochloric acid solution in methanol. The eluate was evaporated to dryness at 40°C under a stream of nitrogen. The residue was reconstituted in methanol and further purified over a strong cation exchange (SCX) resin. During this step AS were converted to their hydrogen form, which is required for an efficient subsequent derivatization. The column eluate was evaporated to dryness. Methylene chloride was then added prior to a second evaporation step to remove any traces of water. AS were then converted to the

TABLE 4 Main Features of the Methods for the Determination of AES/AS in Environmental Samples

Ref.	Chemical	Isolation	Detection	Matrix	Detection limit
Neubecker [65]	AES	SAX—acid hydrolysis—extraction—HBr reaction	GC of alkyl bromides	Influent-effluent-river water	5 µg/L
Fendinger et al. [66]	AS	C2 SPE—SAX—SCX—derivatization	GC of trimethylsilyl ethers	Influent-effluent-river water	5 µg/L
Smedes et al. [67]	AS	None	RP-HPLC fluorescence ion pair with acridinium	Clean solutions	—
Nitschke et al. [68]	AS	Acid hydrolysis—extraction—derivatization	RP-HPLC UV phenyl isocyanate derivatives	Biodegradation liquors	—
Popenoe et al. [69]	AES-AS	C2 SPE	Ion spray LC-MS	Influent-effluent	3 µg/L
Scherler et al. [70]	AES-AS	C18 SPE—TLC—extraction	RP-HPLC Conductivity	Influent-effluent-river water	5 µg/L

corresponding trimethylsilyl ethers (TMSE) by reaction with dimethylformamide   and   N,O-bis(trimethylsilyl)trifluoroacetimide   containing   1%
trichloromethylsilane for 1 hour at 80°C. The mechanism of formation of TMSE,
according to the authors, is sulfate ester cleavage and concomitant silylation.
The TMSEs were then quantified by gas chromatography on a di-
methylpolysiloxane capillary column and using FID. Quantitation was made using a calibration curve made with TMSE derived from normal alcohols. The
applicability of the method to the analysis of AS in samples of influent, effluent,
and river water was demonstrated. The detection limit of the method was 5 µg/L.

Limited information has been published on the HPLC determination of AS.
Because AS do not contain a chromophoric group, a derivatization step is required for their detection and to gain sufficient sensitivity. Smedes et al. [67] reported a reversed-phase chromatographic alkyl homolog separation of AS with
postcolumn derivatization using acridinium chloride as counterion. The mobile
phase consisted of a mixture of acetone and water. The ion-association complex
was on-line extracted into chloroform and its intensity was measured fluorimetrically using an excitation wavelength of 400 nm and an emission wavelength of
470 nm. The procedure has the potential to determine AS in the µg/L range when
combined with appropriate sample concentration techniques. It is, however, unlikely that reversed-phase HPLC with postcolumn extraction and derivatization
can be used for environmental applications because it will not distinguish between various types of anionic surfactants. When applied to environmental samples, the response will be largely dominated by LAS, which is expected to be
present at much higher concentrations than other anionic surfactants such as
AES and AS. Nitschke and Huber [68] described an HPLC method for alkyl sulfates based on the decomposition with sulfuric acid (90°C for 1 hour), extraction
of the resulting alcohols in methylene chloride after addition of sodium chloride
to enhance the extraction efficiency, followed by derivatization of the alcohols
with phenyl isocyanate and reversed-phase chromatography with UV detection.
A gradient elution system based on methanol:water was employed. The method
was applied to the analysis of influent and effluent samples of a laboratory-scale
activated sludge plant. The applicability of the method to the analysis of environmental samples is very doubtful because of the limited clean-up steps.

Popenoe et al. [69] described an HPLC method linked to a mass spectrometer
via an ion spray interface for the simultaneous determination of alkyl sulfate and
alkyl ethoxysulfate in aqueous environmental samples. The combination of
LC/MS enables very selective determination of the individual species both by
alkyl and ethoxylate chain length distribution. Filtered environmental samples
were concentrated on a C2 reversed-phase SPE column. The column was washed
with water, and the AS/AES were then eluted from the column with a
methanol:iso-propanol (80:20, v:v) mixture. The eluate was evaporated to dryness at room temperature using a stream of nitrogen. The residue was reconsti-

tuted in a methanol:water (50:50, v:v) mixture to which deuterated AES (C12, EO1) was added as an internal standard. The HPLC separation was achieved on a C8 reversed-phase column, using a water:acetonitrile:ammonium acetate gradient elution system. From the column effluent, about 40 µL/min was passed to the ionspray source via a splitter device. The mass spectrometer was operated in the negative-ion mode. The method was validated for AES/AS with a chain length distribution from C12 to C15 and an ethoxylate distribution from 0 to 8 representing a total of 36 individual AES/AS species. Recovery of standard additions was generally in excess of 90%. The detection limit was in the low µg/L range for the individual species. This method was used to determine the fate of AES during sewage treatment as reported by Matthijs et al. [1]. Mass spectrometry is the only technique that can deliver this detailed information on the environmental concentration of the individual species, including both the alkyl chain length and EO distribution. It needs, however, to be recognized that MS is a very expensive technique that requires an expert operator and that is not readily available in many laboratories. Therefore this technique should be positioned as a benchmark method for the calibration and validation of other methods, which, although less selective, may be better suited for routine applications.

Scherler et al. [70] presented a method for the simultaneous determination of AES/AS using a combination of thin-layer chromatography (TLC) and HPLC. After addition of 0.5% methanol, the aqueous environmental samples were concentrated on a reversed-phase SPE column. Following elution with methanol the anionic surfactants were separated by preparative TLC on a silica plate using an acetone:tetrahydrofuran mobile phase. This procedure enabled separation of LAS, SAS, and AS/AES. The surfactant zones were then scratched from the plate and the separated surfactants were then isolated from the silica and quantified by separate HPLC analysis on a C18 analytical column using a water:acetonitrile:sodium acetate gradient elution system. AES/AS were quantified by conductivity detection. The detection limit was in the low µg/L range for a 1-L sample. Recovery of standard additions ranged from 70 to 90%. The procedure has been shown to be applicable to the analysis of SAS and LAS as well. For LAS, UV detection is used instead of conductivity detection.

C. Secondary Alkane Sulfonate

SAS are a minor group of anionic surfactants which are mainly used in liquid cleaning formulations. The molecule consists of a mixture of sulfonated paraffins with an alkyl homolog distribution from C13 to C17 and isomers in which the sulfonate group is located at any carbon position except the terminal one. The product comprises mainly monosulfonic acids, but a maximum level of 10% disulfonic acid may be present depending on the synthesis process. In view of its low consumption volume, less attention has been given to the development of specific analytical methods for its determination in environmental samples. In addition, the absence of a chromophoric group necessitates derivatization of the molecule prior to its chromatographic analysis.

Klotz [71] presented a method for the determination of SAS in aqueous and solid environmental matrices based on TLC (see also Table 5). The surfactants present in aqueous environmental samples were first concentrated by solvent sublation into ethyl acetate. The organic extract was then evaporated and the residue was dissolved in methanol. Clean-up of the extract was achieved using anion-exchange chromatography and C18 reversed-phase chromatography. SAS were then separated from other anionic surfactants on a silica TLC plate, using a double development system consisting of methanol and a mixture of acetone:tetrahydrofuran (90:10, v:v) as mobile phases, respectively. Detection was made by fluorescence at 365 nm after derivatization with Prinulin reagent. Quantitation was achieved by an external calibration method. Solid environmental samples such as sewage sludge were Soxhlet-extracted with methanol for 3 hours. The methanol extract was then reduced to a volume of about 5 mL, diluted with 50 mL of water, and this mixture was then extracted with chloroform after addition of distearyl dimethyl ammonium chloride as ion-pairing reagent. The chloroform extract was evaporated to dryness, following which the residue was dissolved in methanol and passed through a strong cation-exchange resin to isolate the SAS from the ion pair. The eluate was then further treated by TLC as described for the aqueous samples. The detection limit was about 1 µg/L for aqueous samples when using a sample volume of 4 L and about 0.1 g/kg for sludge when using a 0.5-g sample. The recovery of the method was generally in excess of 80%. The TLC method requires a series of labor intensive clean-up steps but permits the determination of both mono- and disulfonated SAS. No information is obtained on the alkyl homolog distribution.

Field et al. [72–74] developed a method for the selective determination of SAS in aqueous environmental samples using injection port derivatization gas chromatography coupled to mass spectrometry. Unfiltered samples were concentrated on a C18 disk after addition of C12-SAS as an internal standard. The disks were then dried using a stream of nitrogen and subsequently eluted with chloroform containing 0.025 M tetrabutyl ammonium (TBA) hydrogen sulfate. The latter was added as ion-pair reagent to enhance the elution efficiency from the disk because of the limited solubility of SAS in chloroform. The combined extracts of three subsequent extractions were evaporated to dryness, and the residue was redissolved in chloroform after addition of C18-SAS as internal standard. An aliquot of this solution was then introduced into the GC injector, which was operated at 300°C. In this step SAS were converted into their butyl esters. Separation of the alkyl homologs was achieved using a temperature gradient procedure.

TABLE 5 Main Features of the Methods for the Determination of SAS in Environmental Samples

Ref.	Chemical	Isolation	Detection	Matrix	Detection limit
Klotz [71]	SAS	Liquids: Solvent sublation—SAX—C18 SPE	TLC Fluorescence with Prinulin	Influent-effluent-river water	1 µg/L
		Solids: CH₃OH extr.—CHCl₃ extr.—SCX		Sludge	100 µg/g
Field et al. [72]	SAS	C18 SPE—derivatization with tetrabutyl ammonium hydrogen sulfate	GC-MS Butyl esters	Influent-effluent	20 µg/L
Scherler et al. [70]	SAS	C18 SPE—TLC—extraction	RP-HPLC Conductivity	Influent-effluent-river water	5 µg/L

Detection was made using a mass selective detector with electron impact ionization and operated in selective ion mode. Quantitation was done using external calibration technique; calibration curves were constructed with a commercial mixture of SAS. The distribution of SAS in the particulate and dissolved phases was determined by filtering the environmental samples. Therefore a glass fiber filter was placed on top of the C18 disk and the filter and disk were then treated separately following the procedure described above. The detection limit was reported to be 20 μg/L. Solid samples such as sludges were extracted using the ion-pair technique combined with supercritical fluid extraction (SFE). Injection-port derivatization and subsequent GC analysis was achieved as described above. The ion-pair derivatization represents a simple and efficient technique compared to traditional derivatization methods. In addition, the use of GC results in a separation of the individual alkyl homologs and isomers. Because of the specificity of the mass-selective detection, only a limited sample preparation is required, in this case by the use of a single C18 disk. The only drawback of the reported methodology is that it determines only the monosulfonated SAS. Although the focus of this work was on SAS, the authors have demonstrated that the reported methodology was also applicable to the analysis of LAS in environmental samples.

The combination of TLC and HPLC developed by Scherler et al. [70] and described as part of the AES/AS methodology was proven to be applicable also to the analysis of SAS in aqueous environmental samples.

APPENDIX A: METHYLENE BLUE ACTIVE SUBSTANCES METHODOLOGY FOR THE DETERMINATION OF ANIONIC SURFACTANTS IN BIODEGRADATION TEST LIQUORS [6]

Reagents. Prepare an alkaline buffer solution by dissolving 24 g of sodium hydrogen carbonate and 27 g of sodium carbonate in 1 L of deionized water. Prepare a neutral methylene blue solution by dissolving 0.35 g of methylene blue in 1 L of deionized water. Prepare this solution at least 24 hours before use. Prepare an acid methylene blue solution by dissolving 0.35 g of methylene blue in 500 mL of deionized water. Add 6.5 mL of concentrated sulfuric acid and bring the volume to 1 L with deionized water. Prepare the solution at least 24 hours before use.

Procedure. Filter the aqueous environmental samples directly after sampling, discarding the first 100 mL of filtrate. Transfer a sample volume containing about 20–250 μg of MBAS to a separating funnel and dilute to 100 mL with deionized water. Typically, sample volumes of 10 mL for influent, 50 mL for effluent, and 100 mL for river water are required. Add 10 mL of alkaline buffer solution, 5 mL of neutral methylene blue solution, and 15 mL of chloroform to the separating funnel. Shake this solution gently for 1 minute. Allow the layers to separate and transfer the clear chloroform layer into a second separating funnel containing 110 mL of deionized water and 5 mL of acid methylene blue solution. Shake the solution for 1 minute. Allow the layers to separate, filter the chloroform layer through a piece of cotton wool moistened with chloroform, and transfer the filtrate into a 50-mL graduated flask. Repeat the procedure twice using 10-mL portions of chloroform. Combine the three extracts in the 50-mL graduated flask and dilute to volume with chloroform. Measure the absorbance of the extract at 650 nm in a 1- to 5-cm cell against a chloroform reference. Prepare a reagent blank by repeating the above procedure using 100 mL of deionized water instead of the sample. The absorption maximum of the blank reading should not be higher than 0.02. Subtract the reagent blank from the sample reading and determine the anionic surfactant concentration using a calibration curve (range 0–200 μg) prepared from a suitable LAS standard material.

APPENDIX B: AZURE A ACTIVE SUBSTANCES METHODOLOGY FOR THE DETERMINATION OF ANIONIC SURFACTANTS IN BIODEGRADATION TEST LIQUORS [22]

Reagents. Prepare a sulfuric acid solution (0.1 N) by dissolving 2.8 mL of concentrated sulfuric acid in 1 L deionized water. Prepare the Azure A solution by dissolving 40 mg Azure A in 5 mL 0.1 N sulfuric acid; transfer to a 100-mL volumetric flask and adjust to the mark with deionized water. Prepare a 1000 mg/L stock solution and a 1 mg/L work solution of an LAS reference material.

Procedure. Transfer a sample volume containing about 1–30 μg of anionic active substances in a separating funnel and dilute to 50 mL with deionized water. Typically, sample volumes of 1 mL for influent, 5 mL for effluent, and 50 mL for river water are required. Add 5 mL of 0.1 N sulfuric acid solution, 1 mL of Azure A solution, and 10 mL of chloroform. Shake vigorously for 2 minutes and allow the layers to separate. Drain off the chloroform layer and measure the absorbance of the extract at 618 nm in a 1-cm cell against a chloroform reference. Prepare a reagent blank by repeating the above procedure using 50 mL of deionized water instead of the sample. Subtract the reagent blank from the sample reading and determine the anionic surfactant concentration using a calibration curve (range 0–25 μg) prepared from a suitable LAS standard material.

APPENDIX C: ANALYTICAL METHODOLOGY FOR THE DETERMINATION OF LINEAR ALKYLBENZENE SULFONATE IN ENVIRONMENTAL SAMPLES USING HPLC [8,49,50]

Isolation and Concentration. Precondition a C18 solid phase extraction (SPE) column (6 cm³, Varian or equivalent) by washing with 10 mL of methanol followed by 10 mL of water at a flow rate of about 1 mL/min. Do not let the column dry out once solvated. Pass a representative aliquot of a well-mixed aqueous environmental sample (10 mL for raw and settled influent, 50 mL for effluent, and 250 mL for river water) through the column at a flow rate of 1 mL/min. Wash the column with 5 mL of water and 2 mL of methanol:water (30:70, v:v) to remove any inorganics and highly polar material. Air-dry the column for several minutes and elute LAS from the C18 column with 5 mL of methanol. Pass the eluate through a strong anion exchange column (3 cm³, Varian or equivalent) which has been preconditioned with 10 mL of methanol. Elute LAS from the anion exchange column with 2 mL of methanol:hydrochloric acid (80:20, v:v). Evaporate the eluate to dryness at ca. 40°C under a stream of nitrogen.

Soxhlet-extract a dried and homogenized solid environmental sample (0.25 g for sludge and sediment) with 150 mL of methanol for approximately 4 hours. Adjust the volume of the extract to 200 mL with methanol and pass an aliquot (10 mL for sludge and 100 mL for sediment) through a preconditioned strong anion-exchange column. Elute LAS from the anion exchange column with 2 mL of methanol:hydrochloric acid (80:20, v:v). Dilute the eluate to about 50 mL with suprapur water and adjust to pH 7 with 1 M sodium hydroxide. Pass this solution through a preconditioned C18 SPE column and wash the column with 2 mL of methanol:water (30:70, v:v). Elute LAS from the C18 column with 5 mL of methanol. Evaporate the eluate to dryness at ca. 40°C under a stream of nitrogen.

HPLC Analysis. Resuspend the residue from the clean-up and concentration steps in 0.5 mL of mobile phase and analyze by HPLC. For influent samples dilute this solution 10 times prior to HPLC analysis. Inject 25 µL onto a Chrompack Spherisorb ODS-2 HPLC column (30 cm × 3.9 mm) (or equivalent, e.g., µBondapak C18 column). Run the chromatographic separation isocratically at a flow rate of 1 mL/min using a mobile phase of suprapur water:methanol (16:84, v:v) mixture containing 0.0875 M sodium perchlorate. Operate the fluorescence detector at an excitation wavelength of 232 nm and an emission wavelength of 290 nm. Identify the LAS alkyl homologs on the basis of their retention time by comparison with a chromatogram of a LAS reference sample. Quantify the LAS concentration using a five-point calibration curve made from LAS solutions prepared in the mobile phase at concentrations between 0 and 20 mg/L. Renew the standard work solutions daily from a 1 g/L LAS stock solution prepared in suprapur water.

REFERENCES

1. E. Matthijs, M. S. Holt, A. Kiewiet, and G. B. J. Rijs, Chim. Oggi *14*:9 (1996).
2. G. P. Edwards and M. E. Ginn, Sewage Ind. Wastes *26*:945 (1954).
3. J. Longwell and W. D. Maniece, Analyst *80*:167 (1955).
4. D. C. Abbott, Analyst *87*:286 (1962).
5. International Organisation for Standardization. *Determination of Anionic Surfactants by the Methylene Blue Spectrometric Method*, ISO 7875-1, 1984.
6. Deutsche Einheitsverfahren zur Wasser, Abwasser und Schlammuntersuchung, DIN 38 409 - H23 - 1, 1980.
7. EEC Directive 4311, 1982.
8. E. Matthijs and H. De Henau, Tenside Surf. Deterg. *24*:193 (1987).
9. E. C. Hennes and R. A. Rapaport, Tenside Surf. Deterg. *26*: 141 (1989).
10. H. A. Painter, in *The Handbook of Environmental Chemistry* (N. T. de Oude, ed.), Springer-Verlag, Berlin, 1992, p. 27.
11. R. Wickbold, Tenside Surf. Deterg. *13*:32 (1976).
12. J. D. Fairing and F. R. Short, Anal. Chem. *28*:1827 (1956).
13. H. L. Webster and J. Halliday, Analyst *84*:552 (1959).
14. C. P. Ogden, H. L. Webster, and J. Halliday, Analyst *86*:22 (1961).
15. Q.W. Osburn, J. Am. Oil Chem. Soc. *63*:257 (1986).
16. R. I. Sedlak and K. A. Booman, Soap/Cosmet./Chem. Specialities *4*:44 (1986).
17. W. A. Moore and R. A. Kolbenson, Anal. Chem. *28*:161 (1956).
18. J. van Steveninck and J. C. Riemersma, Anal. Chem. *38*:1250 (1966).
19. D. E. McGuire, F. Kent, L. L. Miller, and G. J. Papenmeier, J. Am. Water Works Assoc. *54*:665 (1962).
20. R. W. G. Cropton and A. S. Joy, Analyst *88*:516 (1963).
21. S. Motomizu, S. Fujiwara, A. Fujiwara, and K. Toei, Anal. Chem. *54*:392 (1982).
22. W. A. M. den Tonkelaar and G. Bergshoeff, Water Res. *3*:31 (1969).
23. P. T. Crisp, J. M. Eckert, and N. A. Gibson, Anal. Chim. Acta *78*:391 (1975).
24. P. T. Crisp, J. M. Eckert, N. A. Gibson, G. F. Kirkbright, and T. S. West, Anal. Chim. Acta *87*:97 (1976).
25. C. G. Taylor and B. Fryer, Analyst *94*:1106 (1969).
26. C. G. Taylor and J. Waters, Analyst *97*:533 (1972).
27. E. A. Setzkorn and A. B. Carel, J. Am. Oil Chem. Soc. *40*:57 (1963).
28. R. D. Swisher, J. Am. Oil Chem. Soc. *43*:137 (1966).
29. W. T. Sullivan and R. D. Swisher, Environ. Sci. Technol. *3*:481 (1969).
30. J. Waters and J. T. Garrigan, Water Res. *17*:1549 (1983).
31. H. Hon-Nami and T. Hanya. J. Chromatogr. *161*:205 (1978).
32. H. Hon-Nami and T. Hanya, Jpn. J. Limnol. *41*:1 (1980).
33. H. Hon-Nami and T. Hanya, Water Res. *14*:1251 (1980).
34. J. McEvoy and W. Giger, Environ. Sci. Technol. *20*:376 (1986).
35. M. L. Trehy, W. E. Gledhill, and R. G. Orth, Anal. Chem. *62*:2581 (1990).
36. C. F. Tabor and L. B. Barber, Environ. Sci. Technol. *30*:161 (1996).
37. J. A. Field, D. J. Miller, T. M. Field, S. B. Hawthorne, and W. Giger, Anal. Chem. *64*:3161 (1992).

38. S. B. Hawthorne, D. J. Miller, D. E. Nivens, and D. C. White, Anal. Chem. *64*:405 (1992).
39. A. Nakae, K. Tsuji, and M. Yamanaka, Anal. Chem. *52*:2275 (1980).
40. A. Nakae, K. Tsuji, and M. Yamanaka. Anal. Chem. *53*:1818 (1981).
41. D. E. Linder and M. C. Allen, J. Am. Oil Chem. Soc. *59*:152 (1982).
42. M. Kikuchi, A. Tokai, and T. Yoshida, Water Res. *20*:643 (1986).
43. H. De Henau, E. Matthijs, and W. D. Hopping, Int. J. Environ. Anal. Chem. *26*:279 (1986).
44. M. S. Holt, E. Matthijs, and J. Waters, Water Res. *23:*749 (1989).
45. M. T. G. Ramon, I. Ribosa, J. Sanchez, and F. Comelles, Tenside Surf. Deterg. *27*:118 (1990).
46. A. Moreno, J. Ferrer, and J. L. Berna, Tenside Surf. Deterg. *27*:312 (1990).
47. H. De Henau, E. Matthijs, and E. Namkung, in *Organic Contaminants in Waste Water, Sludge and Sediment.* (D. Quaghebeur, I. Temmerman, and G. Angeletti, eds.), Elsevier, New York, 1989.
48. HMSO, *Methods for the Examination of Waters and Associated Materials, LAS and Alkylphenol Ethoxylates in Waters, Wastewaters and Sludges by High Performance Liquid Chromatography*, 1993
49. T. C. J. Feijtel, E. Matthijs, A. Rottiers, G. B. J. Rijs, A. Kiewiet, and A. De Nijs, Chemosphere *30*:1053 (1995).
50. M. S. Holt, J. Waters, M. H. I. Comber, R. Armitage, G. Morris, and C. Newbery, Water Res. *29*:2063 (1995).
51. A. Di Corcia, R. Samperi, A. Bellioni, A. Marcomini, M. Zanette, K. Lemnr, and L. Cavalli, Riv. Ital. Delle Sostanze Grasse *71*:467 (1994).
52. J. Sanchez Leal, M. T. Garcia, R. Tomas, J. Ferrer, and C. Bengoechea, Tenside Surf. Deterg. *31*:253 (1994).
53. P. Schöberl, H. Klotz, R. Spilker, and L. Nitschke, Tenside Deterg. Surf. *31*:243 (1994).
54. K. Inaba and K. Amano, Int. J. Environ. Anal. Chem. *34*:203 (1988).
55. E. Matthijs and M. Stalmans, Tenside Deterg. Surf. *30*:29 (1993).
56. A. Marcomini and W. Giger, Anal. Chem. *59*:1709 (1987).
57. A. Marcomini, S. Capri, and W. Giger, J. Chromatogr. *403*:243 (1987).
58. P. H. Brunner, S. Capri, A. Marcomini, and W. Giger, Water Res. *22*:1465 (1988).
59. A. Marcomini, A. Di Corcia, R. Samperi, and S. Capri, J. Chromatogr. *644*:59 (1993).
60. M. A. Castles, B. L. Moore, and S. R. Ward, Anal. Chem. *61*:2534 (1989).
61. J. A. Field, L. B. Barber, E. M. Thurman, B. L. Moore, D. L. Lawrence, and D. A. Peake, Environ. Sci. Technol. *26*:1140 (1992).
62. A. Di Corcia, M. Marchetti, R. Samperi, and A. Marcomini, Anal. Chem. *63*:1179 (1991).
63. A. Di Corcia, R. Samperi, and A. Marcomini, Environ. Sci. Technol. *28*:850 (1994).
64. C. Crescenzi, A. Di Corcia, E. Marchiori, R. Samperi, and A. Marcomini, Water Res. *30*:722 (1996).
65. T. A. Neubecker, Environ. Sci. Technol. *19*:1232 (1985).

66. N. J. Fendinger, W. M. Begley, D. C. McAvoy, and W. S. Eckhoff, Environ. Sci. Technol. *26*:2493 (1992).
67. F. Smedes, J. C. Kraak, C. F. Werkhoven-Goewie, U. A. Brinkman, and R. W. Frei, J. Chromatogr. *247*:123 (1982).
68. I. Nitschke and I. Huber, Fresenius J. Anal. Chem. *346*:453 (1993).
69. D. D. Popenoe, S. J. Morris III, P. S. Horn, and K. T. Norwood, Anal. Chem. *66*:1620 (1994).
70. D. Scherler, M. Schmitt, and H. Waldhoff, Presentation at symposium *Wasch- und Reinigungsmittel* of the German Chemical Society, Garmisch-Partenkirchen, 1995.
71. H. Klotz, Presentation at symposium *Surfactant Monitoring* of the German Chemical Society, Hannover, 1994.
72. J. A. Field, D. J. Miller, T. M. Field, S. B. Hawthorne, and W. Giger, Anal. Chem. *64*:3161 (1992).
73. J. A. Field, T. M. Field, T. Poiger, and W. Giger, Environ. Sci. Technol. *28*:497 (1994).
74. J. A. Field, T. M. Field, T. Poiger, H. Siegrist, and W. Giger, Water Res. *29*:1301 (1995).

4

Molecular Spectroscopy I—Infrared and Raman Spectroscopy

THOMAS M. HANCEWICZ   Department of Advanced Imaging and Measurement, Unilever Research US, Edgewater, New Jersey

I. INTRODUCTION

Vibrational spectroscopy has held a prominent place in the routine analysis of surfactants for many years. There are three main categories of vibrational spectroscopy that provide useful structural information in the analysis of organic and inorganic molecules: mid-infrared (IR), Raman, and near-infrared (NIR) spectroscopies. Preeminent among these techniques is mid-infrared spectroscopy. Traditionally, IR spectroscopy has been the main vibrational spectroscopic technique employed in the surfactant industry for analysis of many types of common materials. Particular importance has been placed on the application of IR to raw material analysis, quality control, and identification of sample fractions extracted from fully formulated products. While there are a multitude of other applications, these three are by far the most important.

The main advantage of IR over other potentially useful techniques such as Raman, NIR, nuclear magnetic resonance (NMR), and mass spectrometry (MS) is that it is fairly easy to perform and somewhat straightforward to interpret given some practice. Most chemists would in fact have had a fair amount of experience using IR instrumentation and interpreting IR spectra through their normal university course of study, by hands-on industrial experience, or both. An introduction to infrared instrumentation and a discussion of all the possible ways infrared spectroscopy can be used to analyze materials is beyond the scope of this chapter. Several very good sources for this type of information on the subject are available [1–11]. A particularly good overview of infrared spectrometry and data-handling considerations for surfactant analysis is given by Clarke [12], and the interested reader is referred to that text. The focus in this chapter will be on the vibrational spectroscopy methodology particularly important for the investigation of anionic surfactants, and consideration will be given to those aspects of vibrational spectroscopy that are particularly appropriate for the analysis of these types of materials.

Raman spectroscopy is another of the vibrational spectroscopic techniques that is extremely useful in the analysis of surfactants, particularly those in which the hydrophile is generally inorganic in nature (sulfate, carbonate, phosphate, etc.). Despite having a much longer history than IR, Raman spectroscopy has suffered from being a technique that is very expensive, difficult to interpret, and delicate to perform. This, however, has changed in recent years with the emergence of low-cost Fourier transform (FT)–Raman systems and also with improvements in modern dispersive systems. A closer look at Raman spectroscopy will be provided later in this chapter.

Near infrared spectroscopy is another useful infrared technique and is used primarily for quality control related to factory-based applications. For this reason, this chapter will only present an overview description of the technique and will leave a detailed review to other sources [3,8,10].

II. THEORY

It is instructive to present an overview of vibrational spectroscopy theory rather than a full discussion of the theoretical physics of the phenomenon. A conceptual approach to the description of the theory of infrared, Raman, and near-infrared spectroscopy is presented here. Hopefully, this will provide a clear picture of the subject matter and a working idea of the complex physical phenomena being described. A theoretical treatment of the ideas presented here can be found elsewhere in a more rigorous form [13–16].

A. Mid-Infrared Spectroscopy

The mid-infrared region of the electromagnetic spectrum extends from approximately 4000 to 200 wavenumbers (cm^{-1}), with nearly all organic and inorganic compounds absorbing radiation somewhere in this region. A practical description of infrared spectroscopy starts with the fundamental idea of absorption of light. In this respect it is very much like UV/VIS spectroscopy, in that light (from an infrared source in this case) is absorbed by a sample placed in its path between the source and a detector. The physical form of the sample may be solid, liquid, or gas, but in each case the result is pretty much the same: some frequencies of the incident IR light will be absorbed by the sample and others will not. The frequencies of light that are absorbed provide a clue to the structure of the material, in particular, the vibrational frequencies of the chemical bonds present in the material. These ultimately identify the bonds that hold the molecule together. Different materials will absorb different frequencies of the incident light depending on the chemical nature of the compounds they are made from and, therefore, the different types of chemical bonds present in the material. The energy transfer between incident light and molecule is accomplished through excitation of electric dipoles between adjacent atoms or groups of atoms. This leads to the fundamental requirement for IR activity: there must be a change in dipole moment for a molecular vibration to be IR active. What this means in simple terms is that there must be a change in charge (electron) distribution in the molecule as it absorbs light. If the charge distribution remains constant, then there is no net absorption of incident light at that particular frequency.

Functional groups in a molecule are the different bonds or groups of associated bonds that give rise to molecular vibrations and, hence, the infrared spectrum. For example, a simple alkyl fatty acid such as stearic acid is composed of a long alkyl chain of C—H bonds (CH_3 and CH_2 functional groups) as well as C=O, C—O, and O—H bonds of the acid group (—COOH carboxylic acid functional group). The individual bonds are referred to as functional groups (C—H, C—O, O—H, C=O), but collections of associated bonds (CH_3, CH_2, and COOH) are also considered to be functional groups. Associated with each functional group are characteristic frequencies of absorption in the IR spectrum. It is the change in charge distribution (dipole moment) within these groups that determines the infrared activity of the compound. A list of simple functional groups of interest in surfactant chemistry is given in Table 1. This table will be expanded in more detail later in the chapter when key anionic surfactant functional groups are discussed on an individual basis.

The combination of absorptions of all functional groups present in a compound produces the complete infrared spectrum. The spectrum is simply a plot of how much light is absorbed at each frequency by the sample, and so the IR spectrum of different materials will be characteristic and distinct for materials made up of different compounds as long as they have different functional groups. For IR spectra the frequency is plotted in wavenumber (cm^{-1}) units to make the numbers easier to work with.

This provides IR spectroscopy with its most powerful characteristic: materials that are different from each other can be distinguished by their IR spectra. Conversely, it also allows unknown materials to be matched against spectra of known compounds in order to determine the identity of the unknown or to at least determine what functional groups are present. This is perhaps an oversimplification since the spectra of very similar compounds (similar functional group composition) often exhibit spectra that are essentially identical (e.g., C_{14} alkyl ester vs. a C_{16} alkyl ester). Differences in physical characteristics of the sample can also alter the appearance of an IR spectrum.

B. Raman Spectroscopy

Somewhat less traditional, but with a longer history in the surfactant industry, is Raman spectroscopy. Although a very popular technique before World War II, it eventually gave way to IR as the dominant vibrational spectroscopic technique subsequent to the war due to the sudden availability of cheap, sensitive mid-IR detectors (a by-product of the war effort). Although very similar to IR spectroscopy in the spectra produced, there are many types of applications in which Raman spectroscopy is superior to IR.

Today, FT-based instruments with very fast data acquisition by dedicated computers dominate the IR industry. Raman instrumentation has also edged in the FT direction, though somewhat more reluctantly due to the popularity and excellent performance of charge coupled device (CCD) detector–based dispersive systems that are capable of outperforming the FT system. These advances in both FT and dispersive Raman instrumentation have begun to favor Raman spectroscopy over IR spectroscopy for many applications.

A Raman spectrum is produced from the same vibrations that give rise to an infrared spectrum. However, Raman is a scattering phenomenon rather than an absorption phenomenon and thus occurs via a different mode of energy transfer between incident light and sample.

TABLE 1 IR Frequencies in Wavenumbers for Major Anionic Surfactant Functional Groups

Functional group	C—H, N—H, O—H stretch region	C=C, C=C stretch region	C—H, N—H, bend, C—N stretch region	P=O, S=O region	C=O, P=O, S=O region	C—H wag/rock region
Alkane C—H	3050–2800		1475–1300			850–700
Alkene H—C=C	3100–2950	1690–1620				700–520
Aromatic H—C=C	3150–3000	1650–1475	1220–975			900–675
Ether C—O—C					1300–1000	620–500
Alcohol C—O—H	3700–3100				1450–1000	
Carbonyl acid	3300–2900	1740–1660			1450–1200	950–860
Carbonyl ester		1750–1720			1325–1100	
Carbonyl ketone		1740–1650			1320–1175	
Amide N—C=O	3500–3050	1720–1475	1690–1580, 1390–1040		1200–1100	
Amine H—N—C	3500–3200					910–640
Sulfate ionic				1150–1085		615–500
Sulfate covalent				1425–1150		650–550
Sulfonate ionic				1250–1010		700–600
Sulfonate covalent				1420–1320	1200–1150	620–525
Phosphate ionic					1120–1030	
Phosphate covalent				1300–1240	1050–975	
Carbonate ionic		1440–1365			950–900	
Ammonium ion	3075–3025		1430–1370			

In infrared spectroscopy there is an absorption of photons related to a change in permanent dipole moment (charge distribution) that leads to a characteristic spectrum. In Raman spectroscopy it is the absorption of photons related to a change in polarizability of chemical bonds that determines activity, but the absorbed light is then reemitted rather than being permanently absorbed. A change in polarizability means that as the bonds absorb incident light, there is an induced charge between the atoms either singularly or in addition to the permanent dipole moment. The change in molecular vibration of this induced charge distribution is referred to as a change in polarizability. The result of this form of energy transfer is that incident photons of one energy are absorbed and reemitted at a different energy, a phenomenon called inelastic scattering. The energy difference between the incident and reemitted light is related to the vibrational energy of the chemical bonds of the molecule and so different functional groups will result in characteristic frequency shifts. The Raman spectrum is produced as scatter intensity versus frequency shift. The frequency units for Raman spectroscopy are also in cm^{-1}.

There is a very large difference in magnitude between infrared absorption and Raman scattering intensities. The Raman effect is so weak that a much higher intensity source is required in order to obtain a spectrum with acceptable signal-to-noise levels. An infrared source is a low-intensity broad band source, while in contrast, a Raman source is a high-powered monochromatic source such as a laser. The Raman experiment is essentially an emission experiment since it is the absolute response at each wavelength that produces the spectrum, not a differential measurement as in IR. A theoretical treatment of Raman spectroscopy is given by Graybeal [16] and is suggested for the interested reader, who is also referred to Hendra et al. [2] and Olsen [10] for a detailed descriptions of the Raman technique with applications.

C. Infrared and Raman Correlation

Since infrared and Raman spectra arise from basically the same molecular vibrations, why do we need both? The answer to that question lies buried in the rather complicated quantum mechanical description of both phenomena. A simplistic explanation is that the two techniques do not really measure the same thing at all, and that there are very real differences in the spectra obtained from a sample using both techniques. The information provided by the two techniques are quite complementary in nature; each provides a different piece to the same puzzle, and the key to the difference is the symmetry properties of the compounds (functional groups) being investigated.

Raman and infrared spectroscopy are considered complementary techniques because of their requirements for activity. As has already been stated, a change in polarizability during a vibrational transition is required for Raman activity, and a change in dipole moment during the same transition is needed in order for the vibrational mode to be infrared active. While a complete description of the Raman effect in not necessary for the purposes of this chapter, a few rules regarding Raman and infrared transition activity are useful when trying to decide which of the two techniques might prove more valuable.

1. The Mutual Exclusion Rule

Without exception, for all molecules with a center of symmetry, vibrational transitions that are allowed in the infrared are forbidden in Raman, and vice versa. Parenthetically, it should be pointed out here that while a transition may be forbidden quantum mechanically, this does not mean that there will be no observable peak in the spectrum. This will only be the case with vibrational modes that behave in an ideal manner. In most cases a forbidden transition implies one that has a very small probability of occurring. The observed result of this is that a typical "forbidden" transition will have a very small peak in the spectrum.

2. Mutually Allowed Transitions

It is generally true, but not always so, that for all molecules that do not have a center of symmetry, the vibrational transitions will be both infrared and Raman active. It is theoretically possible for molecules without a center of symmetry to be either infrared or Raman inactive, so this statement is not absolute. A corollary to this generalization is that when a transition is both infrared and Raman active, those vibrational modes that are symmetric in nature will be strongly Raman active and weakly infrared active, and those that are asymmetric in nature will be strongly infrared active and weakly Raman active.

3. Mutually Forbidden Transitions

In a very small number of molecules there are transitions that are both Raman inactive and infrared inactive since there is no change in either dipole moment or polarizability.

D. Near-Infrared Spectroscopy

As Raman and mid-IR spectra are produced as a result of fundamental vibrational transitions in molecules, NIR spectra result from overtones of fundamental transitions or from combination bands that originate from both fundamentals and overtones. Fundamental vibrations are the very intense absorptions observed in the mid-infrared region ($4000-1000$ cm^{-1}) that result from the direct first-order vibrational transitions. Overtone bands are the results of higher-order vibrational modes arising from anharmonic (nonideal) vibrational distortions of the fundamental modes. Combination bands are linear sums of fundamental bands and/or overtone bands with similar vibrational characteristics. This combination of both overtones and combination bands makes NIR spectra very difficult to interpret in a meaningful way. NIR spectra typically appear as very broad bands of overlapped spectral features. Although difficult to interpret, NIR spectra can be interpreted in a qualitative way through use of correlation charts [17]. Extraction of useful quantitative information from NIR data is not easily accomplished manually since there is no simple way to resolve the overlapped information. However, the application of multivariate mathematics/statistics (chemometric techniques) [18–21] to this problem has greatly increased the amount of useful information that the user is able to generate and interpret.

The role of NIR in surfactant analysis has expanded greatly in recent years. This is mainly due to the "black box" approach that has been employed in fitting NIR systems into the routine quality-control and process-measurement strategy at the factory level. It was not until relatively recent advances in computer-aided mathematical analysis of data that a real understanding of the utility of NIR spectroscopy emerged. While the black box approach worked fine as long as everything went smoothly, as soon as a NIR method began to run into trouble, there was no way to address the problem because no one knew much about interpretation of NIR spectra. Luckily, through recent advances in such interpretation and through prudent use of chemometric techniques for data analysis, a better understanding of how to apply NIR spectroscopy in a robust, meaningful way has emerged [3,8,20,21]. This has led to a better understanding of NIR spectroscopy in terms of how to apply and troubleshoot problems with the technique. Also, key to the rapid advance of NIR methodology has been the development of factory-proven NIR spectrometers coupled with durable high-throughput fiber optics. This has made application of the technique relatively simple for data-collection purposes and has added to the appeal of the technique. NIR has been very successfully applied for raw material identification (discriminant analysis), quantitative analysis, and process monitoring. (A selection of recent work in NIR spectroscopy can be found in Ref. 3.)

III. ACQUIRING SPECTRA

Traditionally, infrared spectra are represented with frequency or wavelength plotted on the x-axis and percent transmission (%T) plotted on the y-axis such that the absorption peaks point downward. This has come about because of the way IR instruments were manufactured, but it is not the ideal representation of spectra for either qualitative or quantitative applications. The preferred, alternate representation is to plot y-axis values as absorbance instead of %T with x-axis values left the same. The infrared active bands are then displayed as upward peaks whose area and peak height are directly proportional to concentration. This is mandatory for quantitative analysis purposes and is somewhat easier to work with for qualitative applications.

As a general rule of thumb, IR spectra should exhibit bands that have a maximum absorbance of 1.0 absorbance units or less for quantitative purposes, but for FT instruments with very large dynamic range, high-resolution A-D converters, this value can be substantially higher (1.5–2.0 Å). If qualitative characterization of a particular band is all that is required, then sloping baselines, noisy spectra, and large absorbance values may be unimportant. Quantitative analysis does, however, demand more in terms of the quality of the data. In particular, the Beer-Lambert relationship becomes severely nonlinear when absorbance values exceed the linear range of the instrument, or when the baseline noise is significantly high. An adequate calibration and validation procedure should always be followed for all quantitative methods in order to determine the linearity of the method. This will also provide an estimate of the error in the analysis.

Typically, 16–32 co-added scans will be sufficient for routine analysis using an FT instrument. This applies to both background and sample scans, with the same number of scans acquired for both to avoid a noise mismatch. In quantitative applications where multivariate analysis is anticipated, it is usually a good idea to acquire as many scans per sample as possible since noise will be the primary limitation. It is always a good idea to design data-acquisition experiments with the eventual application and data-reduction technique in mind. It is aggravating to have to repeat an experiment because the data was acquired at too low a resolution or with an insufficient number of scans.

IV. SAMPLE PREPARATION

Samples of interest in the surfactant industry generally span the range from solid to liquid and everything in between. This often includes very difficult samples such as waxy solids and thick pasty materials (often containing significant amounts of moisture). Therefore, the method of sample preparation and the type of accessory chosen for analysis is extremely important. Raman and NIR spectroscopies are somewhat forgiving in this aspect of analysis since they are not prone to the two major obstacles facing infrared spectroscopy: extremely high absorptivities and opaque cell materials. Both Raman and NIR techniques are capable of dealing with moderate amounts of water in the sample and are somewhat immune to absorption of light by the cell material. In short, both these techniques are only slightly dependent on sample form, and then only when dealing with large-particle-size solid samples. Therefore the main emphasis of this section will be placed on mid-infrared sampling, with Raman and NIR sampling described where appropriate. It cannot be stressed enough that the key to obtaining high-quality spectra is in preparation of the sample.

A. Solids

Solid materials are most often analyzed in the infrared as a ground powder mixed with a solid dispersing agent such as potassium bromide (KBr). The most popular techniques like the pellet method and diffuse reflectance (DRIFTS) method are performed in this manner. Alternately, the sample can be ground and dispersed in a liquid and then analyzed as a thin film between IR transparent plates such as in the mull method.

Raman sampling of solids is much like that for IR analysis. Solid samples are ground and analyzed either as-is or mixed with a solid nonabsorbing dispersing agent. More specialized sampling techniques are also available such as using a rotating NMR tube holder or an x, y moving stage design. A number of different sampling geometries are possible for Raman spectroscopy, but the most common is the 180° back-scatter technique [2].

1.   KBr Disk Method

The most common method for solid sampling in infrared spectroscopy is still probably the KBr disk or pellet method. Given the wide variety of sampling accessories available today, this is largely for historic reasons and not because there is any real advantage provided by the technique.

The method involves mixing about 0.2–0.5% sample by weight in KBr and grinding to produce both a fine powder and complete dispersion. KBr is usually used in this method due to the wide IR wavelength transparency range and low cost; other infrared-transparent materials are also used such as cesium iodide (CsI) and sodium chloride (NaCl), but KBr is the preferred matrix. The powder is then pressed in a metal die to form a translucent-to-clear KBr disk or pellet. The disk is then supported on a standard IR transmission sample holder, which allows a spectrum to be recorded. The reason for grinding the sample in KBr is to reduce the average particle size to well below that of the shortest infrared wavelength being used (4000 cm^{-1}) in order to reduce distortions in the spectra due to dispersion. Advantages of the technique are that it is quick and relatively easy to do with a little practice, and it is inexpensive. The disadvantages are that KBr is hygroscopic and dissolves in water so even a small amount of moisture in either the sample or the KBr make it difficult to produce a quality spectrum. The technique is generally not very reproducible and so cannot be used for quantitative work.

2.   Mull Method

Another solid sampling technique that owes its widespread use to past popularity is the mull method. As with the KBr disk method, the mull technique remains popular as a solid sampling method due to the large number of users who "grew up" using the technique. The idea of the mull method is essentially identical to that of the KBr disk method, but a liquid rather than a solid dispersing agent is used. A small amount of sample is ground first to a fine powder to reduce the average particle size, then a very small amount of mulling liquid is added to the sample and the mixture is ground again to produce a uniform paste of sample in the liquid matrix. The paste is then placed between IR transparent windows (such as KBr, ZnS, or ZnSe), and a transmission spectrum is acquired. Common mulling liquids include mineral oil (Nujol), hexachlorobutadiene (HCBD), and a commercially available perfluorinated hydrocarbon (Fluorolube). Care must be taken when selecting an appropriate mulling liquid for the analysis since each of the mulling liquids themselves absorb strongly in particular regions of the IR spectrum, and consideration must be given to what region of the spectrum is of interest in the analysis. Generally speaking, Nujol is transparent from 1350–400 cm^{-1}, HCBD from 4000–1500 cm^{-1} and <750 cm^{-1}, and Fluorolube from 4000–1350 cm^{-1}. Grinding the sample is also just as important in the mull method as it is in the KBr disk method and for the same reasons.

Advantages of the technique are that water is generally not a problem in small amounts, and it is much easier to prepare a mull than a KBr disk. Disadvantages are that the mulling liquid often obscures useful spectral information of the sample, and refractive dispersion due to differences in the refractive index of the sample and mulling liquid often causes distortions of the final spectrum. The mull method is generally not recommended unless there is some overriding reason for its use.

3. Diffuse Reflectance

DRIFTS is a very useful method for analyzing solid surfactant materials. It continues to gain popularity due to its versatile nature and ease of use and, indeed, has become the method of choice for analyzing most solid materials. Figure 1 shows a schematic of a typical diffuse reflectance accessory.

Sample preparation for DRIFTS analysis usually requires that a small amount of sample (0.1% w/w) be ground in an optically scattering material such as KBr. This is done very much in the same way described for preparing a KBr disk except the powder is not pressed. The loose powder mixture is placed in the

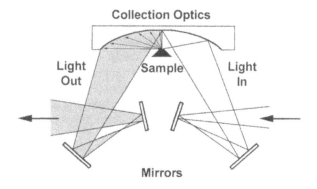

FIG. 1 A typical DRIFTS accessory (Spectra-Tech design).

DRIFTS sample cup and analyzed as-is. For some materials that have a very high scatter efficiency and a low infrared absorptivity, direct acquisition using the neat material may be possible, but in these cases, special design considerations are usually necessary to avoid spectral distortion [22,23]. This is somewhat rare, and, in general, analysis of neat materials is not an accepted sample preparation technique for DRIFTS. In most cases, the sample will absorb almost all the infrared light and the resulting spectrum will appear with flattened or "cutoff" peaks where complete absorption of light occurs. This is unacceptable for any kind of infrared analysis.

In the DRIFTS experiment, the incident beam of IR light is focused at the surface of a powdered sample. The light then goes through a multiple series of microtransmission, refraction, and reflection interactions just below the surface of the powder where light is absorbed, and eventually exits from the powder surface having traversed this convoluted geometry. Only a fraction of the incident light will emerge diffusely scattered, and the rest is specularly reflected from the surface. A proper DRIFTS design will collect as little specular component as possible.

The DRIFTS method is the preferred technique for most solid samples, since little sample preparation is required and addition of an absorbing dispersing agent such as Nujol is not required. This is particularly advantageous if the peaks of interest of the sample overlap those of the dispersing agent.

Quantitative analysis is possible in DRIFTS spectroscopy, but is difficult without the use of mathematical data analysis methods that are capable of working with nonlinear data. For the same reason that %T data must be converted to absorbance data for quantitative work, nonlinear DRIFTS data must be transformed into linear absorption–like data if quantitative measurements are to be performed. The usual way to achieve this is by applying the Kubelka-Munk transform to DRIFTS data prior to quantitative data analysis. In practice, however, the K-M transformation may still result in peak heights/areas exhibiting a large degree of nonlinearity in relation to the concentration. This is because K-M is only valid under a certain set of ideal conditions that make many assumptions about the particle size, sampling depth, absorption cross section, dielectric properties of material, etc. In practice these assumptions are rarely satisfied and the K-M conversion becomes unstable. The best use of K-M is in conjunction with multivariate modeling techniques, such as partial least-squares (PLS) and nonlinear PLS, that are capable of modeling nonlinear behavior to varying degrees [4,21].

4. Raman Sampling of Solids

Sample preparation of solids for Raman spectroscopy is fairly simple and straightforward. Since a Raman spectrometer collects light scattered from the sample, it is just a matter of getting the sample in the laser source focus and making sure it has a reasonably small particle size to reduce refraction and reflection effects. Typically a solid sample is ground in a mortar and pestle, then placed in either a capillary tube or a standard NMR tube for introduction into the spectrometer. In most modern Raman spectrometers, scattered light can be collected in one of several different geometries; 90° scatter, 180° back-scatter, etc. Which geometry is chosen is dependent on a number of factors and will vary depending on the type of experiment and nature of the sample. The 180° back-scatter geometry is preferred in most cases since it provides the highest light-collection efficiency (Fig. 2).

Another solid sampling method for Raman spectroscopy is based on the KBr disk method taken from IR spectroscopy. The powdered sample is compressed as-is into a pellet in a special accessory designed specifically for this purpose. The pellet, still in the press holder, is placed directly into the spectrometer and the spectrum is acquired. A variation on this method is often useful with highly colored samples that may tend to overheat due to laser absorption. The thermal emission from the overheated sample will overwhelm the Raman scatter signal, making the spectrum unusable (and possibly burning the sample as well). A small amount of sample can be ground in a nonabsorbing, highly scattering medium such as KBr, and then compressed in the pellet press described above. The pressed pellet is then placed in the spectrometer and a spectrum acquired. This method will not always work, particularly if the sample of interest has a low Raman scatter efficiency, but in most cases seems to work quite well, and high-quality spectra are obtained.

Another way to address the problem of sample overheating is by modifying a standard NMR tube sample holder so that it will rotate in place in the Raman

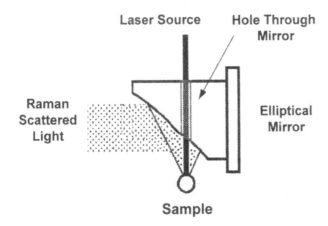

FIG. 2   180° Raman back-scatter reflective collection optics (Nicolet design).

spectrometer. This prevents any local area heating since it is constantly moving at the laser focus as the sample tube rotates. A serious problem with this technique is that the sample tends to drift in and out of laser focus as it rotates, which can cause some problems with data collection and spectral noise.

B. Liquids and Pastes

Liquids, including viscous materials and solutions, can be analyzed in the infrared as thin films between IR transparent plates. Alternately, if the sample is fluid enough, it may be analyzed in a standard liquid cell with either a fixed or variable path length. In either case the spectroscopic sampling mode is by transmission through the sample. An alternative method that is considered a standard method for analyzing liquids is by attenuated total reflection (ATR). Another useful method for fairly viscous samples is to place a thin film of sample on a highly reflecting nonabsorbing surface (aluminum plate or gold-coated slide). The sample can then be analyzed in transflectance mode using a standard specular reflectance accessory or an infrared microscope. ATR and specular reflectance are discussed in more detail in the following sections.

1. Attenuated Total Reflectance

Analysis of liquids and pastelike samples is most easily achieved using the attenuated total reflection technique. There are many different ATR accessories, each designed for a particular type of application [9,11,22,24]. The most common designs are single and multiple reflection designs where the internal reflection element (IRE), also called the ATR crystal, is mounted in a horizontal geometry and sample is applied to the top surface of the IRE. A typical single reflection horizontal ATR accessory is shown in Fig. 3.

The mechanism of energy transfer from IR light to sample, and hence absorption of energy, is not obvious or intuitive in the ATR experiment. A complete treatment

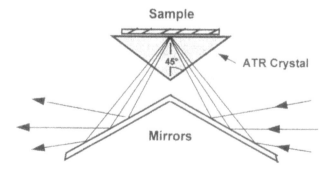

FIG. 3   Single reflection ATR accessory (Harrick design).

of the theory is not given here but is provided by Harrick [11]. When light is internally reflected at an angle less than the critical angle, θ_c, a standing wave called an evanescent wave is formed at the point of reflection. The evanescent wave is electromagnetic in nature and has its electric field pointing perpendicular to the incidence plane of the IRE in the direction of the material of lower refractive index (the sample in all cases). The electric field then interacts with the sample and excites the electric dipoles of the chemical bonds in the sample. This allows transfer of energy between the sample and the IR light in a manner similar to that in the transmission experiment. An additional characteristic of the evanescent wave is that the total absorption by the sample at the point of reflection for a specific wavelength is fixed for a given incidence angle, IRE refractive index, and sample refractive index, and is completely independent of the sample thickness. This makes the phenomenon and the ATR technique extremely reproducible and quantitative so it is ideal for quantitative analysis. In addition, the technique is not seriously affected by particulates or other suspended matter in the sample making the analysis of pastes possible. Aqueous solutions are particularly suited to ATR analysis, and multicomponent analysis of aqueous detergent mixtures has been demonstrated [5].

Infrared light is transmitted into the IRE either by direct pass through the IRE or, as is shown in the diagram, by reflection from gold or aluminum mirrors. The accessory design is such that the light introduced into the crystal strikes the sampling surface of the IRE at an angle less than θ_c, and so is reflected back into the IRE and then to the spectrometer detector. Since θ_c is dependent on the refractive index of both the IRE and sample, the absolute angle varies with a given experiment. Typically an angle of incidence is chosen well below θ_c so that it becomes unlikely that any change in sample properties will cause a problem with the analysis. Typical angles of incidence for ATR experiments are 60°, 45°, and 30°, with 45° being the standard geometry.

ATR crystals are available made with a variety of materials and geometries that provide considerable control over the average effective depth of penetration. Table 2 lists the names and properties of some common types of IRE materials for ATR spectroscopy. ZnSe is the most common material, having a wide wavelength range of transparency and a general inertness to common solvents. AMTIR (Ge-As-Se glass) is another common IRE material that, unlike ZeSe, is particularly resistant to chemical etching by detergents over long-term exposure. AMTIR, however, is substantially more expensive than ZnSe and is more easily scratched and shattered. Germanium (Ge) is also used, but due to its very high refractive index the depth of penetration into the sample is small and renders it difficult to use for anything other than bulk material characterization.

ATR is also useful in studying waxy solids such as fatty acids, esters, and amides. It is sometimes more simple and convenient to use ATR to analyze these materials rather than some of the other solid sampling techniques due the tacky nature of the materials.

TABLE 2 Properties of Typical Internal Reflection Elements for ATR Accessories

IRE material	Transmission range	RI of IRE material	Special properties/cautions
ZnSe	20,000–450	2.4	Attacked by acids/strong alkali and detergents
ZnS	17,000–833	2.2	Attacked by acids; resistant to thermal and mechanical shock
Ge	5,500–600	4.0	Attacked by hot H_2SO_4, aqua regia; brittle; low transmittance range
AMTIR	11,000–625	2.5	Attacked by strong alkalis; resistant to detergents
NaCl	40,000–590	1.49	Susceptible to thermal shock; cleavable
KBr	40,000–340	1.52	Susceptible to thermal shock; cleavable
CsI	40,000–200	1.74	Resistant to mechanical shock; soft
CaF_2	50,000–840	1.39	Very hard; withstands high pressure
AgBr	20,000–300	2.2	Cold flows; reactive
Sapphire	50,000–1780	1.74	Extremely hard; inert
Si	8,300–6,600	3.4	Resistant to thermal and mechanical shock
KRS-5	20,000–250	2.4	Slightly soluble in water; toxic

RI = Refractive index at $1000 \ cm^{-1}$.

2. External Reflection

External reflection in infrared sampling refers to the direct specular reflection of light from the surface of a material. Specular reflection is a coherent "mirrorlike" reflection from a surface with very little scattering. In the traditional sense of the term, external reflection accessories collect specularly reflected light from the surface of a solid sample placed in the path between the source and the detector. Usually a very small portion of the light is absorbed by the sample and most is reflected from the surface. This type of analysis is not usually very useful for IR analysis since the resultant spectra are very difficult to interpret due to derivative shaped bands that appear at wavelengths of very strong IR absorption. There are mathematical methods of transforming these derivative bands into real spectra, but it is not a straightforward process and the corrected spectra may still be somewhat distorted.

Another way of using the specular reflection accessory is as a transflectance accessory. In this mode a thin film of liquid or paste sample is deposited on an IR-reflective surface such as aluminum or gold. The incident beam passes through the sample film, is reflected from the metal surface, and passes through the sample a second time and is detected. This type of absorption-reflection-absorption phenomenon is referred to as transflectance, a concatenation of the words transmission and reflectance. This double-pass of light through the sample is essentially identical to transmission through a thin film of sample, and the resulting spectrum is nearly equivalent to a transmission spectrum. Figure 4 shows a schematic of an external reflection accessory used in transflectance mode.

The technique has the advantage of being very simple to perform and the result is a familiar transmission spectrum. The disadvantages are that it is difficult to perform the technique in a reproducible manner since the sample film is manually applied and very fluid samples will not adhere well to the metal surface, causing irregular films. This can result in extremely thin films being produced which are in general not desired for this type of analysis. Conversely, if the film is too thick, then too much light will be absorbed by the sample causing a saturated appearance of the spectral features.

3. Hyphenated Infrared Chromatographic Techniques

Increased importance has been placed on coupling spectroscopy with various types of chromatographic techniques—the so-called hyphenated techniques. The most common hyphenated vibrational techniques all involve infrared spectroscopy and some form of chromatography. Although GC-IR has been commercially available for many years and has a large user base, the technique does not directly lend itself to applications involving anionic surfactants, for example, since it is quite difficult to nondestructively volatilize and chromatograph ionic species. Newer IR-based chromatographic techniques such as HPLC-IR and GPC-IR, however, do not suffer from this drawback and are very useful for surfactant analysis. The problem has been that commercially available LC-IR systems were generally not available and, if they were available, were quite expensive. This all changed in the late 1980s with the availability of commercial LC-IR products (Lab-Connections, Inc.). Applications using these systems have become commonplace and the number increases dramatically each year as the technique becomes more widespread. Figures 5 and 6 show diagrams of such an LC-IR system [25,26].

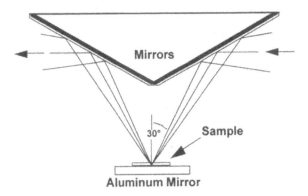

FIG. 4 External reflectance accessory.

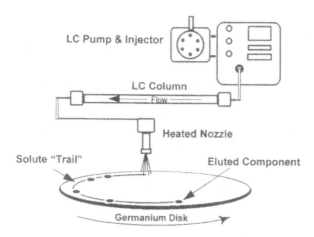

FIG. 5 LC-IR System: chromatograph interface module (Lab-Connections design).

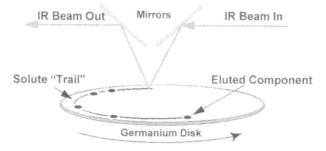

FIG. 6 LC-IR System: IR spectrometer interface module (Lab-Connections design).

The LC-IR interface is a new approach for performing pseudo real-time HPLC data acquisition with infrared detection. One of the unique features of the system is that existing chromatographic and infrared equipment can be used and the two do not need to be located in close proximity to each other. Figure 5 shows a schematic of the chromatographic interface of the Lab Connections LC-IR system. This arrangement would be the same for either a reverse-phase LC system, a normal-phase LC system, or a gel permeation chromatography (GPC) system. Two separate interface modules are used for the analysis. The chromatographic interface uses a liquid deposition, solvent volatilization technique, which produces a thin, narrow film of solute on a germanium (Ge) disk. The disk spins slowly, at a given rotation speed, as the solute trail is deposited in a circular geometry onto the disk. The disk is then removed from the LC interface module and placed in the FTIR interface module. The FTIR portion of the LC-FTIR uses beam-condensing optics to focus the IR beam onto the disk as it traces along the previously deposited solute "trail" at the same rotation speed used at the time of deposition. Figure 6 shows a schematic of the FTIR interface of a Lab Connections LC-IR system. The greatest benefit of this accessory is that it is able to produce high-quality FTIR spectra of every peak in a chromatogram, which provides a means of identifying the components. This is essential for HPLC methods development and component characterization where identification of unknown peaks is paramount. A particularly good example of the utility of the technique has been described by Cuesta Sanchez et al. [25], who showed the application of HPLC-IR to separation and resolution of severely overlapped chromatographic peaks from a mono-alkyl C_{14}–C_{15} di-sodium citrate ester synthesis. The Gram-Schmidt reconstructed chemi-gram and resolved component elution profiles are shown in Figure 7. Gram-Schmidt is a mathematical method of constructing a chromatogram or elution profile, called a chemi-gram, from spectroscopic data. The analysis revealed that there were six actual species present in the single chromatographic peak due to severe interaction between the citrate ester and other components of the solution.

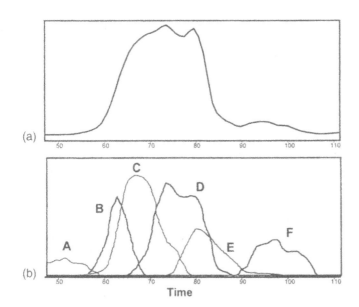

FIG. 7 (a) Elution profile of alkyl citrate unrsolved chromatographic peak (Gram-Schmidt profile). (b) Resolved elution profile from multivariate curve resolution. Profiles A–F are of the resolved pure component profiles.

There are some problems to consider when applying this type of LC-IR methodology. Since the technique uses a solid deposition scheme, the spectra acquired from the Ge disk are essentially a combination of reflectance spectra and transflectance spectra. This is usually not a problem for very thin films of amorphous materials, but when the solute is very crystalline and particulate in nature, there are often slight distortions of the spectra due to specular reflection from the surface of the particles. This is typically not a serious problem, but will cause additive and multiplicative distortion of the spectra when present.

Application of GPC-IR has also proven very useful for analysis of anionic polymers used in surfactant formulations. An example applied to the analysis of polyacrylate copolymers will be discussed in a later section.

4. Microscopes

Infrared and Raman microscopes have become rather standard accessories for vibrational spectroscopic analysis. Both are very expensive and require some amount of expertise to apply effectively, but are still very valuable for particular applications. While IR microscopes have been around for many years, Raman microscopes have just recently come into common use. Fundamentally, IR or Raman microscopes are used primarily to obtain high-quality spectra of very small samples ($2-20$ μm^2) such as fibers (the smallest sample size area for mid-infrared microscopes is about $7-10$ μm^2). This is not particularly necessary for surfactant analysis, but occasionally use of this capability is prudent for characterizing particulates, for example. See Messerschmidt and Harthcock [6] for applications and principles of the IR microscopy.

V. DATA HANDLING

Once a spectrum is acquired using mid-infrared, Raman, or near-infrared, it is often necessary to manually modify the spectra to remove noise, straighten baselines, or generally improve appearance. There are a number of ways a spectrum can be manipulated mathematically with the aid of a computer so that only the pertinent information in the data is retained and other extraneous features are either removed or minimized. Example of extraneous information present in spectra that detract from, and often obscure, the real chemical information are baseline offset, noise, a sloping baseline, water vapor features, spectral features of solvents, etc. While it is sometimes possible to minimize these problematic features before data acquisition, often it is just not possible or cost-effective to do so. Modern data-handling techniques provide tools for minimization of these unwanted features with only slight loss of information. A complete catalogue of data manipulation algorithms would be too lengthy to present here and has been discussed at length in other texts [1,7–9,18]. A brief description of the most common methods is provided here as a quick reference.

A. Baseline Correction

Acquired spectra will frequently have a sloping baseline. Often this slope will be very minor, in which case it is not worth correcting. Occasionally however, a spectrum will have a severe slope due to any of several reasons such as dispersion, poor alignment of the accessory, and improper preparation of the sample. In any event, it is often necessary to correct for this baseline slope. Most spectroscopy software available today has a baseline correction capability. The user can correct interactively using a simple linear baseline or a multipoint baseline for nonlinear distortions, or an auto-correcting process by the computer can be employed. In any case it is important to remember not to overcorrect the baseline. It is fine to correct for legitimate sloping of the baseline due to unavoidable causes, but correction for sloppy sample preparation or a misaligned accessory should never be attempted. If the problem is severe, always try to fix the problem first, then resort to baseline correction when necessary.

B. Subtraction

Spectra subtraction is also a common feature of any spectroscopy software package. It is quite useful when samples are analyzed in a solvent. The spectrum of the solvent can usually be subtracted away revealing the underlying spectrum of the sample. Subtraction is also very useful for compensating for sample compartment purge problems. The CO_2 and water vapor features common to a poorly purged sample compartment can be subtracted from the sample spectrum to produce a clean sample spectrum. This is of particular importance if the carbonyl region is of interest since water vapor features obscure this region of the spectrum ($1800-1300$ cm^{-1}).

C. Smoothing

It is always desirable to acquire spectra using the proper resolution, aperture, and number of scans so that the noise level will be as low as possible and not impact upon an analysis. However, it is inevitable that some spectra acquired will exhibit more noise than is acceptable. This is particularly true for Raman spectra in which in many instances light levels are very low and the number of desired scans is prohibitive. In these cases it is sometimes prudent to apply a smoothing algorithm to reduce the apparent noise. This is done, of course, at the loss of some resolution, but often this is an acceptable trade-off. Of the most popular smoothing techniques, the Savitsky-Golay algorithm is perhaps the most common, with Fourier transform smoothing a close second. Both of these algorithms are acceptable to use, but both ignore the noise in the data, the very thing that it is sought to minimize. This often leads to unacceptable and exaggerated solutions to the problem. A far superior method is called Baysian-Maximum Likelihood Analysis (MLA). MLA is not really a smoothing algorithm, but finds the most likely or statistically probable shape for the spectral peaks and extracts that information from the actual spectrum. The important difference with MLA is that it takes into consideration the statistical distribution of noise in the data using the peak width of the narrowest real spectral peak. The noise distribution and peak characteristics must be specified by the user. This might sound very difficult to do since it may not be obvious how to determine the required information, but in reality there are only a few possible choices for these variables and of those some are more likely than others. It turns out to be much easier to perform than it appears at first. The important thing to consider is that the data is not actually smoothed and if the proper noise and peak characteristics are provided, there is no loss in resolution.

D. Differentiation

Differentiation is most important in NIR spectroscopy, far more so than in either IR or Raman spectroscopy. The purpose of differentiation is to deconvolute the very broad overlapped peaks typical in NIR data. While some applications have been described using differentiation for IR spectra, it is typically not a preferred method since it usually makes the spectra more complicated and, therefore, less obvious to interpret. In NIR applications it is generally true that derivatized data produce better linear models for quantitative analysis than do underivatized data. This is partially due to the ability of the derivative to remove baseline distortions in the spectra. It has been suggested, however, that there are other ways of removing baseline effects that do not distort the spectral information [21,27].

E. Curve-Fitting

Curve-fitting is a method of resolving overlapped peaks in spectra into their separate individual peaks. This is done statistically with the aid of input from the user and requires the user to interactively decide when the best fit to the data has been achieved. While there are times deconvolution may be desired such as for quantitative work using peak height or area calculations, curve-fitting is often applied indiscriminately to "sharpen up" peaks or to provide an apparent increase in peak resolution. In either case, the subjective nature of the technique requires that it be applied very carefully.

VI. SPECTRAL CORRELATION

The most useful method of characterization of any compound is through use of infrared correlation charts. Correlation charts essentially outline, either graphically or in words, the regions of the spectrum associated with particular functional groups. They have been compiled using literature references and personal experience spanning many years for a wide variety of compounds. They are an invaluable tool without which the job of modern spectroscopists would be extremely difficult and time consuming. While similar correlation charts exists for both near-infrared and Raman spectra, discussions of correlation charts for purposes of this chapter will be limited to mid-infrared data due to the more extensive use of that technique in industry and also due to more extensive correlation charts. As a general reference to the reader, correlation charts for mid-IR, Raman, and near-IR spectra are given in Figs. 8–10, respectively. General use of correlation charts is attributed to Colthup [28,29], who compiled an extensive list of NIR, IR, and Raman group frequencies. Also of importance, particularly to work in surfactant identification, are the charts developed by Nyquist [30].

Before considering specific anionic surfactants in detail, a general overview of the alkyl and aromatic regions will be discussed since these features are common to most classes of surfactant compounds. Table 3 lists alkyl and aromatic functional groups commonly found in the spectra of organic materials and surfactant compounds in particular. The assignments listed in the table suggest how correlation charts can be used to estimate structurally related features such as the

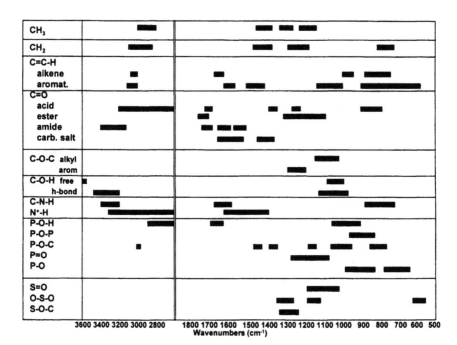

FIG. 8 Mid-infrared spectral correlation chart.

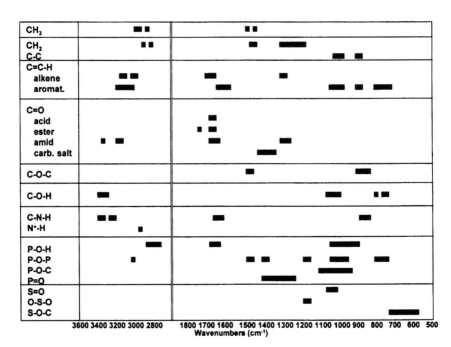

FIG. 9 Raman spectral correlation chart.

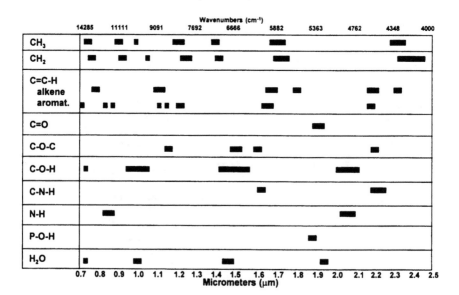

FIG. 10 Near-infrared spectral correlation chart.

TABLE 3 Specific Alkane, Alkene, and Aromatic Functional Group Assignments

Group	Vibration	Wavenumber (cm^{-1})
Alkane groups		
—CH_2—	C—H as. str	2935–2910
	C—H sym. str.	2865–2845
	C—H bend	1475–1440
	Rock	725–720
—CH_3	C—H as. str.	2970–2950
	C—H sym. str.	2890–2860
	C—H sym. bend	1465–1445
	C—H as. bend	1375–1370
Alkene groups		
C=C	Double bond str.	1680–1630
C=C—H	C—H str.	3100–3000
	C—H in-phase wag	995–980 *trans* vinyl
	C—H wag	910–900
	C—H in-phase wag	680–600 *cis* vinyl
Aromatic groups		
Ar—H	C—H str.	3080–3010
	In-plane def.	1290–1000
	Out-of-plane def.	770–730
		710–690 *mono* substitution
		770–735 *ortho* substitution
		960–680 *meta* substitution
		860–800 *para* substitution
C=C	C=C str.	1625–1540
	In-plane ring def.	650–500
	Out-of-plane def.	560–415

as. = Asymmetric; str. = stretch; sym. = symmetric; def. = deformation.

degree of branching in alkanes, the positions of substitution in aromatic compounds, and *cis-trans* isomerism of alkenes.

It is generally very difficult to produce accurate band assignments in the fingerprint region (1350–900 cm^{-1}) since there is usually a large number of severely overlapped peaks in this region. These features are influenced by physical structure of the material, such as its crystallinity, phase behavior, and morphology that magnify the problem of deducing a clear assignment. However, this region often provides valuable information on many aspects of the compounds being studied, and very useful information regarding phase transition temperature [31], crystallinity, and morphology can be obtained by careful study of the features found in this region.

A. Carboxylate Salts

Alkyl and aryl acid salts make up the general surfactant class called acid soaps. These are simply neutralized alkyl acids and as such are characterized by their very specific carbonyl functionality. Soaps belong to the more general class of compounds called carboxylic acid salts and are identified by the appearance of both asymmetric and symmetric $R-CO_2^-$ stretching modes. Both of these modes are affected by the nature of the cation, and the asymmetric mode always occurs at higher frequency than the symmetric mode. As compared to their acid counterparts, soaps have a delocalized CO_2^- group rather than a strict double bond character. What this means is that the charge is spread out more evenly across

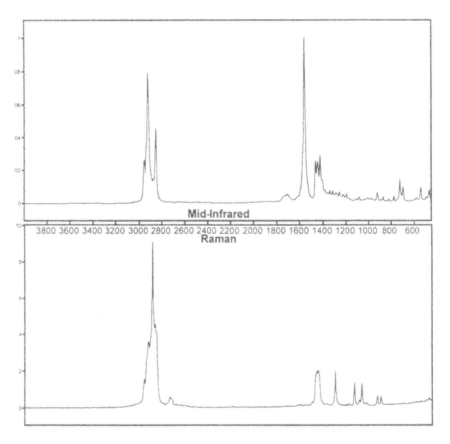

FIG. 11   IR and Raman spectrum of sodium laurate (x-axis is in cm⁻¹).

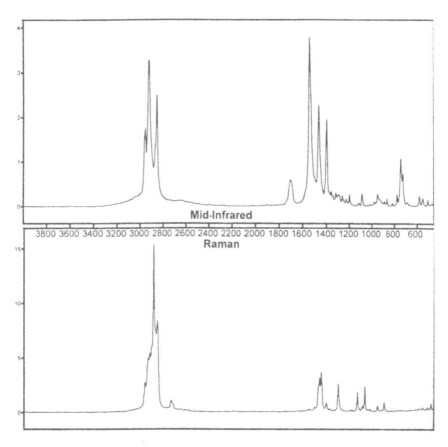

FIG. 12   IR and Raman spectrum of zinc laurate (x-axis is in cm^{-1}).

both the C—O and C=O bonds, forming what is functionally like a hybrid of the two. The two are therefore coupled and provide the characteristic spectral features associated with soaps and all other carboxylate salts. Both asymmetric and symmetric C—O/C=O vibrations occur and are infrared active. The assymetric stretch appears between 1610 and 1530 cm^{-1} and the symmetric stretch appears between 1430 and 1320 cm^{-1}. Medium-to-strong bands due to deformation modes of the carboxylate anion are also seen in the 760–400 cm^{-1} region and are characteristic.

Since the charge distribution among the two C—O functionalities is responsible for the vibrational frequency of the bands, the nature of the counterion

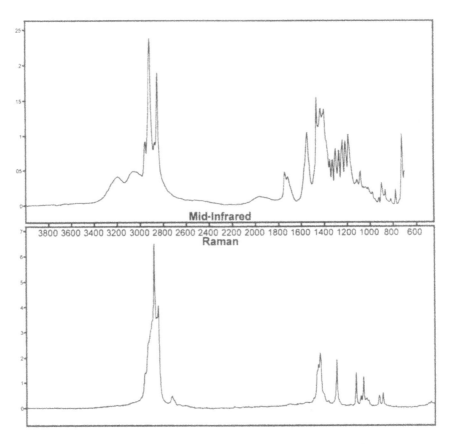

FIG. 13 IR and Raman spectrum of ammonium laurate (x-axis is in cm⁻¹).

will also affect the position and appearance of the carbonyl bands. The stretching vibration of the group depends both on the cation and the organic portion of the salt. Alkali and heavy metal cations, for example, often cause the carbonyl stretching modes to split (one higher and one lower) around the central expected position as seen in the spectra of sodium laurate and zinc laurate in Figs. 11 and 12. If the associated cation derives from an amine, the compound will exhibit NH_2^+ or NH_3^+ stretching absorption bands in the region between 2800 and 2100 cm⁻¹. Absorption bands in this region indicate the presence of acidic protons. Figure 13 shows the IR and Raman spectra of ammonium laurate. Infrared bands near 3200, 3050, and 1425 cm⁻¹ are char-

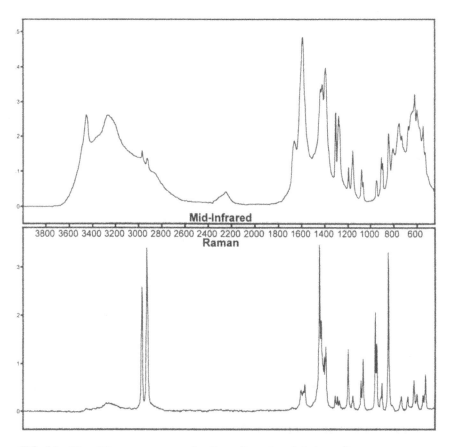

FIG. 14 IR and Raman spectrum of sodium citrate (x-axis is in cm^{-1}).

acteristic of the NH$_4^+$ cation. Figures 11–15 show IR and Raman spectra of a number of carboxylate compounds with different cations. Sodium carbonate (Fig. 15) also belongs to the carboxylate salt group and is an example of an inorganic salt.

Anionic polymers are an important class of additives used in detergent formulations for various reasons and in a broad sense fit under the category of carboxylate salts. The most important of these are the acrylate-based polymers. This includes both polyacrylic acid salts and polyacrylate/maleate block copolymers. This general class of compounds are polycarboxylate polymers that derive their performance characteristics from the highly charged polycarbonyl characters of

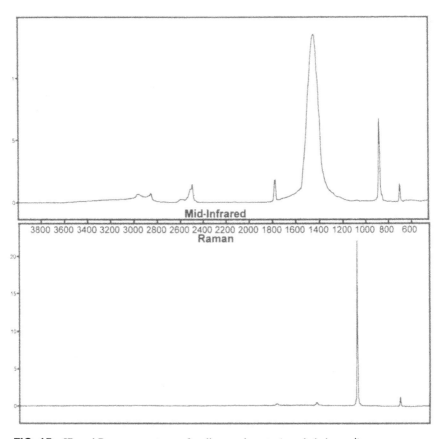

FIG. 15 IR and Raman spectrum of sodium carbonate (x-axis is in cm⁻¹).

the molecule. This class of compounds exhibits characteristic infrared bands that correspond to those discussed previously for carboxylate salt compounds. The primary difference observed is that the polymeric nature of the material tends to broaden out even typically sharp intense features such as carbonyl bands [32]. The result is a spectrum that appears very much like that of the monomer but with much broader, overlapped features. In spite of this, there is still quite a lot of information that can be extracted. This is particularly true if the infrared methodology is coupled with GPC to produce a system capable of providing structural information as a function of molecular weight. An example of this has been demonstrated by Jilani et al. [26], who showed that the block copolymer ratio of acrylate/maleate copolymers could be determined very accurately using

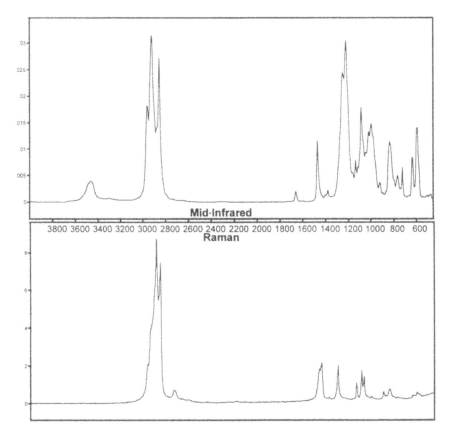

FIG. 16 IR and Raman spectrum of sodium lauryl sulfate (x-axis is in cm⁻¹).

GPC-FTIR. The results reported for this analysis compared favorably to those reported using proton NMR spectroscopy.

B. Sulfates and Sulfonates

Metal alkyl sulfates and metal aryl sulfates have the basic structure: R—O—(O=S=O)—O⁻M⁺, where M is usually either sodium or potassium. These compounds exhibit either one or two strong bands in the region 1230–1185 cm⁻¹ due to the asymmetric SO_3^- stretching mode. One or two additional infrared bands may also occur in the region between 640 cm⁻¹ and 560 cm⁻¹ resulting from the bending and wagging modes of the $O\text{-}SO_3^-$ group. These band positions are fairly stable with respect to cationic substituent and, unlike carboxylate salts, depend only

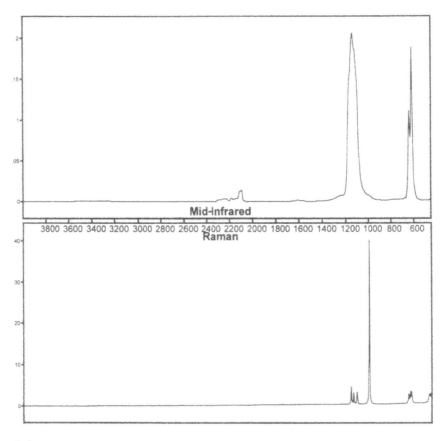

FIG. 17   IR and Raman spectrum of sodium sulfate (x-axis is in cm⁻¹).

slightly on the nature of the cation. Typical spectra for this class of compounds is shown by sodium lauryl sulfate in Fig. 16. Free inorganic sulfate anion by itself shows an intense asymmetric stretching band in the IR region near 1160 cm⁻¹ and a doublet near 650 cm⁻¹. In the Raman spectrum, it shows a single intense symmetric stretching band near 1000 cm⁻¹. These features are considered characteristic and are shown clearly in the spectra of sodium sulfate in Fig. 17. It is usually possible to find sulfate anion in surfactant solutions using this Raman feature (see Fig. 18). Compounds such as sodium alkyl ether sulfates and sodium alkoxy alkyl sulfates will also show a band between 1120 cm⁻¹ and 1080 cm⁻¹ due to the presence of the C—O group either from an ether or an alcohol functionality. Figure 18 shows the IR and Raman spectra of the sodium salt of sulfated glyceryl monolaurate. It is clear from the characteristic Raman band near 1000 cm⁻¹ that free inor-

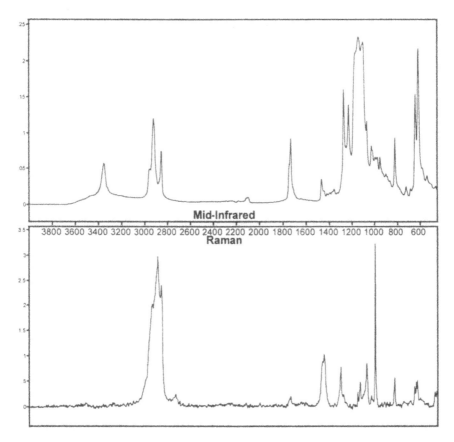

FIG. 18 IR and Raman spectrum of sodium salt of sulfated glyceryl monolaurate (x-axis is in cm⁻¹).

ganic sulfate is present in the sample. In ethoxylated sulfate compounds it is also possible to use the C—O stretching band to estimate the degree of ethoxylation. The intensity of the C—O—C stretch near 1100 cm⁻¹ relative to the CH$_2$ stretching modes between 2945 cm⁻¹ and 835 cm⁻¹ is directly proportional to the extent of ethoxylation.

Metal alkyl sulfonates and metal aryl sulfonates of the form R—O—(S= O)—O⁻M⁺ are similar to their sulfate cousins, but exhibit a relatively strong doublet band near 1200 cm⁻¹. This band is assigned to the normal asymmetric stretching mode of SO$_3^-$. There are actually two asymmetric SO$_3^-$ stretching modes. In some cases, the two modes split and are observed as a doublet. The normal symmetric stretching mode for SO$_3^-$ is observed as the weak band be-

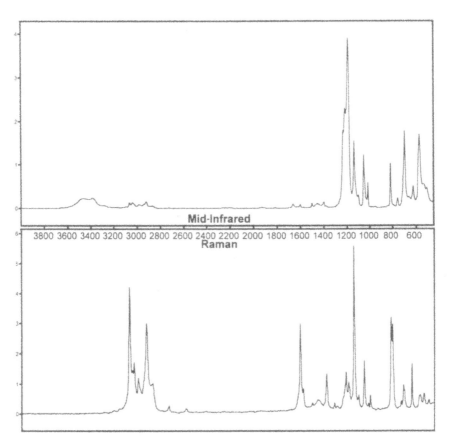

FIG. 19   IR and Raman spectrum of sodium toluenesulfonate (x-axis is in cm⁻¹).

tween 1060 cm⁻¹ and 1020 cm⁻¹. Examples of spectra for these types compounds are shown in Figs. 19 and 20 for aryl and alkyl-aryl derivatives. In addition to the S-O–related functionality, the expected alkyl and aromatic functionalities for these classes of compounds are readily visible in the spectra. It is clear from comparing the IR and Raman spectra of these compounds that the aromatic features are more easily observed in the Raman spectra than in the IR. This would be expected from theory and shows the utility of using both IR and Raman spectroscopy to study these compounds.

An important characteristic of alkyl-aryl sulfonated compounds for performance reasons is the extent of branching in the molecule. Infrared spectroscopy provides a relatively easy method of distinguishing linear from branched alkyl-

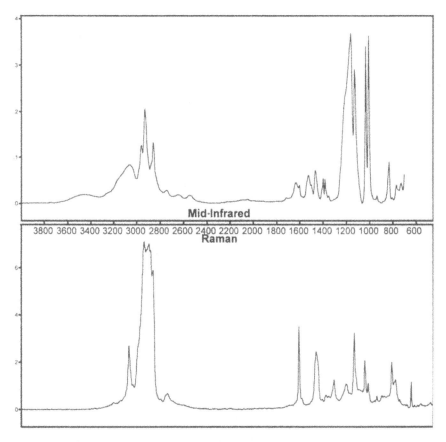

FIG. 20 IR and Raman spectrum of alkyl-ammonium dodecylbenzenesulfonate (x-axis is in cm^{-1}).

aryl sulfonates. The CH$_3$ asymmetric stretch near 2945 cm^{-1} and the symmetric CH$_2$ stretching bands near 2910 cm^{-1} and 2840 cm^{-1} can be used for this determination. The intensity of the asymmetric CH$_3$ relative to the asymmetric CH$_2$ band decreases as the number of CH$_2$ groups in the chain increases. In the linear form of the molecule, the ratio of CH$_3$ to CH$_2$ will be much lower than in the equivalent branched molecule.

Sulfated and sulfonated ester compounds show all the expected functionality of the S–O–related vibrational modes: a very intense band in the region 1230–1200 cm^{-1} and a medium intensity band near 1050 cm^{-1} result from asymmetric and symmetric SO$_3^-$ stretching respectively. They also exhibit the

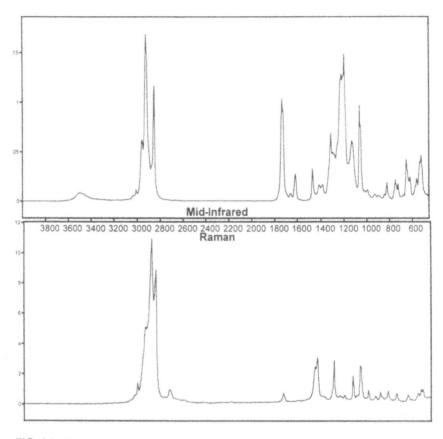

FIG. 21   IR and Raman spectrum of sodium lauryl sulfonacetate (x-axis is in cm⁻¹).

intense band near 1730 cm^{-1} from the ester carbonyl. The isethionates fall into this general class of surfactant and are a special class of sulfonated ester of the general form: R-(C=O)—O—CH$_2$—CH$_2$—SO$_3^-$M$^+$. Two examples of sulfonated esters are sodium lauryl sulfoacetate and sodium stearoyl isethionate, whose IR and Raman spectra are shown in Figs. 21 and 22. Characteristic group frequencies are an intense ester carbonyl between 1730 cm^{-1} and 1720 cm^{-1} and the asymmetric SO$_3^-$ stretching mode at 1190 cm^{-1}. The sulfo-ester shows a medium-intensity 1060 cm^{-1} asymmetric SO$_3^-$ stretch, while the isethionate does not. Both types of compounds show additional bands below 700 cm^{-1}.

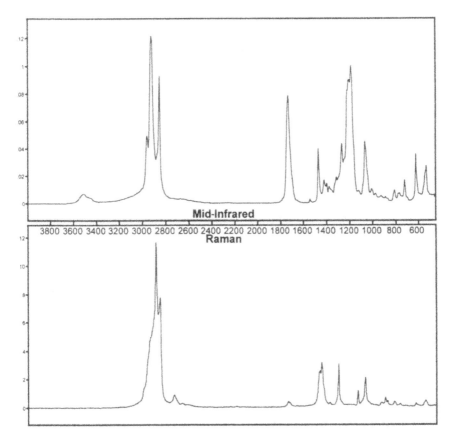

FIG. 22 IR and Raman spectrum of sodium stearoyl isethionate (x-axis is in cm⁻¹).

Polyalkoxy sulfonated ethers make up another class of important anionic surface-active agents. These compounds possess the following basic structure: $R—O—(CH_2—CH_2—O)_n—SO_3^-M^+$. The distinguishing feature of this class of compounds is that of the absorbance near 1100 cm⁻¹ resulting from C—O—C ether stretching mode. This band increases in intensity with an increase in the number of ethoxylate groups compared to band intensities of other functional groups in the molecule. Otherwise, the characteristic group frequencies exhibited by the $CH_2—O—SO_3^-$ group are comparable to those exhibited by the $R—O—SO_3^-M^+$ compounds discussed earlier [17].

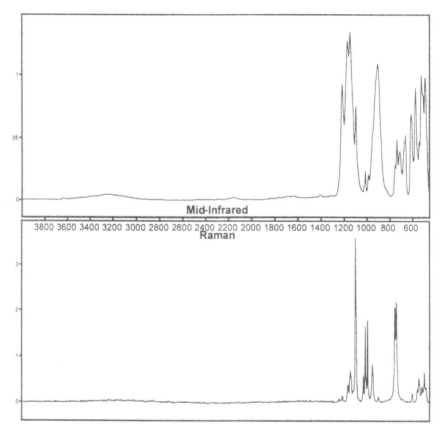

FIG. 23 IR and Raman spectrum of sodium tripolyphosphate (x-axis is in cm⁻¹).

C. Phosphates

The importance of phosphate-based anionic surfactants has decreased in recent years due to the heightened awareness of environmental problems caused by these compounds. Use of these compounds has therefore been restricted in many states within the United States and also in many countries abroad. However, since phosphate-based surfactants are still an important class of surfactant, inclusion in this chapter is warranted.

The main classes of phosphorus compounds for discussion here are the metal phosphates and the alkyl/aryl phosphate classes of compounds that in-

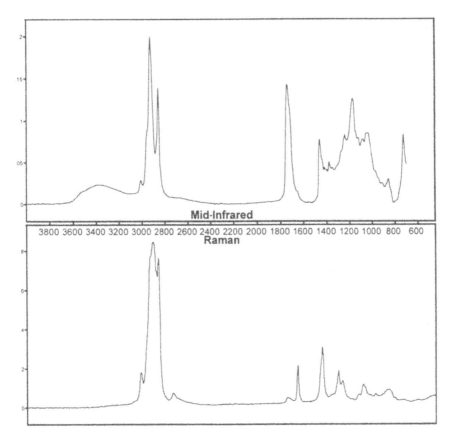

FIG. 24   IR and Raman spectrum of ammonium phosphated castor oil (x-axis is in cm⁻¹).

clude tripolyphosphates, pyrophosphates, metal alkyl phosphates, and ethoxy-lated alkyl and aryl phosphates. All of these compounds contain basic phosphate-related functional groups such as P—O, P—O—P, P=O, P—O—H, etc. The P—O—H groups are very strongly hydrogen bonded, and the O—H stretching mode produces a very broad band, extending from 3700 cm⁻¹ to as low as 1450 cm⁻¹, which may exhibit several peak maxima. It also shows a weak to medium absorbance in the 500 cm⁻¹ region. A band in the region between 1250 cm⁻¹ and 1150 cm⁻¹ is usually due to P=O stretching. The frequency of the P=O stretching vibration is fairly independent of the number of substituents attached to it, but is influenced by the electronegativity of these

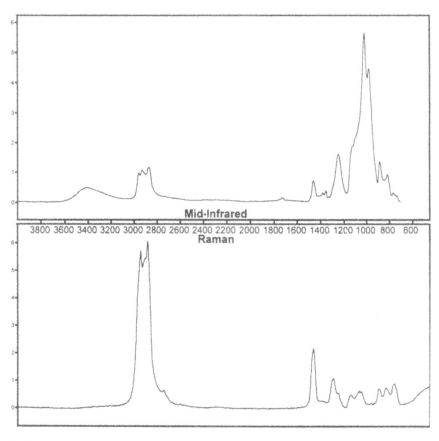

FIG. 25   IR and Raman spectrum of polyoxyethylene (12) octyl phosphate (x-axis is in cm⁻¹).

substituents. The P=O band may sometimes appear as a doublet but if the split is very small, it may appear as a shouldered peak. These compounds also exhibit a strong band in the region 1050 cm⁻¹ to 960 cm⁻¹, which is assigned to a P—O stretching mode.

Pyrophosphates of the form (O=P—O—P=O) have only one P=O band unless the pyrophosphate is an asymmetric compound. This does not usually cause a degeneracy or splitting of the phosphonyl mode since, unlike acid anhydrides, no coupling exist between the two P=O groups. Several examples of spectra of phosphate compounds are provided in Figs. 23–26, which demonstrate the various spectroscopic properties of the types of compounds outlined above.

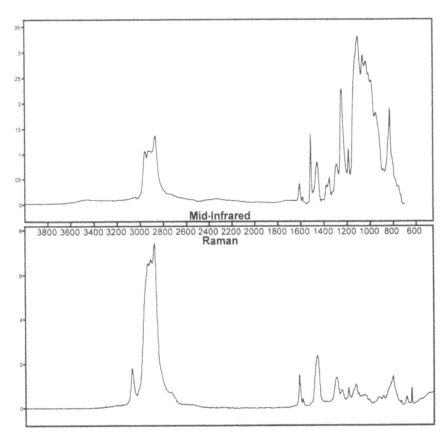

FIG. 26   IR and Raman spectrum of alkyl-phenyl ethoxylated phosphate (x-axis is in cm^{-1}).

VII.   GENERAL REMARKS

This chapter has introduced some of the useful ways vibrational spectroscopy may be applied to the study and characterization of anionic surfactants. Some theory has been discussed along with a description of some of the more useful applications of the technique. A wide variety of sampling and data-handling methods has also been discussed as they pertain to a wide variety of sample types that can be easily analyzed. While it often takes some degree of practice and sometimes many years of experience to gain a full appreciation for the many subtleties of vibrational spectroscopy, this review should suffice as an overview of what is possible.

It should, however, be explicitly stated that vibrational spectroscopy is not an exact technique: rarely, if ever, can the analyst clearly and unambiguously identify a compound using vibrational techniques alone. However, that is not to say that vibrational spectroscopy is not capable of providing very useful structural information. In fact, infrared or Raman spectroscopy is often able to obtain information not forthcoming from any other analytical technique. Vibrational techniques are routinely applied in every aspect of chemistry from the research laboratory to the manufacturing plant. It is fair to say that no other group of analytical techniques is so universally applied to solve such a wide variety of problems. Given the ease and relative low cost of obtaining infrared and Raman spectra, it is well worth applying as a first-line approach in the analysis of most classes of surfactant compounds.

REFERENCES

1. J. R. Ferraro and L. J. Basile, eds., *Fourier Transform Spectroscopy: Applications to Chemical Systems*, Vol. 4, Academic Press, New York, 1979.
2. P. Hendra, C. Jones, and G. Warnes, *Fourier Transform Raman Spectroscopy—Instrumentation and Chemical Applications*, Ellis Horwood, New York, 1991.
3. I. Murray and I. Cowan, *Making Light Work: Advances in Near Infrared Spectroscopy*, VCH, Weinheim, 1992.
4. H. Beaven and R. E. Ricketts, J. Sci. Instr *44*:1048 (1967).
5. M. P. Fuller, G. L. Ritter, and C. S. Draper, Appl. Spectrosc. *42*(2):228 (1988).
6. R. G. Messerschmidt and M. A. Harthcock, *Infrared Microspectroscopy—Theory and Applications*, Marcel Dekker, New York, 1988.
7. S. Johnston, *Fourier Transform Infrared—A Constantly Evolving Technology*, Ellis Horwood, New York, 1991.
8. H. A. Willis and J. H. van der Maas, *Laboratory Methods in Vibrational Spectroscopy*, 3rd ed., Wiley, New York, 1987.
9. P. R. Griffiths and J. A. de Haseth, *Fourier Transform Infrared Spectrometry*, Wiley, New York, 1986.
10. E. D. Olsen, *Modern Optical Methods of Analysis*, McGraw-Hill, New York, 1975.
11. N. J. Harrick, *Internal Reflection Spectroscopy*, Wiley-Interscience, New York, 1967.
12. P. E. Clarke, in *Introduction to Surfactant Analysis* (D. C. Cullum, ed.), Blackie Academic & Professional, New York, 1994, pp. 234–296.
13. M. Davies, ed., *Infra-red Spectroscopy and Molecular Structure*, Elsevier, Amsterdam, 1963.
14. W. P. Richards and P. R. Scott, *Structure and Spectra of Molecules*, Wiley, New York, 1986.
15. M. Orchin and H. H. Jaffe, *Symmetry, Orbitals, and Spectra*, Wiley-Interscience, New York, 1971.
16. J. D. Graybeal, *Molecular Spectroscopy*, McGraw-Hill, New York, 1988.
17. G. Socrates, *Infrared Characteristic Group Frequencies*, Wiley, New York, 1980.
18. W. O. George and H. A. Willis, eds., *Computer Methods in UV, Visible, and IR Spectroscopy*, The Royal Society of Chemistry, Cambridge, 1990.

19. E. R. Malinowski, *Factor Analysis in Chemistry*, 2nd ed., Wiley-Interscience, New York, 1991.
20. D. L. Massart, B. G. M. Vandeginste, S. N. Deming, Y. Michotte, and L. Kaufman, *Chemometrics: A Textbook*, Elsevier, Amsterdam, 1988.
21. H. Martens and T. Naes, *Multivariate Calibration*, Wiley, Chichester, 1989.
22. T. M. Hancewicz, 82nd Annual AOCS Meeting, Chicago, 1991.
23. T. M. Hancewicz, Appl. Spectors. *46*(6):1074 (1992).
24. T. M. Hancewicz, Appl. Spectors. *46*(6),:1073 (1992).
25. F. Cuesta Sanchez, B. G. M. Vandeginste, T. M. Hancewicz, and D. L. Massart, Anal. Chem. *69*:1477 (1997).
26. L. van Gorkom, T. M. Hancewicz, and M. Jilani, 23rd Annual FACSS Meeting, Kansas City, MO, 1996.
27. P. Geladi, D. MacDougall, and H. Martens, Appl. Spectrosc. *39*:491 (1985)
28. N. B. Colthup, L. H. Daly, and S. E. Wiberley, *Introduction to Infrared and Raman Spectroscopy*, 2nd ed., Academic Press, New Yokr, 1975.
29. D. Lin-Vien, N. B. Colthup, W. G. Fateley, and J. G. Grasselli, *The Handbook of Infrared and Raman Characteristic Frequencies of Organic Molecules*, Academic Press, Boston, 1991.
30. R. Nyquist, ed., *The Infrared Spectra Atlas of Surface Active Agents*, Sadtler, Philadephia, 1982.
31. T. M. Hancewicz, Eastern Analytical Symposium, Somerset, NJ, 1995.
32. D. O. Hummel and F. Scholl, *Infrared Analysis of Polymers, Resins and Additives. An Atlas*, Vol. 1, Parts I and II, Wiley, New York, 1971.

5

Molecular Spectroscopy of Anionic Surfactants II—Nuclear Magnetic Resonance Spectroscopy

LEON C. M. VAN GORKOM   Department of Molecular Structure Analysis, Unilever Research US, Edgewater, New Jersey

ARNOLD JENSEN   Department of Analytical Chemistry, Unilever Research US, Edgewater, New Jersey

I. NMR SPECTROSOCPY

A. Introduction

Nuclear magnetic resonance (NMR) spectroscopy is an analytical technique that is used routinely for the identification of organic compounds. Chemists apply ^{1}H and ^{13}C-NMR spectroscopy to determine the structure of synthetic organic molecules and natural products. Although NMR spectroscopy is not a sensitive technique (for ^{1}H-NMR concentrations > 100 μM), the detail and lack of ambiguity of the information makes it the most effective tool for structural identification and elucidation. Improvements in cryo-technology have increased the field strengths of commercially available magnets to 18.8 Tesla. At this field strength, ^{1}H nuclei resonate at 800 MHz, and it is believed that before the turn of the century, a 1 GHz NMR spectrometer will be available for high-resolution NMR spectroscopy. Higher magnetic field strengths result in an increase in spectral resolution and sensitivity, and allow NMR spectroscopy to be applied to lower concentrations of material. Other recent developments, including probe design, cryo-probes that are cooled with liquid nitrogen, and electronics (preamplifiers, receivers, etc.), have also led to significant increases in sensitivity. The technology developed for and the availability of 800 MHz NMR spectrometers have also reduced the price for lower field NMR instruments. In fact, 300 MHz NMR spectrometers are now considered to be inexpensive and are used in open-access types of environments for the analysis of unknowns. Automation has also played an enormous role in the placement of NMR spectrometers in industrial laboratories. Sample changers which hold from 6 to 100 samples allow the unattended acquisition of data on large numbers of samples. The current generation of software provides ease of data acquisition and processing, resulting in a significantly increased throughput for samples.

NMR spectroscopy has great capabilities in the analysis of unknown compounds. Although IR spectroscopy will provide valuable information on the functional groups present in the molecule, a complete analysis is usually limited to a unique spectral match in a data library. Application of mass spectrometry may provide more detailed information using fragmentation patterns, but, as with IR spectroscopy, complete identification is limited to spectral library searches. This does not imply that the structure of an unknown compound can be solved by NMR spectroscopy alone, and, indeed, the information from both IR and mass spectrometry may be used in structural elucidation problems. In general, however, NMR spectroscopy does not rely on spectral matches, and the use of multidimensional NMR experiments unambiguously elucidates the structure of an unknown.

The resonances of the ^{1}H and ^{13}C nuclei, referred to as chemical shifts, are dependent on their chemical environment and, therefore, give information on the structure of the molecule under investigation [1]. As with IR spectroscopy, certain regions in the chemical shifts of ^{1}H and ^{13}C nuclei are indicative of chemical functionalities present in the molecule. The chemical shifts are dependent on the neighboring groups and, especially in larger molecules, are also influenced by the three-dimensional structure through which certain distinct groups may give rise to additional shielding or deshielding of a particular proton or carbon nucleus. Although proton NMR will provide useful information, the chemical shift dispersion is small ($\cong$10 ppm) and the resonance lines cannot always be unambiguously assigned. In the past, chemical shift reagents such as lanthanides were used that would interact with specific residues of the molecule and thus induce significant shifts in the resonance frequencies [2]. Higher magnetic field strengths have increased the chemical shift dispersion, but proton NMR spectroscopy is still not able to provide full analysis. An additional limiting factor is that the chemical shifts are slightly influenced by the solvent.

^{13}C-NMR spectroscopy has a much larger chemical shift range ($\cong$250 ppm) and is more useful for surfactant analysis, but the sensitivity is much lower [3]. The gyromagnetic ratio of ^{13}C is one quarter that of the protons, and the natural abundance of ^{13}C is only 1.1%. The acquisition of a ^{13}C-NMR spectrum, therefore, requires more sample and/or longer measuring times. Higher magnetic fields and better probe design, however, have significantly increased the sensitivity for ^{13}C nuclei.

One-dimensional proton and carbon NMR experiments usually provide sufficient information for the assignment and identification of anionic surfactants. In order to obtain additional information, ^{13}C DEPT (distortionless enhancement by polarization transfer) experiments may be performed. The DEPT, or an analogous experiment, namely, APT (attached proton test), allows the discrimination between CH, CH$_2$, and CH$_3$ groups based on the modulation of the ^{13}C signal due to the coupling with the attached protons [4]. In conjunction with the regular ^{13}C-NMR spectrum, all carbons in the surfactant, including the quaternary ^{13}C nuclei, can be identified.

Another nucleus of interest in NMR spectroscopy of anionic surfactants is ^{31}P (100% natural abundance and high sensitivity). The resonance positions of phosphate and phosphonate groups are dependent on the degree of protonation and, therefore, are extremely pH dependent (i.e., ^{31}P-NMR is used in vivo to determine the inter- and intra-cellular pH in magnetic resonance imaging (MRI) experiments) [5]. The use of ^{31}P-NMR spectroscopy is limited to phosphorus-containing surfactants. The ^{31}P chemical shift is insufficient, however, to completely assign the phosphorylated anionic surfactants, and, consequently, ^{13}C-NMR spectroscopy is used for their identification.

The strength of NMR spectroscopy lies in the application of multidimensional NMR experiments to an analytical problem. In 1981, Jeener performed the first two-dimensional NMR experiment that correlated the chemical shift resonances of protons that were spin-coupled through chemical bonds [6]. Information on nearest neighbors and complete coupled spin systems is obtained by means of two-dimensional COSY (correlated spectroscopy) NMR experiments. Since this initial experiment, numerous two-dimensional experiments have been developed that correlate chemical shifts of nuclei either through bonds (COSY, TOCSY, HMQC, and HMBC) or, in order to obtain three-dimensional structural information, through space (NOESY) [7,8]. The two-dimensional NMR experiments will be described in more detail in Sec. I.B. NMR spectroscopy is not limited, however, to two dimensions and has been extended to three and four dimensions as applied to ^{13}C, ^{15}N isotopically labeled proteins and nucleic acids [9]. The most recent advance in the field of structure analysis of large proteins and nucleic acids is the ability to measure small, incompletely averaged, chemical shift anisotropies (CSA) or dipolar couplings in such molecules [10]. Both the residual CSA and dipolar couplings are dependent on their orientation with respect to the magnetic field and, therefore, provide additional information to the NOESY distance data on the three-dimensional structure of the proteins and nucleic acids.

B. Two-Dimensional NMR Spectroscopy

Several two-dimensional NMR techniques that are often used for the identification and quantification of surfactants are described below. No detailed information is provided on the basic principles of NMR spectroscopy or the experimental details on how to collect NMR data. Treatment is limited to an example of the interpretation of such data (Sec. I.C). With modern NMR spectrometers, however, these experiments are as easily performed as one-dimensional ^{1}H or ^{13}C experiments [11]. For further information on NMR spectroscopy, the reader is referred to textbooks on the subject [7,8,11].

There is a number of ways to determine the identities of anionic surfactants or surfactant mixtures by NMR spectroscopy. First, resonance frequencies of well-isolated peaks can be matched with reference spectra of surfactants. In case of severely overlapping peaks, however, this may not be successful. Similarly, in case of mixtures or impurities, it is uncertain which peaks arise from the same molecule. Application of two-dimensional NMR experiments to such problems may provide additional information on the nature of the surfactants present in mixtures. With modern NMR spectrometers, more detailed information can easily be obtained from multidimensional NMR experiments. Useful two-dimensional NMR experiments for the identification of surfactants are the homonuclear proton correlation (COSY, TOCSY) and heteronuclear proton-carbon correlation (HETCOR, HMQC, and HMBC) spectroscopy [7,8,11]. In the aforementioned experiments, the resonance frequencies of spin-coupled nuclei are correlated and hence give detailed information on the structure of organic molecules. Two-dimensional NMR experiments have an additional time-evolution period during which the correlation between spins is established. A two-dimensional experiment is recorded as a series of one-dimensional NMR experiments (FIDs) in which the time-evolution period is incremented. The data is first Fourier transformed in the directly detected dimension and then in the indirectly detected dimension. Until recently, the acquisition of two-dimensional NMR experiments was complicated and time-consuming. Long phase cycles, ranging from 8 to 32 acquisitions per experiment, had to be used to avoid or reduce artifacts [7,8]. This has changed dramatically over the last few years with the new generation of NMR spectrometers and processing software. Modern spectrometers are usually equipped with gradient capabilities that are used for coherence selection [8,11,12]. This reduces the number of scans to 1 or 2 per experiment and, thereby, the acquisition time. Other advanges of gradients are solvent suppression and reduced T_1 (spin lattice relaxation time) noise in the spectra. Another recent development is the application of linear prediction algorithms in data processing [8,13]. The use of linear prediction reduces the number of actual experiments (or increments) that need to be recorded. For example, a two-dimensional NMR data set may be recorded with only 64 or 128 experiments. After Fourier transformation of the directly detected dimension, interferograms with 64 or 128 data points are obtained. Linear prediction reliably generates up to three times (192 or 384) as many data points before Fourier transformation in the second dimension. Two-dimensional NMR experiments employing proton detection can be performed in 5 to 20 minutes for surfactant solutions of more than 50 mM. Since the two-dimensional NMR experiments are fast and more informative than regular ^{13}C-NMR spectroscopy, they are becoming an alternative to the latter. For quantitative measurements in mixtures, however, one-dimensional NMR spectroscopy with well-resolved resonances must be used (see Sec. I.D).

The resonances of spin-coupled protons (through 2 or 3 bonds) will give rise to cross-peaks in a COSY spectrum and are useful for the identification of spin systems in the molecule under investigation [14]. TOCSY (total correlated spectroscopy) gives similar information and, providing that the mixing time is sufficiently long, shows cross-peaks between all spin-coupled protons within a particular spin system, while COSY only shows cross-peaks of neighboring protons [15]. Still, it has not been confirmed that different isolated proton spin systems belong to the same molecule.

Another experiment is the correlation of ^{13}C nuclei and directly bonded protons by means of HMQC (heteronuclear multiple quantum coherence) spectroscopy [16]. The HMQC experiment employs proton detection and is inherently more sensitive, therefore, than its carbon-detected counterpart HETCOR (heteronuclear correlation). Both experiments provide the same information and are useful in the case of spectral overlap in one-dimensional NMR spectra. Each class of anionic surfactants has its own pattern of cross-peaks. The combination of TOCSY and HMQC produces a powerful two-dimensional technique for the identification of surfactants in mixtures. HMQC-TOCSY does not only correlate the ^{13}C resonance with the ^{1}H resonance of its directly bonded proton but also shows correlations with other protons belonging to the same spin system, for example, protons that are spin coupled with the directly bonded proton [17].

The HMQC does not measure, however, quarternary carbons since no protons are attached to such atoms. In this case, long-range coupling of protons to ^{13}C nuclei must be used. These provide a wealth of information for the determination of the structure of organic compounds and are routinely applied in the identification of complex natural products [18]. In most anionic surfactants, cross-peaks arising from two-, three-, and four-bond ^{13}C-^{1}H couplings may be observed. Since it measures long-range couplings, the HMBC (heteronuclear multiple bond coherence) experiment is the only one that will provide information on whether two isolated spin systems [separated by quarternary, nonprotonated carbons, or heteronuclear atoms (N, O, or P)] belong to the same molecule.

C. Spectral Assignments

Figure 1 shows the ^{1}H- and ^{13}C-NMR spectra of sodium coconut acid *N*-methyl taurate in D_2O. The assignment of these spectra, labeled according to the accompanying structure, was not straightforward and was achieved using the two-dimensional NMR data (COSY, HMQC, and HMBC). This class of surfactants

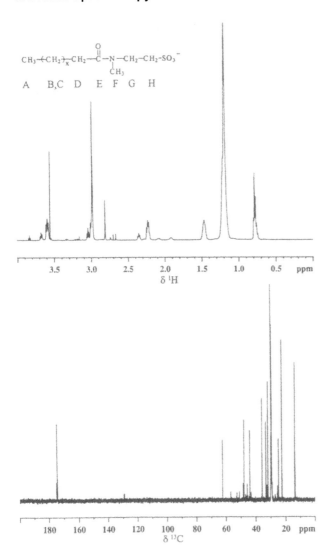

FIG. 1 ^{1}H and ^{13}C-NMR spectrum of sodium coconut acid N-methyl taurate in D_2O.

exists in two rotamer forms because the rotation around the C-N in the methylated amide group is hindered. The existence of rotamers can be deduced from the fact that resonance lines of the N-methyl (H_F) on the amide group and both methylenes (H_D and H_G) on either side of the amide link exist as two lines. Increasing the temperature (up to 90°C) confirmed that these two distinct structures are rotamers and are exchanging into each other. The large peak at 3.55

ppm in the ¹H-NMR spectrum and the one at 63 ppm in the ¹³C-NMR spectrum are not from the taurate and are ascribed to an impurity.

Figure 2 shows the DEPT-135 (CH, CH₃-positive, and CH₂-negative) spectrum of sodium coconut acid *N*-methyl taurate in D₂O. The two positive signals are ascribed to methyl groups since no CH groups are observed in the DEPT-90 spectrum (spectrum not shown).

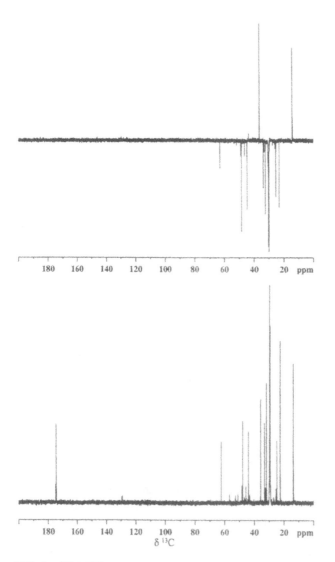

FIG. 2   ¹³C DEPT-135 NMR spectrum of sodium coconut acid *N*-methyl taurate in D₂O.

Figure 3 shows the COSY NMR spectrum of sodium coconut acid N-methyl taurate in D_2O, which was recorded in 10 minutes. In this example, two spin systems can be clearly identified. The first is the alkyl chain spin system in which correlations can be seen between the terminal methyl (H_A, 0.8 ppm) and the methylene backbone (H_B, 1.25 ppm). The methylene backbone (H_B) is coupled to another methylene group (H_C, 1.45 ppm), which in turn is coupled to the methylene moiety adjacent to the carbonyl (H_D, 2.25 and 2.30 ppm). Two specific resonances are observed for the methylene group (H_D) adjacent to the carbonyl due to the two rotamer conformations. No differences in the rest of the alkyl chain are observed

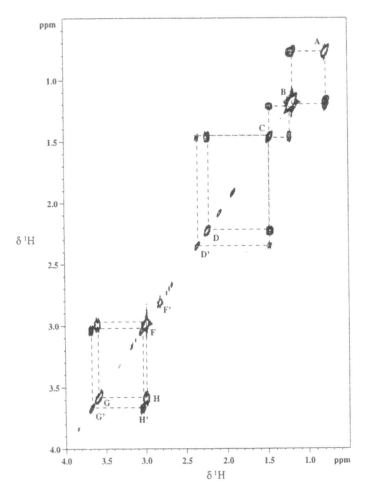

FIG. 3 2D-COSY spectrum of sodium coconut acid N-methyl taurate in D_2O.

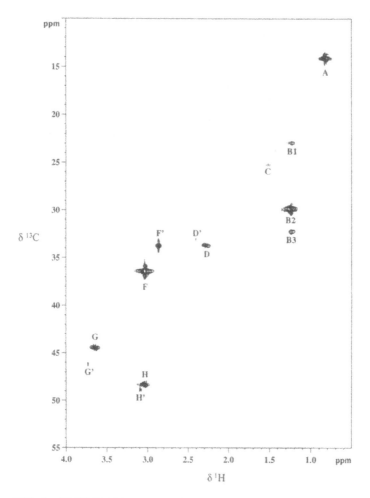

FIG. 4 ^{1}H-^{13}C HMQC spectrum of sodium coconut acid N-methyl taurate in D$_2$O.

for the two rotamers since they are further away from the methylated amide segment with the hindered rotation. The second spin system arises from the methylene groups (H$_G$ and H$_H$) of the taurate. Once again, two distinct spin systems (H$_G$,
3.60 ppm and H$_H$, 3.0 ppm; H$_G$, 3.65 ppm and H$_H$, 3.05 ppm) are observed, indicating the presence of two rotamers. Three distinct singlets that are not spin-coupled to other protons are also evident. The resonances at 2.8 and 3.0 ppm are
ascribed to the N-methyl of the amide group (H$_F$) representing the two rotamers,
while the singlet at 3.55 ppm is ascribed to an impurity. The levels of the two isomers can be determined from the integral ratio in the ^{1}H-NMR spectrum.

Figure 4 shows an HMQC spectrum of sodium coconut acid *N*-methyl taurate in D$_2$O. The HMQC spectrum allows the correlation between ^{13}C and ^{1}H chemical shifts and may be used for identification. These correlations will also be used for the interpretation of the HMBC spectrum. (Fig. 6). The methylene backbone of the alkyl chain with degenerate chemical shifts (H$_B$, 1.25 ppm) in the proton NMR spectrum shows several distinct ^{13}C chemical shifts related to their position in the chain (C$_{B1}$, C$_{B2}$, and C$_{B3}$).

Figure 5 shows the HMQC-TOCSY spectrum of sodium coconut acid *N*-methyl taurate in D$_2$O. The HMQC-TOCSY is particularly useful for the analysis of mix-

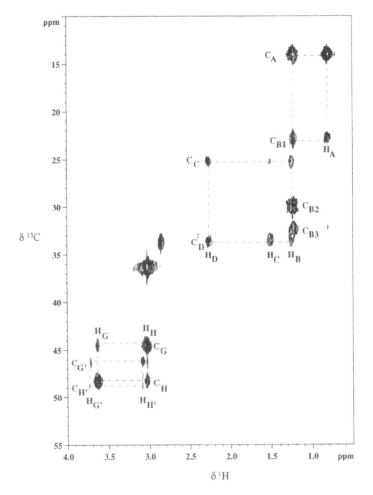

FIG. 5 HMQC-TOCSY NMR spectrum of sodium coconut acid *N*-methyl taurate in D$_2$O.

tures and is shown for illustrative purposes only. Cross-peaks are observed for C_G with H_G and H_H, for C_H with H_G and H_H, and also for protons and carbons belonging to the alkyl chain spin system. Due to the long mixing time (40 ms), the intensities of the cross-peaks between the carbon and its directly bonded proton are seen to be smaller than those of the subsequent protons belonging to the same spin system. (Fig. 4) This is due to the diffusion of magnetization throughout that spin system.

Figure 6 shows the HMBC spectrum of sodium coconut acid N-methyl taurate in D_2O. Cross-peaks indicative of long-range coupling between the protons of the methylene group of each rotamer (H_D, 2.25 and 2.35 ppm) and the car-

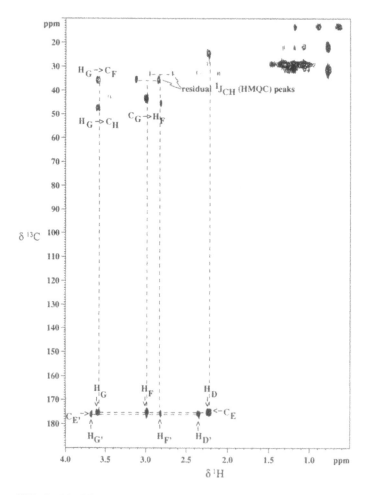

FIG. 6 HMBC spectrum of sodium coconut acid N-methyl taurate in D_2O.

bonyl carbon (C_E) can be seen. Both rotamers have a different carbonyl chemical shift. Long-range couplings between the protons of the N-methyl group of each rotameter (H_F, 3 and 2.8 ppm) with their respective carbonyl carbon as well as between the methylene group next to the amide of each rotameter (H_G, 3.65 and 3.60 ppm) with the corresponding carbonyl carbon (C_E) can also be seen. With all the observed connectivities, it can be concluded without ambiguity that these nuclei are indeed part of the same molecule.

D. New Techniques

In cases in which two-dimensional NMR experiments are insufficient for a complete analysis of the anionic surfactant mixtures, recently developed techniques such as diffusion ordered spectroscopy and LC-NMR may provide better information.

1. Diffusion Ordered Spectroscopy

A technique that may prove successful in the determination of anionic surfactants in mixtures is diffusion ordered spectroscopy (DOSY) [19]. This technique relies on the differences in diffusion coefficients of the anionic surfactants. The basic principle of the experiment is the deliberate dephasing and rephasing of the signal by magnetic field gradients of equal strength. Depending on the time between the gradients, the signal intensity decreases according to the diffusion rate of the molecule under investigation. For example, small molecules will diffuse very quickly out of the plane and, after being dephased by the first gradient, are not in the same position to be rephased by the second gradient pulse, resulting in no observable signal. The DOSY experiment is acquired as a series of one-dimensional NMR spectra in which the entire spectral bandshapes of the individual compounds have been attenuated by varying the gradient strength. The data is analyzed by fitting the decay in the intensities of each resonance by a multiexponential fit. This results in a two-dimensional representation of the DOSY experiment. One dimension corresponds to the spectroscopic data and the other dimension represents the diffusion domain [20]. Recently, statistical analysis of DOSY data by multivariate curve resolution (MCR) using principal factor analysis (PFA), Varimax rotation, and alternating least squares (ALS) optimization generated "pure" spectroscopic profiles of the individual compounds for chemical identification [21], Figure 7a shows the spectrum of a mixture of two anionic polymers, namely, polyacrylic acid (PAA, MW ~ 2kDa) and polyacrylamide (PAM, MW ~ 7kDa) in D_2O. From the regular ^{1}H-NMR spectrum, it was not possible to determine that the sample consisted of PAA and PAM. The data of the DOSY experiment was analyzed using MCR, and the results are shown in Fig. 7b. The "pure" spectra may now be identified as the PAA and PAM polymers, and the signal decay curves representing the diffusion domain indicate that PAM diffuses more slowly than the PAA polymer.

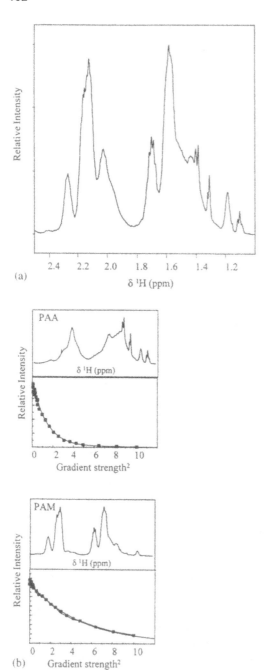

FIG. 7 (a) DOSY spectrum of a mixture of PAA and PAM and (b) the "pure" components with diffusion decay.

2. LC-NMR Spectroscopy

Recently, liquid chromatography has been interfaced with NMR spectroscopy (LC-NMR), which allows the separation of mixtures into individual anionic surfactants and their subsequent detection by means of proton NMR spectroscopy [21]. The application of LC-NMR requires specialized equipment such as an NMR probe with a flow cell. Although LC-NMR has now been established as a routine technique, only a limited number of systems has been sold. In the LC-NMR experiment, the NMR spectrometer can be regarded as just another detector such as UV or RI, and data is continuously acquired on the eluent outflow of the column. NMR acquisition is performed continuously during the separation, and, typically, NMR data is time-averaged over 15–30 seconds. The LC-NMR data is collected as a series of one-dimensional NMR experiments. After Fourier transformation, the data is presented as a two-dimensional plot with the first dimension providing the spectroscopic information for identification and the second dimension representing the elution profile. Although this technique has not been applied to any known surfactant mixtures, it shows great promise in their analysis. An example of the information that has been obtained is the GPC-NMR of an industrial anionic copolymer of acrylic acid and laurylmethyl acrylic acid, Narlex DC-1. Separation by aqueous gel permeation chromatography (GPC) resulted in a broad elution profile. The NMR analysis of this broad peak indicates that there may be a difference in spectral profiles between the leading and trailing edges of the elution peak (Fig. 8a). A more in-depth analysis by multivariate curve resolution generated two "pure"

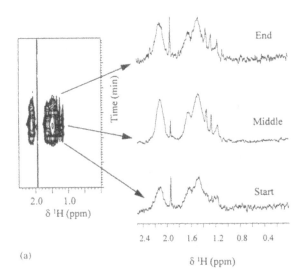

(a)

FIG. 8 (a) GPC-NMR of Narlex DC-1 and (b) "pure" components from MCR analysis with elution profile.

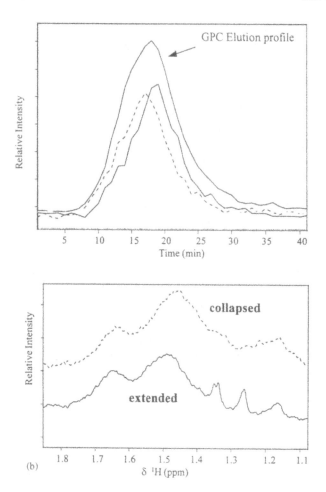

FIG. 8 Continued

spectral profiles, indicating that this polymer exists in two forms, namely, an extended and a collapsed conformation due to the solvent composition used in the polymerization reaction (Fig. 8b) [21]. Figure 8b also shows the elution profiles of both forms of the Narlex DC-1 polymer obtained from MCR data analysis.

E. Quantitative NMR Spectroscopy

NMR may be used to quantify the relative ratios of the individual components in mixtures or to measure the average alkyl chain length or the number of ethylene oxide units present in a surfactant [3]. Examples will be given in the specific spectra for the different classes of anionic surfactants. In the case of anionic surfactants, this is usually done by ^{13}C-NMR spectroscopy, since there is less likely to be signal overlap. In the case of quantitative ^{13}C-NMR spectroscopy, inverse gated decoupled NMR experiments are required. These experiments do not employ irradiation of the protons during the relaxation time in order to avoid the build-up of NOE (nuclear overhauser effects) that may be different for quarternary C, CH, CH$_2$, and CH$_3$ groups. In addition, a long delay [approximately 5 times the spin-lattice relaxation time (T_1)] between successive scans is necessary to allow for full relaxation. This time may be significantly reduced by adding a relaxation agent such as chromium(III) acetylacetonate. Keeping the aforementioned in mind, the integrals of identified peaks of the surfactants are used to determine the relative ratios of the individual components in the mixture. To determine absolute levels of surfactants in samples by NMR spectroscopy, either an internal or an external calibrated standard that has no interfering resonances with those of interest must be used.

F. General Remarks

In general, NMR experiments are straightforward, but small shifts in the resonance frequencies may be observed in different solvents due to polarity, pH, hydrogen bonding, etc. Usually the limiting factor is the solubility of the surfactants in a deuterated solvent, especially in the case of the low-sensitivity ^{13}C-NMR. All the ^{13}C-NMR spectra presented in this chapter were recorded on a Bruker DMX-400 NMR spectrometer equipped with a 5-mm broadband probe.

All of the anionic surfactant samples used to produce the presented spectra were from commercial sources. As such, they may have contained significant levels of impurities such as reaction by-products, reaction starting materials, and solvents. Thus, not all resonance lines may be assigned in all spectra.

All of the anionic surfactants contain some type of cationic counterion. This may be Na or K, in which case they are undetected by ^{1}H or ^{13}C NMR. Alternatively, they may contain an organic counterion such as DEA (diethanolamine) or TEA (triethanolamine). DEA has two resonances in ^{13}C-NMR at 48 and 57 ppm, while the two signals of TEA are much closer together at approximately 55 ppm.

In the remainder of this chapter, the NMR spectroscopy that has been performed on the various classes of anionic surfactants will be discussed. The ^{13}C-NMR spectrum for each of the typical classes of anionic surfactants with the accompanying general structure will be shown.

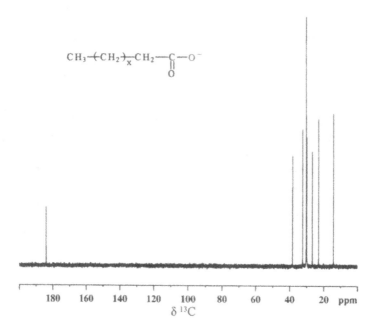

FIG. 9 ^{13}C-NMR spectrum of potassium laurate in D$_2$O.

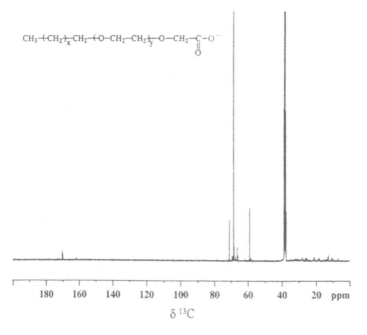

FIG. 10 ^{13}C-NMR spectrum of trideceth-7 carboxylic acid in DMSO-d$_6$.

II. CARBOXYLATES

A. Fatty Acid Salts/Soaps

The assignment of fatty acid salts by means of NMR spectroscopy relies mainly on the resonance frequency of the ^{13}C nucleus of the carbonyl moiety (180–185 ppm). The signals of the fatty acid chain (10–40 ppm) will provide information on the average chain length of the fatty acid. If the data is obtained for quantitative measurements, the average chain length can be determined by adding together the intensities of all signals and then dividing by the intensity of the methyl group ^{13}C signal. Unsaturated fatty acids are identified by resonances at approximately 136 and 130 ppm, indicative of olefinic carbons. Hydroxylated fatty acids such as ricinoleate (12-hydroxy-9-octadecanoic acid, Na salt) show an additional signal at 75 ppm, indicative of a C—O bond. Figure 9 shows the ^{13}C spectrum of potassium laurate in D_2O.

B. Alkyl Ethoxy Carboxylates

This class of surfactants consists of carboxylated fatty ethoxylated alcohols. Figure 10 shows the ^{13}C-NMR spectrum of trideceth-7 carboxylic acid (Incrodet TD-7C,

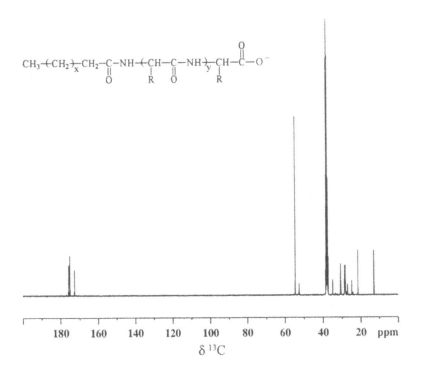

FIG. 11 ^{13}C-NMR spectrum of TEA cocoyl glutamate in DMSO-d_6.

Croda) in DMSO-d_6. The alkyl chain (C_{13}) is heavily branched as deduced from the large number of small resonances in the alkyl region (5–35 ppm). The EO groups are observed at 69 ppm. The peak at 71 ppm is the methylene of the branched alkyl chain next to the ethoxylated part of the molecule, and the peak at 59 ppm is the methylene adjacent to the acid moiety. The acid group is observed at 170 ppm.

C. Amino Carboxylates

This group of surfactants consists of esters of amino acids (succinate, glutamate) with long-chain alcohols. Figure 11 shows the ^{13}C-NMR spectrum of TEA cocoyl glutamate (Alkylglutamate CT12S, Ajinomoto) in DMSO-d_6. Three different carbonyl groups are observed between 174 and 177 ppm. The large resonance at 56 ppm is due to the TEA counterion, and the C-alpha of the amino acid group (adjacent to the nitrogen) is observed at 54 ppm.

III. SULFATED SURFACTANTS

A. Alkyl Sulfates

This group of surfactants consists of sulfates of either primary or secondary alcohols. The alkyl chains are fully saturated and show ^{13}C signals between 5 and 40 ppm. The ^{13}C nucleus next to the sulfate group has an indicative resonance frequency at approximately 70 ppm. The range of the chemical shift of the sulfated methylene is from 67 to 73 ppm for primary and secondary sulfates. Figure 12 shows the ^{13}C-NMR spectrum of an alcohol sulfate (Serdet DSK-40, Huls) in D_2O.

B. Ethoxylated Alkyl Sulfates

This group of surfactants consists of the sulfated ethoxylated alcohols. ^{13}C-NMR spectroscopy is a valuable tool in determining the average ethoxylation of the surfactant. The signals of the EO groups are at a slightly lower field (69–73 ppm) than the methylene carbon adjacent to the sulfate (66–68 ppm). Increased levels of ethylene oxide result in an upfield shift of the resonance position of the sulfated methylene (68 ppm for zero EO, 67–66 ppm for EO levels greater than zero). From the integral ratio of these regions, the average number of EOs per surfactant can be determined (including zero EO). Figure 13 shows a typical spectrum of the ammonium salt of ethoxylated lauryl sulfate (laureth sulfate, Steol CS-330, Stepan) in D_2O.

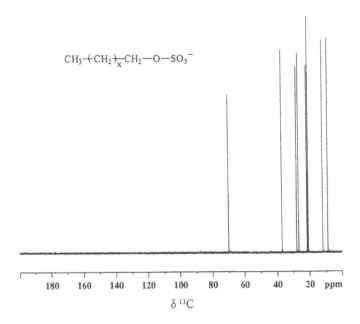

FIG. 12 ^{13}C-NMR spectrum of an alcohol sulfate in D$_2$O.

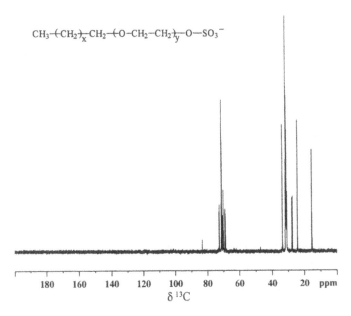

FIG. 13 ^{13}C-NMR spectrum of the ammonium laureth sulfate in D$_2$O.

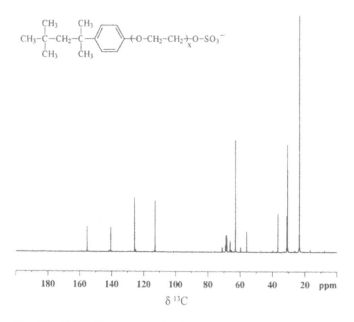

FIG. 14 ^{13}C-NMR spectrum of sodium alkylaryl polyether sulfate in D$_2$O.

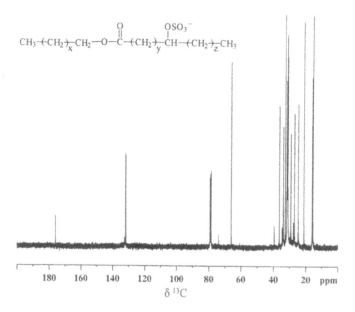

FIG. 15 ^{13}C-NMR spectrum of sulfated butyloleate in CDCl$_3$.

C. Alkyl Phenyl Ethoxy Sulfates

This class of surfactants consists of the sulfated nonionic alkyl phenol EO alcohols. Figure 14 shows the ^{13}C-NMR spectrum of a sodium alkylaryl polyether sulfate (Triton X-770, octylphenyl polyether sulfate, Union Carbide) in D_2O. The resonances at 155, 140, 125, and 112 ppm are indicative of a *para*-substituted aromatic ring. The methyl groups of the branched octyl chain resonate at 31 and 31.5 ppm. The quaternary carbon adjacent to the phenyl group resonates at 38 ppm, while the other quaternary carbon resonates at 32 ppm. The methylene group of the alkyl chain shows a resonance line at 56 ppm. The EO groups are clearly visible and resonate between 60 and 70 ppm. The sulfated methylene shows a peak around 60 ppm. The presence of an additive, namely, 2-propanol, is evident from the resonances at 25 and 64 ppm.

D. Oil Sulfates

This class of surfactants comprises the monoglyceride, diglyceride, and triglyceride derivatives of sulfated fatty acids as well as long-chain alcohol esters of sulfated fatty acids. Figure 15 shows the ^{13}C-NMR spectrum of sulfated butyloleate in $CDCl_3$. The ester carbon is observed at 173 ppm. The alkyl chain shows resonances between 10 and 40 ppm. The sulfated methyne in the oleic acid moiety resonates at 72 ppm, while the methylene of the butyl group adjacent to the ester linkage shows a peak at 64 ppm. The presence of oleic acid is confirmed by the two resonances around 132 ppm. The peak due to the sulfated methyne in the oleic acid is small with respect to the other resonances from the oleic acid, indicating that only low amounts of sulfate have been incorporated.

IV. SULFONATED SURFACTANTS

A. Alkyl Sulfonates

This class of anionic surfactants consists of sulfated alkanes. Figure 16 shows the ^{13}C-NMR spectrum of sodium C_{13}-C_{17} alkane sulfonate (Marlon PS-60, Huls) in D_2O. The alkyl chain is visible at 10–35 ppm, and the methylene group adjacent to the sulfonate group resonates at 60 ppm.

B. Alkyl Glycerol Ether Sulfonates

Figure 17 shows the ^{13}C-NMR spectrum of alkyl glycerol ether sulfonates (AGES) in DMSO-d_6. The alkyl chain is observed between 10 and 35 ppm. The methylene carbon of the alkyl chain adjacent to the glycerol resonates at 67 ppm. The glycerol carbons give rise to three peaks at 74, 70, and 54 ppm. The reso-

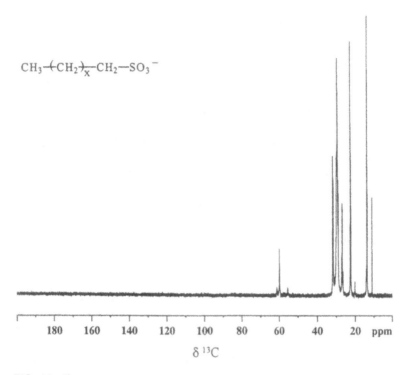

CH$_3$—(CH$_2$)$_x$—CH$_2$—SO$_3^-$

δ ^{13}C

FIG. 16 ^{13}C-NMR spectrum of sodium C$_{13}$-C$_{17}$ alkane sulfonate in D$_2$O.

nance at 54 ppm reflects the sulfonated methylene segment of the glycerol group.

C. Alpha-Olefin Sulfonates

This class of surfactants consists of sulfonated alpha-olefins. During the sulfonation process, the double bond of the alpha-olefin tends to migrate up the chain to the two or three position. In addition, this double bond may be hydrated to form the corresponding 2-, 3-, and 4-hydroxy sulfonates. Figure 18 shows a ^{13}C-NMR spectrum of an alpha-olefin sulfonate with typical impurities. AOS have indicative ^{13}C resonances in the 120- to 140-ppm range depending on the position of the olefinic group. Typically, AOS samples have 10–20% 1-alkene, 30–40% 2-alkene, 10–20% 3-alkene, and 2–5% 4-alkene. The remaining 30% consists of hydrated forms of AOS with resonance lines at approximately 70 ppm. The methylenes directly bonded to the sulfonate group resonate in the 50- to 60-ppm range. Finally, the peaks from 14 to 40 ppm are due to the alkyl chain.

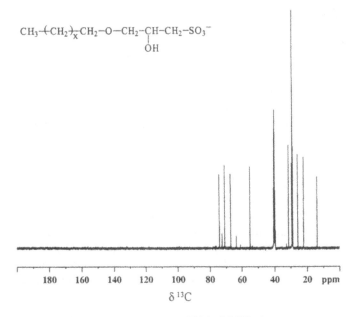

FIG. 17 ^{13}C-NMR spectrum of AGES in DMSO-d$_6$.

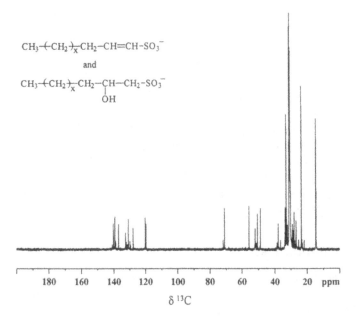

FIG. 18 ^{13}C-NMR spectrum of AOS in D$_2$O.

D.   Alkyl Aryl Sulfonates

1.   Short-Chain Alkylbenzenesulfonates

This category comprises the hydrotropes such as the cumene, toluene, and xylene sulfonates and the sodium salt of *p*-phenolsulfonic acid. Figure 19 shows the ^{13}C-NMR spectrum of sodium cumenesulfonate (Naxonate SC, Reutgers-Neuse Chem. Co.) in D_2O. The signals at 23 and 33 ppm are from the alkyl group's CH_3 and CH, respectively. The *para*-substituted aromatic ring shows four resonance lines at 153 (isopropyl substitution), 140 (sulfonate substitution), 127, and 126 ppm.

2.   Long-Chain Alkylbenzenesulfonates

Figure 20 shows the ^{13}C-NMR spectrum of sodium dodecylbenzenesulfonate (Marlon A-375, Huls) in D_2O. This group of anionic surfactants consists of linear alkylbenzenesulfonates. The resonances between 40 and 50 ppm are from the phenylsulfonate-substituted methyne of the alkyl chain. The resonance frequency of the methyne is dependent on the position of the substitution along the chain. Substitution at the second position gives rise to a signal at approximately 39–40 ppm, and substitution at positions 3, 4, 5, or 6 yields resonances at approximately 47–49 ppm.

Typically, LAS surfactants show an isomer ratio for the 2-, 3-, 4-, and 5-phenylsulfonate substitution of approximately 2:1:1:1. The resonance peaks between 30 and 38 ppm are due the methylene groups adjacent to the substituted methyne.

E.   Diphenyloxide Disulfonates

This class of surfactants consists of two phenyl rings connected by an ether linkage. Both phenyl rings are substituted by a single sulfonate group. One of the phenyl rings is also substituted by an alkyl chain. Figure 21 shows the ^{13}C-NMR spectrum of decyl(sulfophenoxyl) benzene-sulfonic acid (Dowfax 382, Dow Chemical Co.) in D_2O. The number of resonance lines clearly indicates that this surfactant consists of positional isomers and that it may have impurities. The alkyl chain is observed between 10 and 45 ppm, while the aromatic rings are observed between 120 and 160 ppm.

F.   Naphthalenesulfonates

This class of surfactants describes all possible isomers of naphthalene substituted by a single sulfonate. Figure 22 shows the ^{13}C-NMR spectrum of sodium naphthalenesulfonate (Supragil NS-90, Rhone-Poulenc) in D_2O. The aromatic carbons are observed between 120 and 140 ppm.

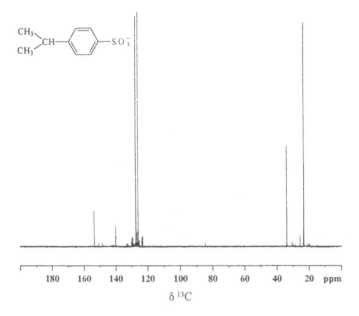

FIG. 19 ^{13}C-NMR spectrum of sodium cumenesulfonate in D$_2$O.

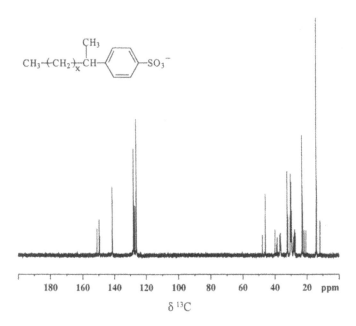

FIG. 20 ^{13}C-NMR spectrum of sodium dodecylbenzenesulfonate in D$_2$O.

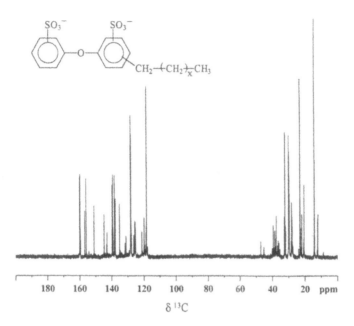

FIG. 21   ¹³C-NMR spectrum of decyl (sulfophenoxyl) benzenesulfonic acid in D_2O.

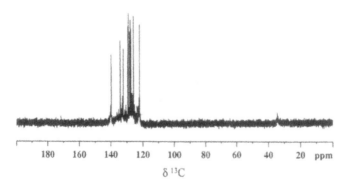

FIG. 22   ¹³C-NMR spectrum of sodium naphthalenesulfonate in D_2O.

G. Alkyl Naphthalenesulfonates

This class of surfactants consists of naphthalenesulfonate in which both aromatic rings have been substituted with alkyl groups. Figure 23 shows the ^{13}C-NMR spectrum of sodium dibutyl naphthalenesulfonate (Eccowet LF, Eastern Color & Chemical Co.) in D_2O. The complexity of the spectrum indicates the presence of possible isomers of the dialkylnaphthalenesulfonate. The butyl groups are observed between 10 and 45 ppm and the aromatic region from 120 to 150 ppm. The large sharp peaks (labeled with asterisks) are believed to be impurities.

H. Alkyl Phenyl Ethoxy Sulfonates

This class of surfactants consists of sulfonated alkylphenol ethoxylates. Figure 24 shows the ^{13}C-NMR spectrum of sodium alkylaryl polyether sulfonate (Triton X-200, Union Carbide) in $CDCl_3$. As observed for its sulfate analog, Triton X-770, the branched octyl chain is evident from the resonances at 31, 31.5, 32, 38, and 56 ppm. The EO groups are observed between 60 and 70 ppm, and the methylene adjacent to the sulfonate is observed at 50 ppm. The aromatic region, 120–160 ppm, shows a characteristic *para*-substitution pattern with the phenyl carbon bonded to the EO group resonating at 156 ppm and the phenyl carbon bonded to the alkyl chain resonating at 142 ppm.

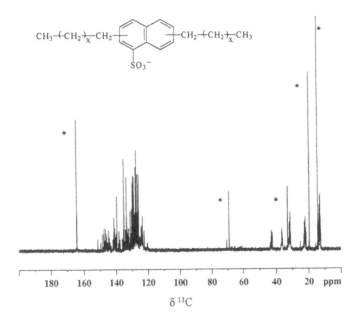

FIG. 23 ^{13}C-NMR spectrum of sodium dibutyl naphthalenesulfonate in D_2O.

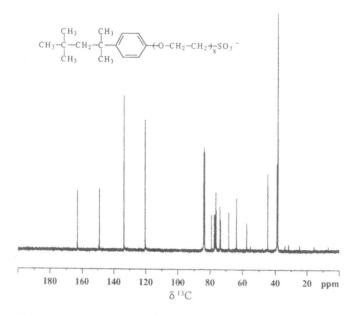

FIG. 24 ¹³C-NMR spectrum of sodium alkylaryl polyether sulfonate in CDCl₃.

I. Sulfoacetates

Figure 25 shows the ¹³C-NMR spectrum of sodium lauryl sulfoacetate (Lathanol LAL, Stepan Chem. Co.) in D₂O. Once again, the alkyl chain is observed between 10 and 35 ppm. The ester carbonyl is observed at 167 ppm. The methylene group between the carbonyl and sulfonate is observed at 55 ppm. The methylene segment of the alkyl chain adjacent to the oxygen of the ester linkage resonates at 66 ppm.

J. Sulfo Methyl Esters

Figure 26 shows the ¹³C-NMR spectrum of methyl-2-sulfolaurate (Alpha Step MC-48, Stepan Chem. Co.) in D₂O. The ester group can be identified by the resonance at 171 ppm. The methoxy moiety resonates at 66 ppm and the sulfonated CH group shows a peak at 53 ppm. The alkyl chain is observed between 15 and 35 ppm.

K. Alkoyl Isethionates

Alkoyl isethionate shows alkyl peaks between 10 and 35 ppm, and an ester carbonyl is observed at approximately 173 ppm. The incorporated isethionate has

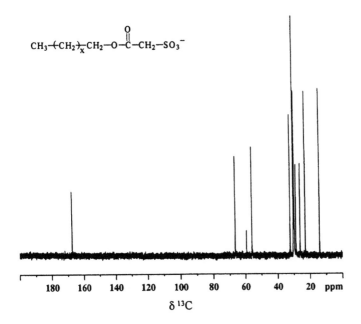

FIG. 25 ^{13}C-NMR spectrum of sodium lauryl sulfoacetate in D$_2$O.

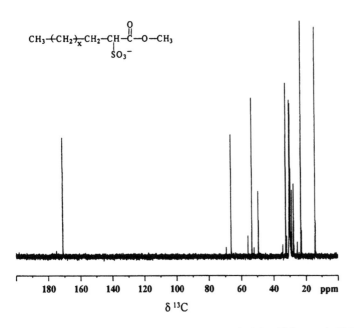

FIG. 26 ^{13}C-NMR spectrum of sodium methyl-2-sulfolaurate in D$_2$O.

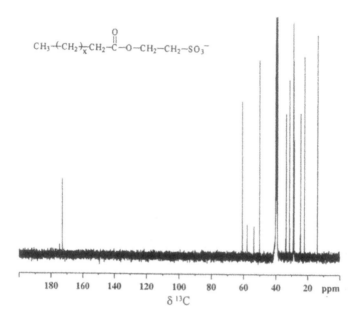

FIG. 27 ^{13}C-NMR spectrum of coconut oil acid ester of isethionate in DMSO-d$_6$.

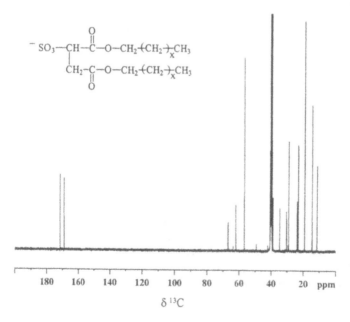

FIG. 28 ^{13}C-NMR spectrum of sodium di-2-ethylhexyl sulfosuccinate in DMSO-d$_6$.

two signals at approximately 60 and 50 ppm. Figure 27 shows the ^{13}C-NMR spectrum of coconut oil acid ester of isethionate in DMSO-d$_6$. This class of surfactants may show residual amounts of sodium isethionate. This residue is easily identified by two resonances of equal intensities at approximately 57 (O—CH$_2$) and 53 ppm (—CH$_2$—SO$_3$), which are observed between the peaks of the alkoyl isethionate.

L. Sulfosuccinates

This class of anionic surfactants comprises the mono- and di-esters of succinic acid and long-chain alcohols (ethoxylated alcohols etc.). Figure 28 shows the ^{13}C-NMR spectrum of sodium di-2-ethylhexyl sulfosuccinate (Serwet WH-170, Huls) in DMSO-d$_6$. The ester group next to the sulfate methyne is observed at 171 ppm, while the other ester moiety resonates at 168 ppm. The sulfonated CH of the succinic backbone can be observed at 62 ppm, and the methylene group of the succinic backbone resonates at 34 ppm. The methylene groups of the alcohols adjacent to the ester linkages were observed at approximately 67 ppm. The branched alkyl chain has resonance lines between 10 and 40 ppm. The large peaks at 57 and 19 ppm are assigned to an impurity of ethanol.

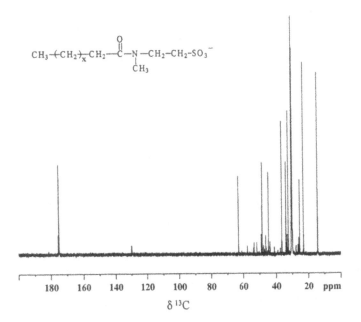

FIG. 29 ^{13}C-NMR spectrum of sodium coconut acid-*N*-methyl taurate in D$_2$O.

M.   Alkoyl Taurates

Figure 29 shows the ^{13}C-NMR spectrum of sodium coconut acid *N*-methyl tau-rate (Geropon TC-42, Rhone-Poulenc) in D_2O. The amide carbonyl resonates at 175 ppm. The *N*-methyl, *N*—CH_2, and the CH_2—SO_3 have resonances at 36, 44, and 48 ppm, respectively. A small amount of an unsaturated alkyl chain is ob-served at approximately 130 ppm. Although not obvious from the ^{13}C-NMR spectrum, this class of surfactants exists as rotamers due to the hindered rotation around the methylated amide bond (see Sec. I.C).

V.   AMINO SURFACTANTS

A.   Aminoacetates and Aminopropionates

Although this class of surfactants is amphoteric, such surfactants may be ob-served in anionic fractions of detergent formulations. Figure 30 shows the ^{13}C-NMR spectrum of di-sodium laurimino dipropionate (Monateric 1188M, Mona) in D_2O. The alkyl chain carbons are observed between 10 and 35 ppm. The methylene of the alkyl chain adjacent to the amino group resonates at 53 ppm. The methylene segment of the propionate moiety adjacent to the amino group resonates slightly upfield at 49 ppm, and the methylene group next to the car-bonyl resonates at 33 ppm. The carbonyl is observed at 180 ppm. A simple dis-tinction between mono-and di-propionates (acetates) can be made from the ratio of the signal intensities from the propionate and the alkyl chain. In this example, the signal intensity of the methylene of the alkyl chain (at 53 ppm) is half the size of the propionate methylenes (at 49 ppm). Due to the symmetry of the mole-cule, both propionates are magnetically equivalent.

B.   Amphoacetates and Amphopropionates

These surfactants also belong to the amphoteric class but may be observed in an-ionic surfactant fractions. Figure 31 shows the ^{13}C-NMR spectra of lauroam-phoacetate (Miranol HMA-32, Rhone Poulenc) in D_2O. The upfield line at 175 ppm is from the amide carbonyl, while the acid carbonyl resonates at 177 ppm. The incorporated glycolic acid shows one resonance at 58 ppm. The alcohol group has two resonances at 58.5 (CH_2—O) and 56 ppm (N—CH_2). The resonance at 53 ppm is ascribed to the methylene carbon next to the tertiary amino group. The two resonances at 35 and 36 ppm are ascribed to the methylene groups adja-cent to the amide group with the upfield resonance (35 ppm) belonging to the fatty acid moiety. Finally, the resonance line at 61 ppm is assigned to an impurity of glycolic acid.

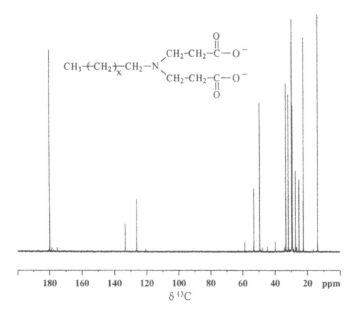

FIG. 30 ^{13}C-NMR spectrum of di-sodium laurimino dipropionate in D$_2$O.

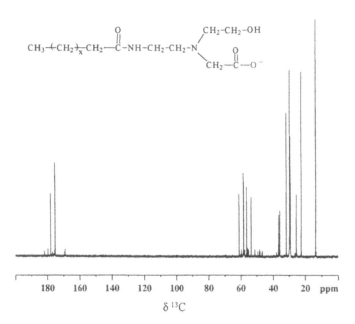

FIG. 31 ^{13}C-NMR spectrum of lauroamphoacetate in D$_2$O.

C. Alkoyl Sarcosinates

Figure 32 shows the ^{13}C-NMR spectrum of sodium lauroyl sarcosinate (Mapro-syl 30, Stepan Chem. Co.) in D_2O. The methylene segment next to the carbonyl of the amide link resonates at 33 ppm. The N-methyl group resonates at 52 ppm, and the methylene group between the amide and carbonyl group has a chemical shift of 54 ppm. The amide carbonyl resonates at a higher field (175 ppm) than the acid carbonyl (176 ppm).

D. Sulfosuccinamates

Figure 33 shows the ^{13}C-NMR spectrum of tetra sodium dicarboxyethyl stearyl/oleyl sulfosuccinamate (Monawet SNO-35, Mona Industries) in D_2O. The resonances in the 170- to 180- ppm region are assigned to the carbonyl groups. Small amounts of unsaturated alkyl chains are evident from the lines at 130–131 ppm. The resonance at 66 ppm is assigned to the sulfated methyne, while the carboxylated methyne resonates at 65 ppm. The resonances at approximately 40 ppm are ascribed to the alkyl chain methylene groups substituted onto the nitrogen, whereas the methylene adjacent to the acid group resonates at 38 ppm. Finally, the methylene group next to the amide moiety

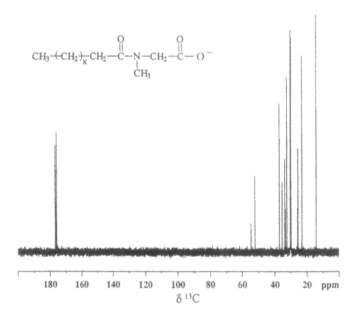

FIG. 32 ^{13}C-NMR spectrum of sodium lauroyl sarcosinate in D_2O.

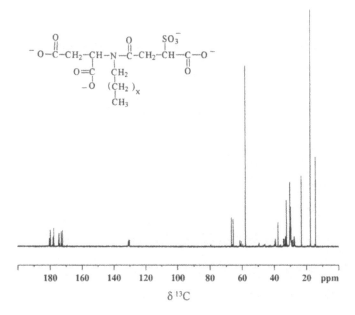

FIG. 33 ^{13}C-NMR spectrum of tetra sodium dicarboxyethyl stearyl/oleyl sulfosuccinamate in D$_2$O.

resonates at 33 ppm. The large peaks at 57 and 17 ppm are assigned to an impurity of ethanol.

VI. PHOSPHORYLATED ANIONIC SURFACTANTS

A. Phosphate Esters of Fatty Alcohols

This class of anionic surfactants consists of the mono- and di-esters of phosphoric acid. Figure 34 shows the ^{13}C-NMR spectrum of the 2-ethylhexyl ester of phosphoric acid (Servoxyl VPTZ-100, Huls) in CDCl$_3$. The methylene segment of the alkyl group adjacent to the phosphate resonates at 69 ppm, and the rest of the alkyl chain shows resonance lines between 10 and 40 ppm.

B. Phosphate Esters of Fatty Ethoxy Alcohols

This class of surfactants consists of the phosphate esters of ethoxylated alcohols. Figure 35 shows the ^{13}C-NMR spectrum of polyoxypropylene (5) polyoxyethylene (10) cetyl ether phosphate (Crodafos SG, Croda) in DMSO-d$_6$. The alkyl chain is observed between 10 and 35 ppm. The ethoxylated part of the surfactant is observed between 60 and 75 ppm.

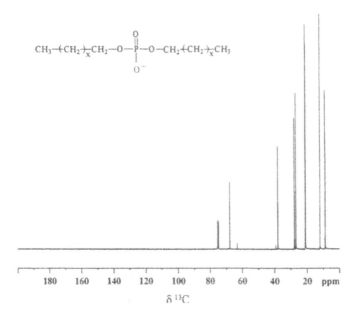

FIG. 34   ¹³C-NMR spectrum of 2-ethylhexyl ester of phosphoric acid in CDCl₃.

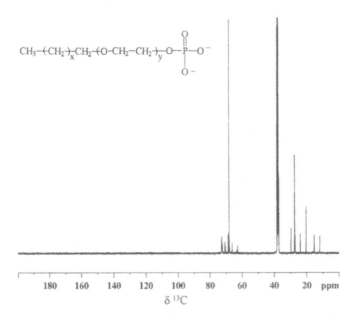

FIG. 35   ¹³C-NMR spectrum of polyoxypropylene (5) polyoxyethylene (10) cetyl ether phosphate in DMSO-d₆.

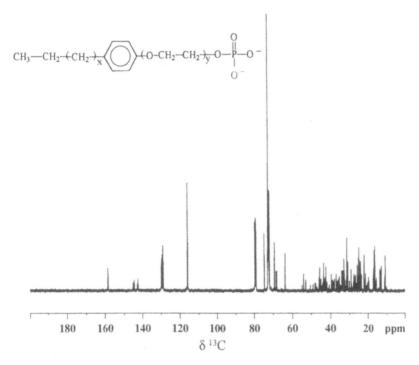

FIG. 36 ¹³C-NMR spectrum of poloxyethylene branched nonyl phenyl ether phosphate in CDCl₃.

C. Phosphate Esters of Alkyl Phenyl Ethoxylates

This class of surfactants comprises the phosphate esters of alkyl phenyl ethoxylates. Figure 36 shows the ¹³C-NMR spectrum of poloxyethylene branched nonylphenyl ether phosphate (Rhodafac PE-510, Rhone-Poulenc) in CDCl₃. The number of resonances in the aliphatic region (10–50 ppm) suggests a mix of branched alkyl chains. The phenyl group shows resonances between 115 and 160 ppm. The EO groups are clearly visible between 60 and 80 ppm.

REFERENCES

1. A. E. Derome, *Modern NMR Techniques for Chemistry Research*, Pergamon Press, Oxford, UK, 1987.
2. H. Konig, in *Anionic Surfactants: Chemical Analysis* (J. Cross, ed.), Marcel Dekker, New York, 1977, pp. 141–192.
3. E. Breitmaier and W. Voelter, *Carbon-13 NMR Spectroscopy: High Resolution Methods and Applications in Organic Chemistry and Biochemistry*, VCH, Germany, 1989.

4.   D. M. Doddrell, D. T. Pegg, and M. R. Bendall, J. Magn. Reson. *48*:323 (1982).
5.   D. G. Gorenstein, in *Phosphorus-31 NMR: Principles and Applications* (D. G. Gorenstein, ed.), Academic Press, Orlando, 1984, pp. 7–33.
6.   J. Jeener, Ampere International Summer School, Basko Polje, 1971.
7.   F. J. M. Van De Ven, *Multidimensional NMR in Liquids: Basic Principles and Experimental Methods*, VCH, New York, 1995.
8.   W. R. Croasmun, *Two-Dimensional NMR Spectroscopy, Applications for Chemists and Biochemists*, VCH, New York, 1994.
9.   G. M. Clore and A. M. Gronenborn, Prog. NMR Spectrosc. *23*:43 (1991).
10. J. Prestegard and J. Tolman, 38th Experimental Nuclear Magnetic Resonance Spectroscopy Conference (ENC), Orlando, FL, 1997.
11. S. Braun, H.-O. Kalinowski, and S. Berger, *100 and More Basic-NMR Experiments: A Practical Course*, VCH, New York, 1995.
12. P. B. Barker and R. J. Freeman, J. Magn. Reson. *64*:334 (1985).
13. G. Zhu and A. Bax, J. Magn. Reson. *98*:192 (1992).
14. W. P. Aue, E. Bartholdi, and R. R. Ernst, J. Chem. Phys. *64*:2229 (1975).
15. A. Bax and D. G. Davis, J. Magn. Reson. *65*:355 (1985).
16. A. Bax, R. H. Griffin, and B. L. Hawkins, J. Magn. Reson. *55*:301 (1983).
17. L. Lerner and A. Bax, J. Magn. Reson. *69*:375 (1986).
18. A. Bax and M. F. Summers, J. Am. Chem. Soc. *108*:2093 (1986).
19. K. F. Morris and C. S. Johnson Jr., J. Am. Chem. Soc. *114*:3139 (1992).
20. K. F. Morris and C. S. Johnson Jr., J. Am. Chem. Soc *115*:4291 (1993).
21. L. C. M. Van Gorkom, S. Alkan, A. Jensen, M. Jilani, and T. Hancewicz, 38th Experimental Nuclear Magnetic Resonance Spectroscopy Conference (ENC), Orlando, FL, 1997.

6

Mass Spectrometry of Anionic Surfactants

HENRY T. KALINOSKI   Unichema North America, Chicago, Illinois

I. INTRODUCTION

A number of significant advances have been developed since the analytical chemistry of anionic surfactants was last described in this series [1], particularly in mass spectrometry. These advances have enabled mass spectrometry to become a primary, routine tool in the identification and characterization of anionic surfactants and mixtures containing the materials. This chapter will describe these developments, giving some details of the principles of operation and characteristics of the techniques. There will also be specific descriptions of the use of the techniques in the characterization of anionic surfactants. These applications will be illustrated with figures and mass spectra to highlight the appearance and characteristics of the data. These illustrations will indicate the strengths and weaknesses of each approach and allow comparison among the various approaches. In many instances, more than one approach to mass spectrometry will yield the information required to address a particular analytical question. Therefore, there will be little in the way of recommendations regarding the "best" approach to any particular problem.

Many of the developments that have allowed greater application of mass spectrometry to the characterization of anionic surfactants have also been applied in the analysis of cationic surfactants. These developments in mass spectrometry were the focus of a chapter in a previous volume in this series [2] on the analytical chemistry of cationic surfactants. As this previous chapter described in great detail the basics of mass spectrometry and the fundamentals of the newer techniques, these topics will only be described in enough detail to familiarize the reader with the basic concepts. Much more effort is devoted to the specifics of applications to anionic surfactants. An effort has also been made to incorporate sufficient references to allow practitioners and researchers access to appropriate documentation permitting deeper exploration and understanding of any particular aspect of the field. Some reviews and reference works [3–5] are included in the bibliography, which include a greater scope and depth regarding the powerful technique of mass spectrometry.

The intent of this chapter is to introduce recent equipment, technique, and application developments which have broadened the use of mass spectrometry in the study of anionic surfactants. It is intended to give an overview of a dynamic and growing area of application. The level of detail will only be sufficient to allow those currently familiar with anionic surfactants to understand the expanding role modern mass spectrometry can play in addressing identification and characterization problems. It will also allow those familiar with mass spectrometry to gain an understanding of the nature, complexity, and challenges posed by the study of anionic surfactants in a range of environments and circumstances. This chapter contains only enough specific experimental detail to illustrate the aspects of instrumental operations and sample preparations that would be required to give appropriate context to the use of mass spectrometry as a problem-solving tool. As with any analytical technique, mass spectrometry is only one approach to yield information on a specific problem. The information now available through mass spectrometry must be combined and compared with the information available through the other techniques described in this volume to address any specific problem thoroughly.

Information for this chapter was acquired through a combination of literature research and the personal experiences of the author. For the most part, the literature references will be relatively recent, with many significant changes and advances in mass spectrometry occurring in the last decade. Where appropriate, references to earlier fundamental work which made later advances possible are also included. One primary source of information was the proceedings from the annual conferences of the American Society of Mass Spectrometry (ASMS). Most significant advances and developments in this field are described and advanced through this meeting. This chapter and its associated bibliography is by no means an exhaustive or complete list of all references and applications of mass spectrometry of anionic surfactants. It should provide a good introduction to the topic and the groundwork for a more detailed examination of any aspect of the topic. The applications described, and the associated figures, were produced either directly by the author or by his colleagues and coworkers. The illustrations and figures will include information on conditions used to produce the particular spectrum. Other conditions or approaches may yield similar or comparable results.

The chapter is structured around particular techniques in mass spectrometry, not around the various types of anionic surfactants. The emphasis is on developments and approaches that facilitate or extend the capabilities of mass spectrometry. Although mass spectrometry is now more than ever used as a detector for chromatography, illustrations and examples of such applications are probably better described and included in the chromatography chapter of this volume. Each of the approaches described have been or can be used on any of the major anionic surfactant types. In some cases, particular examples were chosen to illustrate the value of the mass spectrometry technique or some characteristic aspect of the data for that type of surfactant. The areas of application include, either directly or indirectly, references to applications to all of the major anionic surfactant classes. Amphoteric surfactants, bearing both anionic and cationic characteristics, will not be covered. One class that will not be addressed in any great detail are the salts of long-chain carboxylic acids (soaps). While these materials are anionic surfactants and have been addressed using techniques of mass spectrometry, the primary approach for analysis by mass spectrometry—ester formation followed by gas chromatography-mass spectrometry (GC/MS) [6]—is of much more limited utility for the other classes of anionic surfactants. The nature and chemistry of the derivatives formed strongly or completely dominate the mass spectra of these compounds and are better described and compared in a dedicated volume on the topic. Many of the newer techniques and approaches for mass spectrometry were developed to address the limitations and complications of the derivatization-GC/MS methods. This is not to say that GC/MS itself will not be described to some extent. But again this description will be directed toward the advantages and characteristics of the approach that aid and enhance the mass spectrometry aspects of the analysis.

There will be some thoughts and information regarding recent trends, the newest developments, and the likely future directions of mass spectrometry as applied to anionic surfactants. As with any review containing indicators of future developments, it will be worthwhile to reexamine this topic and chapter a few years hence to determine the validity and value of these thoughts.

II. BASICS OF MASS SPECTROMETRY

A. General

In its simplest definition, mass spectrometry (MS) is the measure of the intensity of an ion creating a signal versus its mass-to-charge ratio (m/z). The visual display of this information (on paper, film, a glass plate, or a computer screen) is known as a mass spectrum. From this measure, information on the molecular mass of a compound forming an ion and some understanding of the composition and structure of the ion can be gained. To obtain this information, all that is required is formation of a material into an ion, followed by separation of the ions from one another based on their molecular masses and production of a recordable signal from each ion. The instrumentation required to accomplish such a process can be quite complex and expensive, based on how many means can be used to form ions, what approaches are used to introduce the ions into the mass separator, the accuracy and precision of the mass separation device, and the efficiency and sensitivity of the entire system. A great variety of sample introduction systems, ion sources, mass analyzers, detectors, and recording devices have been developed, used, and commercialized for this purpose. Most of the pieces in a great number of combinations and configurations have been used to produce and analyze ions from anionic surfactants and their mixtures. At one time, the size, cost, complexity, and characteristics of the technique and its applications were so great that most locations or facilities were fortunate to have a single instrument requiring one or more highly specialized and dedicated individuals to operate it and only in useful operation for small amount of time. The advances described in the remainder of this chapter and a number of other advances have now made it possible for mass spectrometry to be quite inexpensive, compact, computer-controlled and regulated, capable of long periods of unattended operation, and able to be understood and used by individuals with a limited amount of specific training and education. Many current and recent users of mass spectrometry received little formal training in the technique and would not consider themselves mass spectrometrists. This is not to say that large, expensive, and complex systems no longer exist, but rather the technique is now also available to a great variety of researchers, scientists, and practitioners. It is from this wide availability that increases in understanding and application are currently being developed.

The one recent advance in technology that has had the greatest impact on the development, utility, and proliferation of mass spectrometry will not be discussed here at all. The advances in computer technology and capability have influenced all aspects of the science and practice of mass spectrometry. Computers and their associated peripheral devices are inextricably linked with modern mass spectrometers, and most current systems could not have been designed, built, or operated without the computers. It must not be forgotten, however, that it is the formation of the ions and their detection that are the fundamental components of a mass spectrometry experiment. The computers can be exceptionally useful in preparing the instrument and experiment, controlling the stability for long periods of time, and making outputs highly readible and useful. The real product of a mass spectrometry experiment, however, is data on m/z versus intensity.

For the later stages of ion analysis (separation and detection) to be conducted, the material being analyzed must be in the gas phase. To yield the greatest sensitivity with the minimum amount of side reactions, all other "uninteresting" species must be eliminated from the analysis system. For this reason, analyzers and detectors is mass spectrometers are operated in a vacuum, usually from 10^{-3} to 10^{-9} torr. At these pressures, little is present in the mass spectrometer other than the species of analytical interest. When other materials are present, such as in chemical ionization [7] or collision-induced dissociation [3], they are specifically introduced to affect some change. The need for high vacuum is one factor increasing the price and complexity of mass spectrometer systems. A number of different analyzers, using electrical, magnetic, radiofrequency (RF), or gravitational fields (either alone or in combination), have been developed for mass spectrometry. There have been great advances in the styles, designs, complexity, combinations, and performance of mass spectrometer analyzers. Once an ion is formed and introduced into the analyzer, however, its behavior is similar to other ions of similar mass and charge. As there is little unique or characteristic regarding analyzers with respect to ions from anionic surfactants, developments in this area will not be described in great detail. One area regarding analyzers that will get some attention is the technique of tandem mass spectrometry [3]. In this technique, ions are formed and separated based on mass. The mass spectrometer instrument itself is then used as a reaction vessel with a mass-selected species further reacted. The products of this reaction are then analyzed on a second mass analyzer within the same instrument. A range of information on the structure of an ion or the nature of components in a mixture can be gained from this technique. More details on this technique are given in later application sections.

The initial steps in the mass spectrometry experiment are introduction of the sample into the system and initial ion formation. It is in these areas that the most significant advances, at least regarding anionic surfactants, have occurred. As such, most of the information in this chapter will be on these advances and their applications to anionic surfactants. The specific techniques to be covered are fast atom bombardment [8], thermospray [9], and electrospray [10]. One could argue that each of these approaches is an ionization method, but that limits some of the real advantages these techniques bring to mass spectrometry. In each case, the technique also acts as a sample introduction system, and each has allowed analysis of materials considered unsuitable for mass spectrometric analysis prior to these developments. It must be acknowledged that none of these approaches was developed for the purpose or with the intention of analyzing anionic surfactants. As such, credit should be given to those who applied and implemented these techniques, in this case to anionic surfactants, as well as to those who first described and developed the procedures. A range of other introduction and ionization techniques have been applied to the mass spectrometry of anionic surfactants. These include "classical" electron and chemical ionization, field ionization/field desorption [11], laser ionization [12], and the particle-beam interface for mass spectrometry [13]. For a variety of reasons, some described later, each of these is less frequently practiced than the other techniques mentioned. Significant information and important directions for future work, however, were gained through the use of these techniques, so they are included in this work.

B. Electron and Chemical Ionization

The techniques of electron and chemical ionization are referred to above as classical. These are the basic processes that have been used over most of the history of mass spectrometry to form gas-phase ions. In both cases, an analyte molecule is introduced in the gas phase into the ion source of the mass spectrometer. The gas-phase molecule interacts with either an electron or a plasma of charged reagent ions. In either process, a fraction of the analyte molecules is charged and the ions drawn from the source for analysis. As the initial ionization process has a fairly well-defined and narrow energy spread, the energy distribution of the resulting product ions is also fairly narrow and well-defined. The separation in the mass spectrometer is based, in part, on the energy of the ions moving through a field. This narrow energy distribution allows for high resolution and discrimination in the analyzer field, permitting highly accurate assignment of mass-to-charge ratio. The electron ionization (EI) process forms predominantly positive radical ions from organic molecules by stripping a single electron from the molecule. In chemical ionization (CI), the reaction with the plasma transfers a charged species (proton, ammonium ion, chloride ion) to an organic analyte molecule. This adduct ion is then analyzed in the mass spectrometer. In both EI and CI, the charge (z) on the analyte ions is almost always one, so the m/z value is a measure of the mass of the ion. The EI process forms almost exclusively positive ions, while CI can be used to produce either positive or negative ions. Both processes can be highly energetic, far more than is usually required to form ions from most organic molecules. This excess energy is transferred to the analyte molecule and is dissipated primarily through bond cleavages and rearrangements of the analyte ions, particularly in EI. These processes lead to fragmentation of the analyte ions and can, in many instances, produce only fragment ions and no species representative of the intact analyte molecule. The CI process usually transfers less energy to an analyte molecule and a greater percentage of ions characteristic of the intact species is generally found in CI mass spectra. Fragments in CI are primarily formed through loss of small neutral molecules (H_2O, NH_3, HCl, CH_3OH, etc.) so some information on functional groups can be obtained. Both EI and CI are routinely used in combination with gas chromatography (GC/MS). More details on GC/MS, particularly in relation to the analysis of anionic surfactants, are given later. An understanding of the possible fragmentation routes and gas-phase chemistry of ions formed in mass spectrometry can yield information on the structure of an analyte. These processes and an understanding of the underlying chemistry form the basis for a large percentage of the applications of mass spectrometry.

The physical and chemical properties of anionic surfactant materials have, for the most part, precluded the use of EI and CI for their analysis. Both approaches are limited to materials that can be transferred to the gas phase of a mass spectrometer without significant degradation prior to ionization. This is accomplished either directly in the high vacuum of the mass spectrometer or with heating of the material in the high vacuum. Most anionic surfactant materials exhibit no vapor pressure, even at the high vacuum of a mass spectrometer. Heating of anionic surfactant materials usually results in thermal degradation well prior to any significant transfer of ions to the gas phase. The EI mass spectra of alkylbenzenesulfonates have been produced [14] through direct introduction of the materials into an EI ion source. This can allow for characterization of isolated materials but is a less useful tool for many of the circumstances found in surfactant analysis. Most often when EI or CI mass spectra are required for anionic surfactants, the materials were either modified to increase their volatility for analysis or were introduced using an apparatus specifically designed for the purpose.

One such apparatus is the particle-beam interface for liquid chromatography/mass spectrometry [13,15]. As with the thermospray interface to be described in greater detail later, the particle-beam interface was designed specifically to allow direct introduction of the effluent of a flowing liquid system into a spectroscopic system. Unlike other inlet techniques, it was designed to use standard EI and CI sources. It cannot, however, operate at the 1 mL/min rate used for standard bore (4.6 mm i.d.) LC columns. A liquid stream, at a flow rate of 0.1–0.5 mL/min, is nebulized in a slightly heated chamber to affect desolvation. This chamber is differentially pumped to remove evaporated solvent and the nebulizing gas (usually helium). This serves to produce a "beam" of solid particles or low-volatility liquid drops for injection into the EI or CI source of the mass spectrometer. The use of EI and CI ionization permits greater structural information to be obtained directly on the injected sample when compared with other inlet and ionization techniques. There have been some descriptions of the use of the particle-beam technique in the analysis of anionic surfactants [16,17]. The results indicated that intact ionic species were introduced to the mass spectrometer ion source. The ion source temperature had significant impact on the ability to produce ions related to the intact surfactant species. Higher source temperatures, resulted in lower abundance of intact anions. Alkyl sulfates, alkyl ethoxysulfates, and alkylbenzenesulfonates were all found to yield useful EI and CI mass spectra, in both positive and negative ion detection modes. The sulfate surfactants showed weak signals for the intact molecules while the sulfonate surfactants gave stronger ions for the intact parent species.

One major drawback of the particle-beam inlet system is its lack of sensitivity, often requiring a large amount of sample to produce a useful spectrum. In addition, not all materials are amenable to the approach, as indicated by the variable responses of the anionic surfactants. These characteristics, combined with the somewhat complicated operation of the technique and the development of other, more powerful techniques for coupling liquid systems to mass spectrometry, have limited the development and application of the particle-beam interface.

III. FAST ATOM BOMBARDMENT

A. General

The limitations of the EI and CI methods provided the continuing drive to develop alternative ionization methods for mass spectrometry. Techniques were developed that extended the utility of mass spectrometry to less volatile and more labile organic molecules. These included electrohydrodynamic (EHD) ionization [18], surface ionization [19], and the previously mentioned field desorption [11]. These were found to yield limited information, to be not universally applicable, and to be rather difficult to perform regularly. One of the techniques in surface analysis is known as secondary ion mass spectrometry (SIMS) [20]. SIMS involves bombardment of a solid surface, which may be covered with an organic material, by a beam of energetic ions. These ions, the primary beam, may be derived from an inert gas (argon or xenon) or from a metal (gold, bismuth, gallium). The interaction of the charged primary ions with the surface caused ions to be ejected from the surface. These secondary ions from the surface, and any material present on the surface, are sampled for mass spectrometry. The SIMS process is highly energetic, and secondary ion yields, particularly from organic materials, are relatively low and short-lived.

An alternative to SIMS was described by Barber et al. [8] in 1981. The approach, now known as fast atom bombardment (FAB), uses energetic inert gas atoms (argon or xenon) to bombard a surface. The novel aspect of Barber's approach, however, was the use of a low volatility, viscous liquid matrix to introduce the sample to the ion source of the mass spectrometer. This liquid matrix, usually glycerol, permits the surface exposed to the energetic atom beam to be constantly replenished with sample with a signal lifetime ranging from minutes to over an hour. This increased signal lifetime enhances the ability to perform more advanced mass spectrometric techniques, enables acquisition of accurate mass data, and facilitates a range of tandem mass spectrometric experiments. Sample preparation for FAB is primarily dissolving the analyte in the liquid matrix. As the technique relies on mass transfer in the liquid matrix and surface activity of analyte materials, a number of studies [21–25] describe the direct relationship between analyte surface activity and ease of ionization in FAB. As the mechanism of technique and its applications have been extensively described [2,20,26–28], only basic principles of operation and applications to anionic surfactants are given here.

B. Mechanism

The drop of viscous liquid matrix is introduced to the ion source vacuum chamber of a mass spectrometer on a direct insertion probe. The probe tip is made of conducting material allowing the probe to be grounded, avoiding charge buildup on the probe. The tip should also be relatively inert, to avoid reaction between the probe surface and the analyte species. The energetic atoms (usually between 6 and 10 kV for gases) bombard the sample target, interacting with the sample-matrix mixture resulting in a sputtering of ionic and neutral materials from the sample target. Energetic ions (usually cesium, at energies of up to 35 kV) can also be used as the primary particles. The resulting mixture of ions and neutrals is sampled and the ions focused by the optics of the mass spectrometer for subsequent mass analysis. Both positive and negative ions can be formed and sampled in the FAB process, depending on the nature of the sample and the tuning of the instrument. The charge (z) on the ions produced is almost exclusively one, with m/z then being a direct measure of ionic mass. The primary beam has a broad energy spread, imparting a broad energy distribution in the analyte ions. The great abundance (or high intensity) of secondary ions compensates for the broad distribution, allowing significant discrimination in the mass spectrometer ion source. This source discrimination permits high-resolution mass analysis and accurate mass assignment.

The FAB process was originally classified as a "soft" ionization technique, similar to CI, due to the tendency to form proton attachment $(MH)^+$ or proton abstraction $(M-H)^-$ ions. The ease with which the FAB process produces gas-phase ions from materials (such as anionic surfactants) present as ions in the matrix solution suggested that the technique only serves to desorb ions from the matrix [29,30]. Further studies indicated energies of the order of 30 eV could be transferred to the analyte ions [31] allowing for some bond cleavage and fragmentation. Fragmentation in FAB is similar to that found in CI, with the loss of small neutral molecules from polar functional groups. Other cleavages and rearrangements, including charge-remote fragmentations, have also been observed [32–34].

The matrix in FAB plays a significant role in the ionization of analyte species. The analyte must show some solubility in the matrix but should not react with it. The matrix in FAB often contributes ions to the mass spectral background, which may interfere with spectral interpretation. These background ions and species arising from the bombardment of the matrix may change with time [35], further complicating interpretation. Matrix ions may form adducts with the analyte, and the presence of other materials in the matrix, particularly alkali metal ions, can lead to the formation of adduct ions with analytes. In the analysis of anionic surfactants, glycerol, *m*-nitrobenzyl alcohol, and triethanolamine have been used effectively in the author's laboratory. A more complete range of matrix materials and their physical properties has been described in references [36,37].

One limitation to the original FAB technique was that it is a "batch" process. A single sample in a discrete drop of matrix is analyzed at one time. The high matrix concentration limits sensitivity and contributes to high background. The approach is not ideal for mixtures, as the presence of one component or the matrix can impede, inhibit, or complicate the analysis of a second component. Further, the batch approach does not permit on-line coupling to separation techniques such as liquid chromatography. To address these issues, the "static" direct probe method was modified to allow a dynamic, flowing system [38–40].

In the "flow-FAB" technique, the direct insertion probe contains a fused silica capillary tube, much like a GC/MS or direct-liquid-introduction LC/MS interface. The capillary (75–100 µm ID) allows introduction of a solvent, generally 5% glycerol in water, directly onto the FAB target in the mass spectrometer ion source. Bombardment of the target and flowing, evaporating liquid is conducted in the same manner used for static FAB. Many probes are constructed with some absorbent material near the tip to aid removal and evaporation of solvent and matrix, providing stable flow rates, ion signals, and vacuum system operation. The system can be used in a flow-injection mode [41], with samples injected into a flowing carrier solvent, or in conjunction with a liquid separation technique, such as HPLC [38,40] or capillary electrophoresis [42–44]. To allow sufficient evaporation of flow-FAB solvent without overwhelming the mass spectrometer vacuum source, flow-FAB operates at microliter per minute flow rates. Separation techniques directly coupled using flow-FAB must also operate at these rates or the column effluent must be split prior to introduction to the flow-FAB probe. These flow rates are compatible with capillary or packed capillary HPLC or capillary electrophoresis.

Fast atom bombardment provides a relatively simple and straightforward means for producing ions for mass spectrometry from highly intractable materials. The characteristics that make anionic surfactants otherwise difficult to address by mass spectrometry render them ideal candidates for analysis using FAB. The technique is probably the benchmark in mass spectrometry of anionic surfactants against which other approaches should be measured.

C. Application to Anionic Surfactants

Owing to the strong compatibility of anionic surfactants with FAB ionization, the technique has found great utility in the analysis of systems containing anionic surfactants [36,45–56]. All of the major types of anionic surfactants have been addressed using FAB ionization with the typical applications being analysis of surfactants directly, characterization of commercial mixtures and finished products, and analysis of environmental samples containing anionic surfactants and their degradation products. Both positive and negative ion detection have been used with FAB ionization, with characteristic ions produced in both modes.

In positive ion FAB/MS, mass spectra usually contain an abundant ion corresponding to the anion neutralized with one cation and ionized by a second cation $(A + C_2)^+$. In these cases, the cations are typically sodium, the predominant cation found in commercial surfactants. Neutralization and ionization by protons, ammonium or potassium ions are also found, as are ions containing a mixture of cations. There can also be cluster ions where the initial species is accompanied by neutral surfactant molecules giving rise to a series of ions of the form $(A_n + C_{n+1})^+$. These clusters can often extend out to the limit of the mass range of the mass spectrometer.

For negative ion FAB/MS, the mass spectra are characterized by intense signals corresponding to the intact anions (A^-). As with the positive ions, cluster ions of the anion and neutral surfactant $(A_n + C_{n+1})^-$ are also formed. Clusters between analyte ions and matrix components, often common in FAB mass spectra, are seldom seen in either positive or negative ion mass spectra of anionic surfactants.

The typical background in a FAB mass spectrum, positive or negative ion, is characterized by regular signals (a peak at every mass) and ions related to the matrix, including matrix clusters. As anionic surfactants ionize particularly well in FAB, these background characteristics are suppressed or nearly eliminated.

As commercial anionic surfactants are almost always mixtures of various chain lengths and/or degrees of ethoxylation, the FAB mass spectra always contain a great mixture of ions corresponding with these individual components. As all the related anionic surfactant components in a single class have similar surface activities, ionization potentials and gas-phase ion affinities, the ions present in the FAB mass spectrum of a mixture are roughly indicative of the relative concentrations of the components in the mixture. This can be confirmed by comparing the relative abundances of related ions in the positive and negative ion mass spectra of the same sample. The relationship between relative ion abundance and relative mixture composition does not necessarily hold true for mixtures of surfactant classes, where ionization of one type or species may inhibit or supress ionization of another class.

Fragment ions are not often abundant in FAB mass spectra but are almost always present. Fragments are found in both positive and negative ion detection but are generally more abundant in positive ion mode. The type and intensity of fragment ions produced depends on a number of factors including analyte type and structure, matrix, bombarding species and energy, distance between atom source and sample, type of analyzer, exposure time to bombarding beam, and other factors. Typically the fragmentation is characteristic of the loss of small neutral molecules (H_2O, NH_3, HCl, etc.). As many common anionic surfactants do not have small, polar functional groups, this type of fragmentation is somewhat limited.

As structural information on anionic surfactants is quite limited in a single-stage FAB/MS analysis, tandem mass spectrometry is often used to increase the amount of structural information. The value of this approach was recognized very early on in the analysis of anionic surfactants using FAB [45,46]. This approach was greatly facilitated by the generally long sample lifetime and abundant ion current produced in FAB. These early efforts coupling FAB and tandem mass spectrometry addressed many of the common commercial surfactant types (ABS, LAS, alkyl sulfates, alkylethoxy sulfates, alpha-olefin sulfonates, sulfosuccinates, soaps) including amphoteric surfactants like sarcosinates and hydrotropes such as xylenesulfonate [46]. A similar approach was used for fluorine-containing anionic surfactants [47]. Much of the understanding of the structure and decomposition of anionic surfactant ions in mass spectrometry, in both positive and negative ion mode, was gained in this early work using FAB ionization.

Based on the ability to produce strong signals from small amounts of samples for anionic surfactants, FAB found use early on in the evaluation and analysis of environmental samples [48,49]. In these applications, materials isolated from environmental samples were shown to be readily addressed using FAB/MS with relatively little sample clean-up. Information allowing identification of surfactants present in environmental samples could be made directly from FAB mass spectra without sample derivatization or chromatography. Often little interference from other materials present in environmental samples was found. Better sensitivity or specificity is reported for mass spectrometry–based approaches compared with other characterization methods.

The characteristics of FAB mass spectra are seen in the negative ion FAB mass spectrum of a commercial alkylbenzenesulfonate surfactant in Fig. 1. The series of ions at m/z 297, 311, 325, and 339 corresponds to the intact anions of the C_{10} through C_{13} ABS homologs. The general peak at every mass background is present, as are low abundance fragment ions. The characteristic fragments in this spectrum are due to loss of molecular hydrogen (H_2) from the anions, producing ions at m/z 295, 309, 323, and 337 and the alkyl fragments

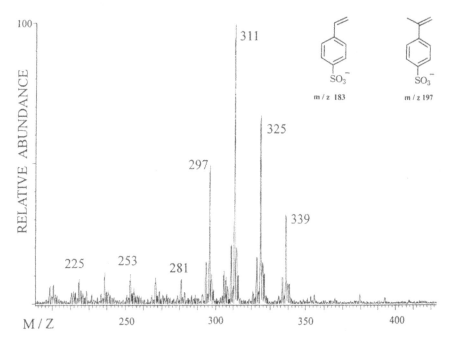

FIG. 1 Negative ion fast atom bombardment mass spectrum of a commercial alkylbenzenesulfonate (ABS) surfactant mixture. The mass spectrum was produced on a VG Trio-3 triple-quadrupole instrument from a glycerol matrix using a VG cesium-ion gun operated at about 30 kV. The inset shows the structures of the m/z 183 fragment ion produced from linear alkylbenzenesulfonate (LAS) surfactants and the m/z 197 ion found as a fragment of branched chained ABS materials. (The mass spectrum was produced by P. J. W. Schuyl of the Unilever Research Laboratory at Vlaardingen, The Netherlands.)

at m/z 211, 225, 239, 253, 267, and 281. For all fragments the charge is supplied by the sulfonate group, so that fragmentation is remote from the charge site. As the scan range for this spectrum began at m/z 200, it cannot be seen whether the spectrum contains an ion at either m/z 183 or 197. These ions were found to be indicative of the alkyl-chain structure in the alkylbenzenesulfonate surfactants [41,46]. The branched-chain (ABS) surfactants show an ion at m/z 197, corresponding to a propenylbenzenesulfonate anion (Fig. 1, inset) arising from beta-cleavage of the alkyl chain. The linear alkylbenzenesulfonates (LAS) display an ion at m/z 183 of the structure shown in the inset in Fig. 1. The presence of these ions in the mass spectra or tandem mass spectra of alkylbenzenesulfonate surfactants can be used to determine the type of surfactant present in a sample [41].

An example of a more complete FAB mass spectrum of an anionic surfactant is shown in Fig. 2. The negative ion FAB mass spectrum of an alkylglyceryl ether sulfonate (AGES) surfactant is characterized by the predominant monosulfonate anion at m/z 323 for the C_{12} alkyl chain and little in the way of fragment ions. The characteristic low level of FAB background ions is also present. Along with the m/z 323 ion are analogous anions for the C_{14} and C_{16} chains at

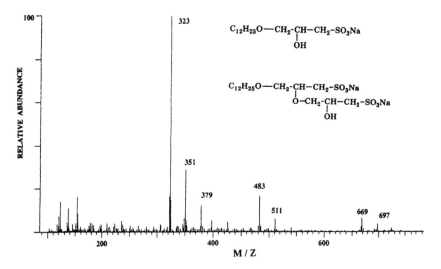

FIG. 2 Negative ion fast atom bombardment mass spectrum of an alkylglyceryl ether sulfonate (AGES) surfactant. The mass spectrum was produced on a Finnigan-MAT TSQ70 triple-quadrupole instrument from a glycerol matrix using a Finnigan fast atom source with xenon as the bombarding gas at about 8 kV. The inset shows the structures of the AGES monosulfonate and the disulfonate AGES "dimer." (FAB mass spectra of AGES were also provided by A. Pereira of Unilever Research US, Edgewater, NJ.)

m/z 351 and 379. Other, higher mass ions are of significant interest in this spectrum. The AGES surfactant was originally produced for use as a synthetic detergent that could be formulated in a toilet bar that would be mild to the skin [57,58]. The material also imparted other desirable characteristics to the formulation. The synthetic route to this surfactant produces not only the monosulfonate (shown in Fig. 2) but also a disulfonate "dimer," also shown in Fig. 2. The presence and relative amount of this dimer would influence the performance and formulation characteristics of the surfactant. FAB mass spectrometry provides an ideal means to identify and characterize this surfactant and formulations containing the material. The ion series at m/z 483, 511, and 539 corresponds to the singly charged anions of the disulfonate, again C_{12}, C_{14}, and C_{16} chain lengths. The second sulfonate group is neutralized by sodium. The relative abundances of these ions roughly correlates to the amount of each species present in the surfactant.

Other significant ions in the spectrum are clusters of the primary anions. The series beginning at m/z 669 corresponds to two monosulfonate anions and a sodium cation $(A_2 + C)^-$. There is also a small signal at m/z 425, corresponding to the m/z 323 anion and a glycerol molecule. Structurally significant fragments are limited, with the m/z 153 ion corresponding to loss of the alkyl group as an alkane. This species could arise from all of the primary monosulfonate anions.

FAB mass spectrometry can also be used to analyze formulations containing anionic surfactants directly. Figures 3 and 4 show the negative and positive ion FAB mass spectra of a personal wash bar formulation containing synthetic anionic surfactants. The anions of AGES monomer at m/z 323, 351, 379, and 407 are found along with the dimer ions at m/z 483, 511, and 539 in the negative ion mass spectrum (Fig. 3). Corresponding sodium cationized species for the monomer are found in the positive ion FAB spectrum (Fig. 4) at m/z 369, 397, and 425. A series of ions corresponding to the dimer is found starting at m/z 529. Evidence of other formulation components are also seen in these spectra. In Fig. 3, there is an ion at m/z 270, and the positive ion spectrum in Fig. 4 shows ions at m/z 294 and 316. These three ions all correspond to lauryl sarcosinate (Fig. 4, inset). With an anion molecular weight of 270 Daltons, the positive ion at m/z 294 corresponds to the anion with a proton and a sodium and the m/z 316 ion with the anion and two sodium ions. These ions were previously seen in the FAB mass spectrometry of the neat sarcosinate surfactant [46]. One other component present in this formulation displays the ions at m/z 199 (negative, 3) and m/z 245 (positive, 4). These correspond to C_{12} fatty acid or soap. This may be present as a separate component or as a minor component in the sarcosinate surfactant. Cluster ions between all components in the mixture are found at higher masses.

It should be noted that the relative abundances of the various components

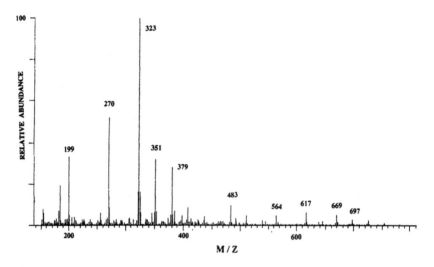

FIG. 3 Negative ion fast atom bombardment mass spectrum of a sample of a personal wash bar found to contain the AGES surfactant along with other surfactant materials. The mass spectrum was produced on a Finnigan-MAT TSQ70 triple-quadrupole instrument from a glycerol matrix using a Finnigan fast atom source with xenon as the bombarding gas at about 8 kV.

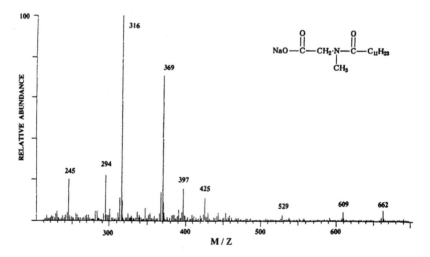

FIG. 4 Positive ion fast atom bombardment mass spectrum of a sample of a personal wash bar found to contain the AGES surfactant along with other surfactant materials. The mass spectrum was produced on a Finnigan-MAT TSQ70 triple-quadrupole instrument from a glycerol matrix using a Finnigan fast atom source with xenon as the bombarding gas at about 8 kV. The inset shows the structure of sodium lauryl sarcosinate also found as a constituent of this formulation.

do not directly correspond between the positive and negative ion spectra. At least one reason for this is that the sarcosinate is an amphoteric surfactant, which may allow it to ionize more effectively than the AGES in the positive ion mode. Further, the synergistic, suppression, or inhibition effects of the mixture of surfactants may behave differently in the positive and negative ion modes. Overall, this direct mixture approach must be viewed as a qualitative rather than quantitative tool. Coupling of FAB mass spectrometry with chromatography or combination of the information gained from FAB with other data on the system may permit a more quantitative evaluation of a full formulation.

IV. THERMOSPRAY

A. General

There has been a long quest to develop an interface between liquid chromatography (LC) and mass spectrometry analogous to the interface between GC and MS [4]. Among the reasons for this is that many more organic materials are amenable to separation and analysis using LC than can be addressed using GC. Further, compounds analyzed using LC are often able to be addressed without derivatization or sample clean-up prior to chromatography. Efforts in LC are limited by the lack of a "universal" detector able to respond predictibly to a wide range of materials with approximately equal efficiency. The characteristics required of this "ideal detector" were described [4,59] to include high sensitivity, linear, universal, predictable response, reliability and convenience, unaffected by mobile phase and environmental conditions, and providing qualitative information on the detected component. In many respects, the mass spectrometer meets these qualifications. LC also meets the requirements of an "ideal inlet to mass spectrometry" [4], minimum sample preparation, separation, and sequential delivery of samples. It is also amenable to unattended, computer-controlled operation. Unfortunately the operational conditions of LC, particularly the high flow rate of relatively low volatility solvent (primarily water), exceed the capability of even the most sophisticated MS vacuum systems. Any coupling between LC, particularly the dominant, packed column form, and MS must address this great difference.

While FAB ionization permitted mass spectrometry to be applied to many of the types and classes of materials amenable to LC separations (low volatility, polar, ionic, thermally labile), there were still limitations to its use. Among these were the static, batch nature of the analysis and the requirement for the viscous matrix. Both of these circumstances limit the applicability of the technique for use in a dynamic, flowing system.

One technique was developed specifically to couple standard bore (4.6 mm i.d., 1 mL/min flow) LC with high-vacuum mass spectrometry—the thermospray interface [8,60]. A number of studies [4] have probed the operational mechanisms, optimized the performance, and detailed the applications of the technique. These studies demonstrated the broad applicability of the technique as well as its severe limitations. Even with these limitations, as defined by the requirements for both LC and MS described above, thermospray is the most successful interface between these two techniques ever developed.

B. Mechanism

In addition to being developed primarily as an LC/MS interface, thermospray is unique among the approaches described in this chapter in being both an inlet technique and an ionization mechanism. These two processes can be used independently from one another, offering greater flexibility than some alternatives. Thermospray was developed to address reverse-phase LC separations using aqueous-based mobile phases and ionic buffers. A wide range of LC conditions, including gradient elution, can be accommodated. The technique requires specific hardware, including additional pumping capacity above that available on most mass spectrometers. It can be coupled directly with other LC detectors, such as UV absorption, as long as the mass spectrometer is the final detector.

The initial thermospray process is the same whether it is being used as an inlet or an ionization technique. The flowing mobile phase, containing the dissolved analyte and buffers, is heated in a vacuum region to vaporize and nebulize the mixture for spraying into the ionization region of the mass spectrometer. When used as an inlet, this vapor cloud of analyte, solvent, and buffer is exposed to a high-voltage discharge or a glowing filament. These processes initiate gas-phase ion-molecule chemistry among the species present in the vapor, producing ions for analysis. The overall mechanism is somewhat analogous to chemical ionization with the mobile-phase components being used as the reagents. The developers of the technique found that ions could be produced even in the absence of the discharge or filament [8,60], leading to the exploration and development of thermospray ionization. It appears that the combination of nebulization with heat, the high mobile-phase velocity and the ionic components present in the mobile-phase leads to the production of gas-phase ions for mass analysis.

Both the inlet and ionization modes of thermospray produce sample-ion adduct species, with the structure and composition dependent on the materials present during the thermospray process. Both positive and negative ions are formed, with almost equal facility. By and large, the charge (z) on the ions is one. Overall, this is a form of "soft" ionization with little excess energy being imparted to analyte ions. The amount of energy transferred during the ionization process can be regulated through control of voltages on ion source optics. This can be optimized to yield primarily intact molecular adducts, containing information on analyte molecular weight, or a high fraction of fragment ions, containing information on analyte structure. As with CI, the loss of small neutral molecules from the polar or ionic analytes is the primary fragmentation route.

The procedure does have its drawbacks. The process and resulting mass spectra are complicated as thermal degradation does occur, apparently both before and after ion formation. Further complicating the mass spectra are ions corresponding to buffers, clusters of mobile phase, and analytes complexed with mobile phase and buffer ions. Not all analyte species yield useful, interpretable, or intense mass spectra. It is not straightforward to predict a priori which analytes will be amenable to successful thermospray analysis. Although the process can accommodate, in fact can rely on, ionic buffers, these buffers must be volatile at the thermospray operating conditions. This may require modification of the LC separation conditions, not always possible or acceptable to the chromatographer or researcher.

These are oversimplifications of the thermospray process, and more details are available in other references [4,8,60]. Many of the limitations of the approach have been overcome by other techniques described in this chapter. Overall, the approach does offer significant advantages to analysts and researchers and should be considered for the analysis of anionic surfactants.

C. Application to Anionic Surfactants

There has been a number of applications of thermospray mass spectrometry to anionic surfactant analysis reported [61–66]. Typical surfactant types (ABS, AES, secondary alkane sulfonates, and alkylphenol ethoxysulfates) have been addressed in these analyses. In all cases, the benefits of mass spectrometry (specificity, sensitivity, structural information) in the characterization of typical complex surfactant mixtures are primary driving forces in the development of the application. The use of tandem mass spectrometry was described in some of the applications, including the analysis of fluorine-containing surfactants [63]. Tandem mass spectrometry often improved the analyses, as it allowed direct analysis of mixtures without the need for chromatographic separations [61,63,65].

Overall, the use of discharge ionization with thermospray appears to yield better results in the analysis of anionic surfactants. A high discharge voltage, however, can lead to excessive background signal, loss of signal-to-noise, and a decrease in sensitivity. When directly compared [61], the use of FAB ionization appears preferred over thermospray for analysis of anionic surfactants.

Figures 5 and 6 show representative positive and negative ion thermospray mass spectra, again for the anionic AGES surfactant. The positive ion mass spectrum (Fig. 5) reveals some of the characteristics and difficulties of the technique,

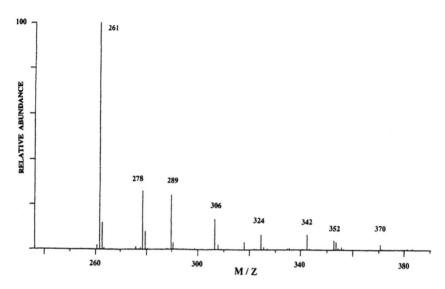

FIG. 5 Positive ion thermospray mass spectrum of an alkylglyceryl ether sulfonate (AGES) surfactant. The mass spectrum was produced on a Finnigan-MAT TSQ70 triple-quadrupole instrument using a Finnigan thermospray source. Sample was dissolved in a 60:40 methanol:0.1 N ammonium acetate mobile phase. The thermospray vaporizer was operated at 45°C and the inlet jet at about 380°C.

as the signals from the intact molecular species (*m/z* 342 and 370 for C_{12} and C_{14} chain lengths) are of relatively low abundance. The spectrum is dominated by ions related to protonated (*m/z* 261 and 289) and ammonium-adduct (*m/z* 278 and 306) ions of alkylglycerol. This material may be present in the spectrum as a thermal decomposition product of the surfactant or may be present in the surfactant as a synthetic by-product. Use of chromatography prior to mass spectrometry would confirm the origin of the free alkylglycerol component. If present as a synthetic by-product, this spectrum indicates that the alkylglycerol ionizes much more readily than the AGES surfactant. In this respect, positive ion thermospray is not an effective approach for the characterization of such surfactants. Other ions present (*m/z* 324 and 352) correspond to water loss from the ammonium adduct ions of the hydroxyl-substituted surfactant.

The mass spectrum of AGES surfactant in Fig. 6 is an example of the more appropriate use of negative ion detection for thermospray mass spectrometry of this type of material. The base peak (*m/z* 323) corresponds to the intact anion of the C_{12} chain-length surfactant. No evidence of an alkylglycerol component is found. It may not be present in the sample or may not yield useful ions in the negative ion mode. Again, chromatography may be of some help. This

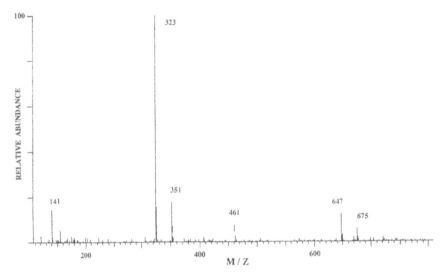

FIG. 6 Negative ion thermospray mass spectrum of the same alkylglyceryl ether sulfonate (AGES) surfactant as shown in Fig. 5. The mass spectrum was produced on a Finnigan-MAT TSQ70 triple-quadrupole instrument using a Finnigan thermospray source, under the conditions used for Fig. 5.

ambiguity and apparently selective ionization is a drawback of the thermospray technique.

The mass spectrum also contains a signal corresponding to the disulfonate AGES dimer at m/z 461, with a proton present on one sulfonate group. The relative abundance of the AGES dimer signal does not appear to be representative of the actual amount of AGES dimer in the sample, as determined by other techniques (including FAB mass spectrometry). In one more instance, chromatography may offer additional information. The relative ionization efficiency of the monosulfonate over the disulfonate limits the ability to quantitate each species directly from the thermospray spectrum.

There is one other characteristic feature of thermospray spectra, cluster ions, present in Figure 6. The ions at m/z 647 and 675 correspond to clusters containing two surfactant anions and a proton to neutralize one anion (chain lengths C_{12}–C_{12} and C_{12}–C_{14}, respectively; a cluster of C_{14}–C_{14} is also present at m/z 703). Smaller ions corresponding to sodium-bound clusters are present at m/z 669 and 697. Not present in this spectrum are ions corresponding to analytes clustered with solvent. This cluster formation serves to further complicate the interpretation of thermospray mass spectra, particularly when no chromatographic separation is conducted prior to mass spectrometry.

There is often little useful structural information on anionic surfactants present in thermospray mass spectra, particularly negative ion spectra. Tandem mass spectrometry can be used to increase the amount of structural information available. Figure 7 shows the collision-induced dissociation (CID) mass spectrum of the m/z 323 ion from AGES (from Fig. 6). Only two fragments are found, and no residual parent ion signal remains in this low-energy tandem quadrupole (triple-quad) mass spectrum. The m/z 81 ion corresponds to the HSO_3^- anion and m/z 137 corresponds to loss of the C_{12} alkyl chain as the alcohol. A likely, stable structure of the m/z 137 ion is given in the figure. From these data, the identity of the surfactant as a sulfonate can be determined and the presence of a C_{12} chain, bound as an ether, can be inferred. Other forms of analysis (chromatography, IR, NMR) should be used in combination with thermospray MS/MS to determine the identity of an unknown surfactant fully.

Significantly absent in the MS/MS spectrum of m/z 323 (Fig. 7) are ions related to the intact alkyl chain or structurally related to C_{12} alkylglycerol. This suggests that fragmentation to yield structurally significant ions from the alkyl chain is not favored in negative ion MS/MS. Further, routes that would lead to ions found in the positive ion thermospray mass spectrum of AGES (Fig. 5) are

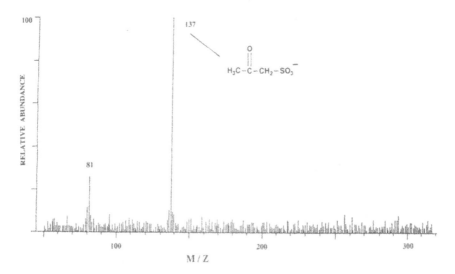

FIG. 7 Negative ion collision-induced dissociation mass spectrum (CID-MS/MS) of the m/z 323 ion from the thermospray ionization of the AGES surfactant shown in Fig. 6. The MS/MS spectrum was produced on a Finnigan-MAT TSQ70 triple-quadrupole instrument using argon as the collision gas and a collision energy of 19 eV. The inset shows a likely structure of the m/z 137 ion formed by loss of the C_{12} alkyl chain and rearrangement of the resulting product ion.

unavailable to the anion. To understand more fully the structures and fragmentation behavior of thermospray-produced surfactant ions, both positive and negative tandem mass spectrometry should be conducted.

Thermospray mass spectrometry offers a very powerful approach to the rapid
characterization of the products that contain anionic surfactants. Figure 8 shows
one negative ion thermospray mass spectrum from an LC/MS analysis of a commercial hard-surface cleaner. Liquid chromatography on a polymethacrylate gel
column was used to separate compound classes and thermospray mass spectrometry used to identify the various constituents. The mass spectrum in Fig. 8 indicates that this fraction contains C_{10} through C_{14} alkylbenzenesulfonate. The
presence of the *m/z* 197 ion suggests that the surfactant is the branched ABS material, with *m/z* 197 having the propenylbenzene sulfonate structure previously
described. There are a number of low-intensity cluster ions also present in the
spectrum, but they offer little useful information to aid characterization of this

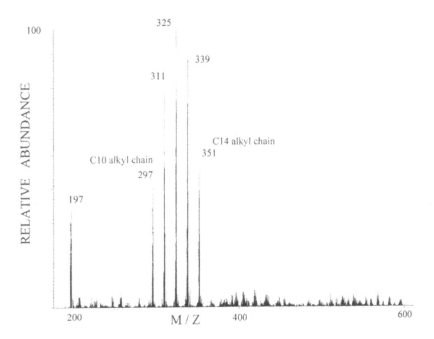

FIG. 8   Negative ion thermospray mass spectrum of a fraction of a commercial hard-surface cleaner separated using LC. The mass spectrum was produced on a Hewlett-Packard
MS Engine using a Hewlett-Packard thermospray source in the filament on mode. The
sample was separated in a 50:50 acetonitrile:0.1 M ammonium acetate mobile phase. The
thermospray probe was operated at 96°C and the ion source at 250°C. (The mass spectrum
was produced by J. Finn of Lever-Ponds Canada, Toronto, Ontario.)

surfactant. Other components of this hard-surface cleaner identified using the LC-thermospray approach include alkyl sulfates, soap, and nonionic alcohol ethoxylates. The thermospray spectrum in Fig. 8 can be compared with the FAB spectrum of a commercial ABS surfactant shown in Fig. 1. The thermospray spectrum does not show the high background signal seen in the FAB spectrum, nor does it show the hydrogen-loss ions or alkyl fragments that characterize the FAB spectrum. The higher mass cluster ions found in the thermospray spectrum are not seen in the FAB spectrum.

Although other approaches may offer advantages, the use of thermospray mass spectrometry is still a powerful and useful tool in the characterization of samples containing surfactant mixtures. The ability to couple directly with standard-bore LC columns and to use LC conditions very familiar to chromatographers offers great advantage for routine screening of complex surfactant mixtures. The combination with tandem mass spectrometry further extends the range, flexibility, and amount of information available through the thermospray technique.

V. ELECTROSPRAY

A. General

Even with the proliferation of new and refined approaches for producing gas-phase species for spectroscopic analysis, all were found to be limited in one way or another. A primary limitation of the various approaches was the need to accommodate the high vacuum of the mass spectrometer ion source region. Building on the work of Dole and coworkers [67–71] and later Iribarne and Thomson [72,73] in the production of charged droplets from liquid solutions, John Fenn and coworkers developed an ionization mechanism for mass spectrometry that functions at atmospheric pressure [5,10]. These developments allowed the sample inlet and ionization to be decoupled from the high vacuum of the mass spectrometer. This technique, called electrospray by Fenn [10], has since been more thoroughly developed and applied more broadly as atmospheric pressure ionization [74,75] to a wide range of interesting and useful applications. Of greatest significance to the analysis of anionic surfactants is that the approach was seen early on as an ideal method for the analysis of "large organic molecules that are too complex, too fragile, or too nonvolatile for ionization by more conventional methods" [10].

B. Mechanism

Very briefly, electrospray is affected by exposing a flowing liquid sample to a high electric field at the end of a capillary or needle. The end of the needle is specifically prepared to support stable operation and is maintained at elevated voltage relative to ground. The resulting field gradient between the needle tip and a lower voltage nozzle induces a spray of ionized droplets, which can be sampled by the inlet of a mass spectrometer. Analyte ions corresponding to proton attachment or cation or anion adducts are generally formed. Polar mobile phase components, ions dissolved in the mobile phase, or other analyte molecules are generally responsible for the charged species. Materials present as preformed ions in solution, such as anionic surfactants, ionize particularly well in electrospray. A number of systems have been developed to optimize and control the ionization process [5,10,74,75], some using gas flows to aid nebulization or droplet evaporation. These systems have also sought to maximize ion-sampling efficiency, perform gas-phase chemistry, address some particular application, or generally improve system performance. Primarily for ease of initial development, early efforts at electrospray were conducted on quadrupole analyzers. This also allowed accommodation of the generally low translational energies of electrospray-produced ions. Interest in the technique and its wide applicability has seen adaptations to a range of analyzers, including ion traps and ion cyclotron resonance instruments [75].

As with flow-FAB, small-bore capillary tubing and low (μL/min) flow rates are used in electrospray. Similar to thermospray, auxiliary pumping is required to affect the transition from atmospheric pressure to the vacuum of the analyzer. Modifications, such as pneumatic assistance, have been made to accommodate somewhat higher flow rates (to 50 μL/min) [76]. Overall these parameters are consistent with microscale LC, capillary electrophoresis, and other low-flow liquid separation techniques [74–76]. Higher flow rate systems have been addressed with atmospheric pressure ionization (API) by splitting the effluent of a separation column prior to electrospray [77].

Primarily aqueous-based solvents have been used for electrospray, although Dole's original work used a benzene:acetone solution [67–71]. Addition of alcohols or other solvents lowers the surface tension of the aqueous systems, resulting in a more uniform production of small droplets. The specific composition used is often determined by the analysis or separation being addressed by electrospray, but pure water is seldom used. Auxiliary nebulization or the use of a "sheath" flow to add a modifying solvent [78] has expanded the range of liquids amenable to electrospray.

Ionization in electrospray occurs from a surface in a region of high electric field. In this manner, the technique is comparable to field desorption [11]. The analyte is present in a condensed liquid phase, so electrospray is somewhat analogous to electrohydrodynamic ionization (EHD) [14]. Overall, the mechanism is not fully understood and is still subject to some debate [10,75]. Whatever the exact process, the technique is very mild, producing ions of intact molecular species with little or no fragmentation. Both positive and negative ions can be formed, although operating conditions in positive ion detection mode are somewhat more stable. Species ionized in electrospray can carry a large number of charges, with $z = 50$ or more [79,80]. Combined with the lack of fragmentation, this characteristic of electrospray (multiple ionization of intact species) has permitted mass spectra of large biomolecules (molecular weights > 50,000) to be routinely obtained on conventional analyzers with m/z ranges of 2000 and less [79,81,82]. The technique can be extremely sensitive for compounds that ionize particularly well, with attomole (10^{-18} M) detection limits found for some materials. Coupled with its relative ease of use, these characteristics have greatly expanded the utility of mass spectrometry into fields, particularly biochemistry, where the technique had no use only a few years ago.

There is an atmospheric pressure ionization approach which uses the electrospray process more as a sample inlet than as an ionization technique. In atmospheric pressure chemical ionization (APCI), the nebulized droplets from the electrospray needle are exposed to a discharge or glowing filament. In this respect it is somewhat similar to thermospray and CI. This approach allows higher flow rates to be accommodated, and the analyte does not necessarily need to be ionized in solution or by other ionic materials also present in the mobile phase.

With all the benefits and positive aspects of the technique, it might be difficult to describe drawbacks of electrospray. It is a very low flow rate mechanism, so that ready coupling to standard-bore LC is difficult. Changes to conventional and well-established separations may need to be made to accommodate electrospray mass spectrometry. The technique produces almost exclusively intact molecular species, with little structural information on analytes available. Changes can be made to mass spectrometer ion optics so that electrospray-formed ions can be made to undergo ion-molecule chemistry, increasing the amount of structural information available. More frequently, additional stages of mass analysis, in a tandem quadrupole, tandem sector, ion trap, or ion cyclotron resonance analyzer, are used to obtain structural information on analytes. The technique does operate at high voltage (3–7 kV), so precautions are generally built into systems to avoid exposure to dangerous conditions.

It must be noted that some types of surfactants strongly interfere with electrospray ionization [83] and can completely quench the signals from some analytes. This is more of an issue when the surfactant is not the analyte of interest but is present as part of a separation technique (micellar chromatography) or from analyte preparation/separation (biological, environmental, forensic samples). This signal suppression appears to be directly related to charge and surface activity of the surfactant.

Overall, electrospray, or more generally atmospheric pressure ionization, is a sensitive, effective, rugged, and reliable means to produce ions for mass spectrometry from an exceptionally wide range of compounds. Many materials that are found difficult or intractable, even with FAB or thermospray, can be readily addressed using API. It is the flexibility and sensitivity that renders the technique ideal in the mass spectrometric study of anionic surfactants.

C. Application to Anionic Surfactants

For many of the same reasons and in many of the same areas as the other forms of mass spectrometry, applications to anionic surfactant systems are now being made using electrospray ionization [84–97]. The systems being described are primarily surfactants present in fully formulated consumer products or in complex environmental samples. When present in environmental samples, the good sensitivity of electrospray enables low levels of detection and permits analyses to be performed with few preparation or concentration steps being performed. Electrospray spectra of anionic surfactants are most often produced from aqueous-based (usually water-alcohol) systems, but a nonaqueous system using methanol with benzene or carbon tetrachloride and negative ion detection has been reported [92]. Electrospray mass spectrometry is also coupled with tandem mass spectrometry in order to increase the amount of structural information available on the ions produced. This includes coupling with an ion-trap analyzer to increase the sensitivity of the systems over tandem quadrupole instruments [85] and to perform multiple-stage tandem mass spectrometry for structure elucidation [86]. Sample solutions were diluted by a factor of 100 over that used for the triple quadrupole system to permit analysis on the ion trap. Such a system may be particularly effective when working with environmental systems, where sample amount may be limited and the effort may be focused on the identification of unknown degradation products.

Electrospray mass spectra of anionic surfactants are generally comprised only of ions corresponding to the intact molecules, either as anions (in negative ion mode) or as the anion clustered with two cations (in positive ion mode) to neutralize and ionize the material. No fragment ions are found in either positive or negative ion electrospray mass spectra. Although spectra can be produced in either polarity, positive ion electrospray systems are reportedly more stable, more robust, and more sensitive. Detection limits, with positive ion detection, for anionic surfactants in the femtogram (10^{-15} g) range are indicated.

The negative ion electrospray mass spectrum of a commercial alkylbenzenesulfonate surfactant in Fig. 9 is characteristic of what is produced using this approach. The ions present in the spectrum (m/z 297 through 339) correspond to the intact anions of the C_{10}–C_{13} alkylbenzenesulfonates. The small ions present at m/z 255 and m/z 367 correspond to the C_7 and C_{15} alkyl chain lengths, respectively, and there is also a small signal at m/z 283 for the C_9 component. No other components, clusters, or fragments are found. This spectrum can be compared to mass spectra of similar ABS mixtures obtained using FAB (Fig. 1) and thermospray (Fig. 8). The ability to produce a mass spectrum free from complicating

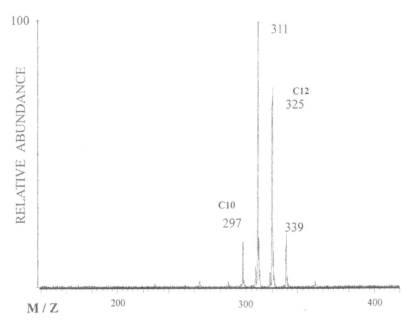

FIG. 9 Negative ion electrospray mass spectrum of a commercial alkylbenzenesulfonate surfactant. This mass spectrum was produced on the same VG Trio-3 instrument used to produce the FAB mass spectrum shown in Fig. 1 using a VG Plasma-spray source. (The mass spectrum was produced by P. J. W. Schuyl of the Unilever Research Laboratory at Vlaardingen, The Netherlands.)

clusters, adducts, and fragments can enhance the ability to interpret the spectrum, to establish relationships between sample components, and to perform other mass spectrometric experiments.

The electrospray mass spectrum of a more complex anionic surfactant, an alkyl ethoxysulfate, is shown in Fig. 10. In this case, the surfactant is comprised of the C_{12} and C_{14} alkyl chain lengths, and ions containing up to 12 moles of ethylene oxide (44 Daltons) can be detected. Also present in this spectrum are ions corresponding to C_8, C_{10}, and C_{16} chain lengths (m/z 209, 237, and 281), not detected when other forms of ionization are used on such materials.

The real power of the technique is exhibited in the spectra shown in Figs. 11 and 12. Each of these are negative ion electrospray mass spectra of fully formulated detergent products. The samples were diluted in mobile phase and injected into the electrospray source without further clean-up or separation. The surfactant systems represented are a simple alkylethoxy sulfate (with C_{12} and C_{13} alkyl chain lengths) plus xylenesulfonate (m/z 185) in an "ultra" concentrated liquid hand-dishwashing detergent (Fig. 11, top), a more complex alkylethoxy sulfate

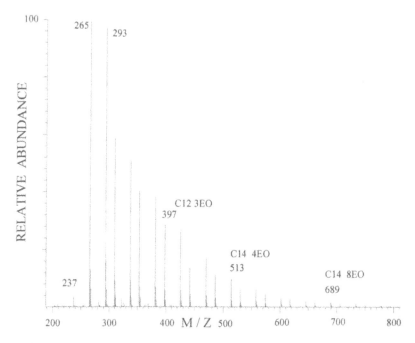

FIG. 10 Negative ion electrospray mass spectrum of a commercial ammonium alkyl-EO sulfate surfactant mixture. The mass spectrum was produced on a Finnigan TSQ700 triple quadrupole instrument using a Finnigan electrospray source. The sample was introduced using a 67:33 methanol:0.01% ammonium acetate in water mobile phase at 5 μL/min in flow-injection mode. The source was operated at 3.2 kV with the capillary at 150°C and a 20 μL/min air sheath. (The mass spectrum was produced by B. J. Shay of Unilever Research US, Edgewater, NJ.)

(with C_{12}–C_{15} chain lengths), fatty acids (m/z 171, 199, 227, 255) and xylenesulfonate in a standard liquid laundry detergent (Fig. 11, bottom), and a mixture of alkylethoxy sulfate (C_{12} and C_{14} chain lengths) and alkylbenzenesulfonate (C_{10} through C_{13} chain lengths) with xylenesulfonate in a cold-water wash (Fig. 12). In all cases, the "clean" spectra produced by electrospray mass spectrometry allow ready interpretation of these complex mixtures. It would need to be determined whether the relative abundances of the ions, especially for the structurally distinct materials, are directly indicative of the relative concentrations of components in the formulations.

The strength of the electrospray method is the relative ease with which spectra can be obtained, even on complex mixtures, once the system has been set up and optimized. These complex commercial surfactant mixtures (in diluted form) can even be used as tuning solutions for the electrospray mass spectrometer sys-

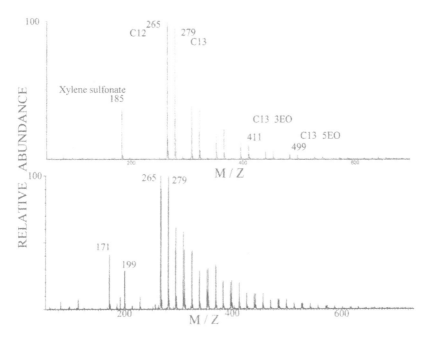

FIG. 11 Negative ion electrospray mass spectra of fully formulated commercial detergent products, an "ultra" concentrated hand-dishwashing liquid detergent (top) and a standard concentration liquid laundry detergent (bottom). Mass spectra were produced on a VG-Micromass Platform II instrument using a VG Plasma-spray ion source operated at 60°C. Samples were dissolved in a 50:50 acetonitrile:0.2% ammonium hydroxide mobile phase. (The mass spectra were produced by L. O. Hargiss of Novartis Pharmaceutical, Summit, NJ.)

tems [98]. This apparent ease of operation can lead practitioners, skilled and novice alike, to become complacent with the mechanisms operating within the instruments. There can still be significant amounts of chemistry occurring in the mobile phase and at the electrospray tip (it can act as an electrochemical cell), which can modify or perturb the resulting mass spectra produced. Not all materials readily produce electrospray spectra, which can result in mixture components being overlooked and ignored. The situation also exists in which some components, particularly anionic surfactants, can inhibit or suppress the ionization of other components [83] in a mixture. The system, even with assisted nebulization or other APCI approaches, is still a very low flow rate method, and coupling to separation techniques can be somewhat complicated.

Overall, it appears that the current trend is toward the use of electrospray (or more generally APCI) for nearly all applications involving polar and low-volatil-

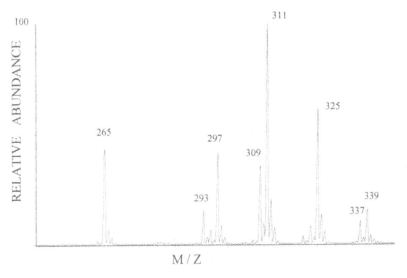

FIG. 12 A portion of the negative ion electrospray mass spectrum of a fully formulated cold-water hand-wash laundry product. The mass spectrum was obtained under the conditions described for Fig. 11. (The mass spectrum was produced by L. O. Hargiss of Novartis Pharmaceutical, Summit, NJ.)

ity anlaytes. This is now being coupled with powerful, compact, benchtop analyzers, such as the ion traps, to produce a versatile, flexible, and information-rich system available to a wide range of scientists and researchers, including surfactant scientists.

VI. OTHER APPROACHES

A. Field Desorption

One of the earlier approaches to the analysis of nonvolatile and polar materials by mass spectrometry involved placing a solution containing the analyte on a specially prepared wire emitter and evaporating the solvent. The emitter, containing about a microgram of sample, was inserted into the high vacuum ion source. The emitter was exposed to a high-voltage electric field, creating a very high field gradient between the emitter and the mass spectrometer ion optics. This field gradient caused materials on the emitter to be desorbed and ionized. The ions formed are sampled by the mass spectrometer. In cases where the field alone does not induce sample desorption, a current can be passed through the wire, heating the sample. This field desorption (FD) technique [11] was found to be quite mild, producing primarily intact molecular species with little fragmentation. The technique was also quite sensitive, with as little as 10^{-13} g of material required to produce useful spectra. It was found that different classes of compounds, in this case surfactants, would desorb when different heater currents were applied to the emitter [99]. This permitted highly selective analysis of complex surfactant mixtures.

The FD technique has been found to have serious drawbacks, even more apparent when compared with other recent developments. It is far more difficult to optimize and use FD than the other current methods. Like electrospray, a high-voltage source and field is used in FD, but FD is much more amenable to use on magnetic sector, as opposed to quadrupole, instruments. The primary difficulty, however, is that FD produces only short-lived, highly variable currents of analyte ions. These analyte ion currents are also very weak, and specially designed analyzers need to be used to gain useful information [99].

Field desorption mass spectra of anionic surfactants have been produced [100,101], with the positive ion spectra consisting primarily of the surfactant ionized by attachment of a sodium cation $(M + Na)^+$. There were also higher molecular weight clusters of surfactants and sodium adducts. A more extensive study [99] used FD in combination with tandem mass spectrometry to analyze a mixture of various surfactants. In this study, anionic surfactants desorbed at the highest emitter currents (45–60 mA). The use of collisionally activated dissociation (CAD) tandem mass spectrometry [102,103] allowed structural characterization of anionic surfactants to be made. The primary fragment from the positive sodium ion adduct of a sodium alkylsulfonate was at m/z 126 $(Na_2SO_3)^+$, corresponding to loss of the alkyl substituent and allowing ready identification of alkyl chain length. There is also a series of fragments, also containing the sodium sulfonate function, 14 Daltons apart, corresponding to fragmentation of the alkyl chain. The work also explored the FD/CAD spectra of alkylbenzenesulfonates, producing similar ions yielding information on surfactant structure.

The FD/CAD approach was used in an environmental study of surfactant mixtures in surface waters [104]. Again, the use of various emitter currents allowed differentiation of the various surfactant classes and CAD allowed identification of specific surfactants. Alkylbenzenesulfonates were identified in the environmental samples, with microgram quantities of material required to produce spectra. The difficulty in producing spectra from the anionic surfactants was likely due to salts from the sample preparation procedure. Although not conducted in the referenced work, it was indicated that the ability to obtain negative ion spectra would likely have aided work on the anionic surfactants. An earlier effort [105] indicated that negative ion spectra of alkylbenzenesulfonate surfactants from detergents could be selectively produced down to a concentration of 10^{-5} M. The presence of cationic surfactants would affect the intensity and desorption behavior of alkylbenzenesulfonate anions. The mass spectra of anionic surfactants produced by FAB and FD were compared in a number of studies [106–108] with FD producing abundant intact molecular species and some structure-specific fragment ions found in the FAB spectra.

Overall, the FD approach is likely too difficult and fraught with inherent complications to be of routine use in the characterization of anionic surfactants. An understanding of the underlying principles of FD, however, will permit better understanding of the mechanisms of other ionization techniques.

B. Laser Ionization

A laser can also be used as the energy source for ionization in mass spectrometry (photoionization), and this approach has been used in the analysis of surfactants [109–113]. Some advantages of laser ionization include high sensitivity, minimum sample preparation, small sample requirements, rapid analysis, and elimination of the need for a liquid matrix. This lack of the liquid matrix allows laser ionization to be conducted on mass analyzers requiring very high vacuum, such as Fourier-transform ion cyclotron resonance (FT-ICR) instruments [112,113]. Use of the FT-ICR system allows for very high-resolution, accurate mass measurement of ions. The lack of a matrix requirement can also permit desorption/ionization directly from a sample of interest, such as a cloth [114]. This avoids the need to remove or extract the materials of interest from a test environment and overcomes secondary sample preparation effects such as selective precipitation of analytes onto a FAB or laser ionization target.

The mass spectra resulting from laser ionization can contain almost exclusively intact molecular species, or the abundance of fragment ions can be increased through use of higher power on the ionizing laser. Both molecular weight and structural information can be produced using this technique. The fragments found are similar to those seen with other "soft" ionization techniques.

The FT-ICR systems using laser ionization are among the most expensive instruments available for mass spectrometry, limiting applications on materials such as anionic surfactants. More recently, laser ionization is being coupled with one of the simpler and much less expensive mass analyzers, the time-of-flight (TOF) system [115,116]. This may lead to a greater use of the technique for characterization of anionic surfactants. In such a system, the analytical sample is mixed with a solid matrix material to aid in absorption of light energy and in desorption of analyte ions. This matrix-assisted laser desorption ionization (MALDI) overcomes some of the difficulties of standard laser desorption and leads to reproducible ionization and desorption for mass spectrometry. A rapid analysis, with limited sample preparation, can be achieved if an appropriate matrix is chosen for the analyte of interest, including anionic surfactants [117].

VII. GAS CHROMATOGRAPHY/MASS SPECTROMETRY

Mass spectrometry is most effective when a single pure component is present in the instrument for ionization and analysis. Most problems utilizing mass spectrometry, however, involve mixtures of materials. Some form of separation is required to gain the greatest amount of information from the mass spectrometer. When materials are thermally stable and volatile, or can readily be made volatile, gas chromatography/mass spectrometry (GC/MS) provides an ideal combination to obtain separation, identification, and perhaps quantitation on components in a mixture. High-resolution, capillary gas chromatography is also about the most efficient separation technique for amenable compounds, with the ability to separate structural isomers of organic compounds from one another completely. It can also be a very sensitive technique, with nanograms of material able to yield a signal for quantitation well above the level of background noise. The two techniques are highly complementary in nature, and a variety of interfaces coupling the techniques have been developed [118–121]. The most common interface is the direct inlet for capillary open-tubular GC. In this approach, the fused-silica separation column is terminated in close proximity to the mass spectrometer ion source. Much of the column effluent is swept into the ion source for analysis, and the helium GC carrier gas is readily pumped away by the instrument's vacuum system. Electron and chemical ionization are most commonly used in GC/MS with large libraries of EI mass spectra of volatile compounds available from commercial sources. The basics of GC/MS and its application are given in Refs. 118–121.

Problems involving anionic surfactants could well utilize the advantages provided by GC/MS. The efficient, high-resolution, selective separations would be quite useful in mixture characterization. The sensitivity and selectivity of detection would be of great value in the common environmental problems that usually include anionic surfactants [122–125]. Unfortunately, the use of GC/MS for the analysis of anionic surfactant mixtures is not straightforward. Anionic surfactants are formally charged and of very low volatility, the antithesis of an ideal GC candidate. Further, procedures to render many materials, such as those containing organic acids or polar hydroxyl groups, amenable to GC analysis are not readily applicable to anionic surfactants. The surfactants are often present as mixtures and are not readily isolated from materials that interfere with derivatization reactions. Further, the functional groups formed in derivatization of anionic surfactants are generally not very stable, and the derivatives decompose or react prior to the completion of an analysis.

GC/MS analysis has, however, been very effectively used in the analysis of anionic surfactants, primarily for environmental problems. A complete description of the activities, methods, approaches, and outcomes is probably better given in a work on the environmental analysis for anionic surfactants. A number of recent references [126–139] are included for those seeking more information on the use of mass spectrometry in environmental activities.

The most commonly encountered anionic surfactants in an environmental sample are the alkylbenzenesulfonates, both linear (LAS) and branched (ABS). Along with LAS, the related, alicyclic dialkyltetralinsulfonates (DATS) are also found. Many of the references given describe approaches using mass spectrometry in the identification, characterization, and quantitation of these materials and related biodegradation products. A variety of derivatization schemes are given and the derivatization-GC/MS approaches used are found to be sufficiently sensitive and selective to permit detection at environmental levels (µg/L, ppb) [126]. Mass spectrometry and tandem mass spectrometry often provide required specificity for determination and quantitation of materials that are unable to be separated, even by highly efficient GC [127].

One interesting approach to both the isolation of anionic surfactants from environmental samples and their subsequent analysis by GC/MS involved the use of tetrabutylammonium cation to form ion pairs with the anionic surfactants [128]. Additionally interesting was that the ion pairs were quantitatively extracted from sewage sludge using supercritical carbon dioxide. Both secondary alkanesulfonates (SAS) and LAS surfactants were extracted at greater than 90% recovery. The ion pairs formed butyl esters of the sulfonates in the hot injection port of the GC/MS instrument, allowing GC/MS determination without class fractionation. The approach was later applied as a single-step procedure using a solid-phase extraction (SPE) step to isolate the surfactants [129].

The physical and chemical properties that limit the use of GC/MS in the analysis of anionic surfactants also minimize the utility of supercritical fluid chromatography (SFC) and its coupling with mass spectrometry (SFC/MS) [140]. In SFC, high-pressure carbon dioxide, above its critical temperature of about 31°C, is used as the mobile phase for chromatography. Fused-silica open-tubular columns similar to GC columns are used for high-resolution separations. The interface with mass spectrometry can be accomplished using a number of methods, including a direct capillary inlet [140]. As an unheated, LC-type injector is used for SFC, the injection-port derivatization method previously described could not be used for SFC. Other derivatization methods for anionic surfactants could be used for sample preparation for SFC and SFC/MS. This approach was used for SFC/MS of a derivatized surfactant on an ion-trap mass spectrometer [141]. Sufficient data was obtained to elucidate the structure of the unknown surfactant.

Until other techniques can couple high-resolution separations with sensitivity and specificity in detection to compare with GC/MS, the additional effort to render materials such as anionic surfactants amenable to GC/MS will continue to be made. This seems to be particularly true in the environmental field, where GC/MS is well established and has a long and distinguished track record.

VIII. CONCLUDING REMARKS

Mass spectrometry has made great contributions to the study and understanding of the chemistry of anionic surfactants and systems containing them. The technique has earned its place in the surfactant scientist's problem-solving toolbox. The type and amount of information available through the various forms of mass spectrometry are available through no other analytical approach. The recent advances in inlet and ionization capabilities place the technique well within the reach of most involved in the study of surfactant systems. These advances allow the surfactant scientist to concentrate on the system and the problem and not so much on the technique and its practice.

There is not yet one ideal approach for the analysis of anionic surfactants by mass spectrometry. Each method has its own strengths and drawbacks, hopefully described and illustrated here. A rule-of-thumb would be to use the technique that is currently and most readily available. Although it may not be ideal, some information can be gained. That information can be coupled with information gained from other techniques or from later mass spectrometric analyses using more powerful or more specific techniques.

The understanding of each technique should lead to an improved understanding of other techniques in mass spectrometry. The knowledge gained in the development and application of field desorption and thermospray, for example, led to a better understanding of the principles and operations of electrospray and APCI. Knowledge of the fundamentals and shortcomings of electrospray and FAB can lead to a new generation of techniques, approaches, and applications.

Although FAB had been the benchmark for mass spectrometry ionization for low-volatility materials and thermospray was the first very widely used LC/MS interface and ionization technique, both of these techniques now appear to be losing importance to electrospray and MALDI. The information gained from those older techniques should not be forgotten in pursuit of the newest, latest, and greatest approaches. There is still much to be gained, particularly in applications involving anionic surfactants, in the understanding and use of all these techniques of mass spectrometry.

REFERENCES

1. J. Cross, ed., *Anionic Surfactants: Chemical Analysis*, Marcel Dekker, New York, 1977.
2. H. T. Kalinoski, in *Cationic Surfactants: Analytical and Biological Evaluation* (J. Cross and E. J. Singer, eds.), Marcel Dekker, New York, 1994.
3. K. L. Busch, G. L. Glish, and S. A. McLuckey, *Mass Spectrometry/Mass Spectrometry*, VCH Publishers, New York, 1988.
4. A. L. Yergey, C. G. Edmonds, I. A. S. Lewis, and M. L. Vestal, *Liquid Chromatography/Mass Spectrometry: Techniques and Applications*, Plenum Press, New York, 1990.

5. J. B. Fenn, M. Mann, C. K. Meng, S. F. Wong, and C. M. Whitehouse, Mass Spectrom. Rev. *9*:37 (1990).
6. European Oleochemicals and Allied Products Group, Preparation of Methyl Esters of Fatty Acids, ISO 5509-1978, Avenue E. Van Nieuwnhuyse 4, Brussels.
7. A. G. Harrison, *Chemical Ionization Mass Spectrometry*, CRC Press, Boca Raton, FL, 1983.
8. M. Barber, R. S. Bordoli, D. Sedgwick, and A. N. Tyler, J. Chem. Soc., Chem. Commun.: 325 (1981).
9. C. R. Blakley, J. C. Carmody, and M. L. Vestal, J. Am. Chem. Soc. *102*:5931 (1980).
10. M. Yamashita and J. B. Fenn, J. Phys. Chem. *88*:4451 (1984).
11. H. D. Beckey, *Principles of Field Ionization and Field Desorption*, Pergamon, Oxford, 1977.
12. M. Karas and F. Hillenkamp, Anal. Chem. *60*:2299 (1988).
13. R. C. Willoughby and R. F. Browner, Anal. Chem. *56*:2625 (1984).
14. P. Agozzino, L. Ceraulo, M. Ferrugia, E. Caponetti, F. Intravaia, and R. Triolo, J. Colloid Interface Sci. *114*:26–31 (1986).
15. R. C. Willoughby and F. Poeppel, Proceedings of the 35th Annual Conference on Mass Spectrometry and Allied Topics, American Society for Mass Spectrometry, Denver, 1987, p. 289.
16. J. N. Alexander and C. J. Quinn, Proceedings of the 39th ASMS Conference on Mass Spectrometry and Allied Topics, American Society for Mass Spectrometry, Nashville, 1991.
17. B. H. Solka, Proceedings of the 40th ASMS Conference on Mass Spectrometry and Allied Topics, American Society for Mass Spectrometry, Washington, DC, 1992, p. 1464.
18. J. H. Callahan, K. Hool, J. D. Reynolds, and K. D. Cook, Int. J. Mass Spectrom. Ion Proc. *75*:291 (1987).
19. U. K. Rasulev, E. G. Nazarov, K. S. Tursunov, and T. V. Arguneeva, Zh. Anal. Khim. *46*:1802 (1991).
20. P. A. Lyon, ed., *Desorption Mass Spectrometry: Are SIMS and FAB the Same?*, ACS Symposium Series 291, American Chemical Society, Washington, DC, 1985.
21. W. V. Ligon and S. B. Dorn, Int. J. Mass Spectrom. Ion Physics *61*:113 (1984).
22. W. V. Ligon and S. B. Dorn, Int. J. Mass Spectrom. Ion Physics *57*:75 (1984).
23. M. Barber, R. S. Bordoli, G. Elliott, D. Sedgwick, and A. N. Tyler, J. Chem. Soc., Faraday Trans. 1 *79*:249 (1983).
24. A. J. DeStefano and T. Keough, Proceedings of the 31st Annual Conference on Mass Spectrometry and Allied Topics, American Society for Mass Spectrometry, Boston, 1983, p. 130.
25. J. D. Wernery and D. A. Peake, Rapid Commun. Mass Spectrom. *3*:396 (1989).
26. K. L. Busch and R. G. Cooks, Science *218*:247 (1982).
27. H. R. Schulten and R. P. Lattimer, Mass Spectrom. Rev. *3*:231 (1984).
28. A. Dell and G. W. Taylor, Mass Spectrom. Rev. *3*:357 (1984).
29. R. J. Day, S. E. Unger, and R. G. Cooks, Anal. Chem. *52*:557A (1980).
30. M. Barber, R. S. Bordoli, G. Elliott, D. Sedgwick, and A. N. Tyler, Anal. Chem. *54*:645A (1982).

31. M. Rabrenovic, T. Ast, and J. H. Beynon, Int. J. Mass Spectrom. Ion Physics. *61*:31 (1984).

32. A. M. Buko, L. R. Phillips, and B. A. Fraser, Biomed. Mass Spectrom. *10*:408 (1983).

33. J. W. Dallinga, N. M. M. Nibbering, J. van der Greef, and M. C. Ten Noever de Braun, Org. Mass Spectrom. *19*:10 (1984).

34. G. W. Garner, D. B. Gordon, L. W. Tetler, and R. D. Sedgwick, Org. Mass Spectrom. *18*:486 (1983).

35. G. Székely and J. Allison, J. Am. Soc. Mass Spectrom. *8*:337 (1997).

36. R. L. Cochran, Appl. Spectrosc. Rev. *22*:137 (1986).

37. J. L. Gower, Biomed. Mass Spectrom. *12*:191 (1985).

38. Y. Ito, T. Takeuchi, D. Ishi, and M. Goto, J. Chromatogr. *346*:161 (1985).

39. R. M. Caprioli, T. Fan, and J. D. Cottrell, Anal. Chem. *58*:2949 (1986).

40. R. M. Caprioli, Anal. Chem. *62*:477A (1990).

41. A. J. Borgerding and R. A. Hites, Anal. Chem. *64*:1449 (1992).

42. R. D. Minard, D. Chin-Fatt, P. Curry, Jr., and A. G. Ewing, Proceedings of the 36th ASMS Conference on Mass Spectrometry and Allied Topics, American Society for Mass Spectrometry, San Francisco, 1988, p. 950.

43. R. M. Caprioli, W. T. Moore, K. B. Wilson, and S. Moring, J. Liq. Chromatogr. *480*:247 (1989).

44. J. S. M. de Wit, L. J. Deterding, M. A. Moseley, K. B. Tomer, and J. W. Jorgensen, Rapid Commun. Mass Spectrom. *2*:100 (1988).

45. K. B. Tomer, F. W. Crow and M. L. Gross, Int. J. Mass Spectrom. Ion Phys. *46*:375 (1983).

46. P. A. Lyon, W. L. Stebbings, F. W. Crow, K. B. Tomer, D. L. Lippstreu, and M. L. Gross, Anal. Chem. *56*:8 (1984).

47. P. A. Lyon, K. B. Tomer, and M. L. Gross, Anal. Chem. *57*:2984 (1985).

48. F. Ventura, J. Caixach, A. Figueras, I. Espadler, D. Fraisse, and J. Rivera, Wat. Res. *23*:1191 (1989).

49. M. Valls, J. M. Bayona, and J. Albaigés, I. J. Environ. Anal. Chem. *39*:329 (1990).

50. S. L. Hunt, F. E. Behr, L. D. Winter, P. A. Lyon, R. L. Cerny, K. B. Tomer, and M. L. Gross, Anal. Chem. *59*:2653 (1987).

51. F. Ventura, J. Caixach, J. Romero, I. Espadler, and J. Rivera, Water Sci. Technol. *25*:257 (1992).

52. F. Ventura, A. Figueras, J. Caixach, J. Rivera, and D. Fraisse, Org. Mass Spectrom. *23*:558 (1988).

53. J. Romera, F. Ventura, J. Caixach, J. Rivera, and R. Guerrero, Environ. Toxicol. Water Qual. *8*:383 (1993).

54. R. M. Facino, M. Carini, P. Minghetti, G. Moneti, E. Arlandini, and S. Melis, Biomed. Environ. Mass Spectrom. *18*:673 (1989).

55. J. Rivera, J. Caixach, A. Figueras, D. Fraisse, and F. Ventura, Biomed. Environ. Mass Spectrom. *16*:403 (1988).

56. J. A. Field, L. B. Barber II, E. M. Thurman, B. L. Moore, D. L. Lawrence, and D. A. Peake, Environ. Sci. Technol. *26*:1140 (1992).

57. V. Mills and E. O. Korpi, Nonsmearing Detergent Bar, U.S. patent 2,988,511 (1961).

58. D. D. White and E. O. Korpi, Alkyl glyceryl ether sulfonate mixtures and processes for preparing the same, U.S. patent 3,024,273 (1962).
59. L. R. Snyder and J. J. Kirkland, *Introduction to Modern Liquid Chromatography*, John Wiley and Sons, New York, 1974.
60. C. R. Blakley, J. C. Carmody, and M. L. Vestal, Anal. Chem. *52*:1636 (1980).
61. M. A. Mabud, P. A. Dreifuss, W. E. Killinger, and M. W. Smith, Proceedings of the 37th ASMS Conference on Mass Spectrometry and Allied Topics, American Society for Mass Spectrometry, Miami Beach, 1989, p. 929.
62. R. E. A. Escott and D. W. Chandler, J. Chromatogr. Sci. *27*:134 (1989).
63. H. F. Schroeder, Vom Wasser *77*:277 (1991).
64. J. D. Reynolds, D. Stubbs, and M. Barringer, Proceedings of the 39th ASMS Conference on Mass Spectrometry and Allied Topics, American Society for Mass Spectrometry, Nashville, 1991, 1354.
65. H. F. Schroeder, Korresp. Abwasser *39*:387 (1992).
66. H. F. Schroeder, Vom Wasser *79*:193 (1992).
67. M. Dole, L. L. Mack, R. L. Hines, R. C. Mobley, L. D. Ferguson, and M. B. Alice, J. Chem. Phys. *49*:2240 (1968).
68. L. L. Mack, P. Kralik, A. Rheude, and M. Dole, J. Chem. Phys. *52*:4977 (1970).
69. G. A. Clegg and M. Dole, Biopolymers *10*:821 (1971).
70. D. Teer and M. Dole, J. Polym. Sci. *13*:985 (1975).
71. M. Dole, H. L. Cox, Jr., and J. Gieniec, Adv. Chem. Ser. *125*:73 (1973).
72. J. V. Iribarne and B. A. Thomson, J. Chem. Phys. *64*:2287 (1976).
73. B. A. Thomson and J. V. Iribarne, J. Chem. Phys. *71*:4451 (1979).
74. R. D. Smith, J. A. Olivares, N. T. Nguyen, and H. R. Udseth, Anal. Chem. *60*:436 (1988).
75. E. C. Huang, T. Wachs, J. J. Conboy, and J. D. Henion, Anal. Chem. *62*:713A (1990).
76. H. R. Udseth, J. A. Loo, and R. D. Smith, Anal. Chem. *61*:228 (1989).
77. J. J. Conboy, J. D. Henion, M. W. Martin, and J. A. Zweigenbaum, Anal. Chem. *62*:800 (1990).
78. R. D. Smith, C. J. Barinaga, and H. R. Udseth, Anal. Chem. *60*:1948 (1988).
79. S. F. Wong, C. K. Meng, and J. B. Fenn, J. Phys. Chem. *92*:546 (1988).
80. J. B. Fenn, M. Mann, C. K. Meng, S. F. Wong, and C. M. Whitehouse, Science *246*:64 (1989).
81. J. A. Loo, H. R. Udseth, and R. D. Smith, Rapid Commun. Mass Spectrom. *2*:207 (1988).
82. T. R. Covey, R. F. Bonner, B. I. Shushan, and J. D. Henion, Rapid Commun. Mass Spectrom. *2*:249 (1988).
83. K. L. Rundlett, and D. W. Armstrong, Anal. Chem. *68*:3493 (1996).
84. M. F. W. Nielen, J. Chromatogr., A *712*:269 (1995).
85. R. J. Strife, J. Schwartz, M. Bier, and J. Zhou, Proceedings of the 43rd ASMS Conference on Mass Spectrometry and Allied Topics, American Society for Mass Spectrometry, Atlanta, 1995, 160.
86. R. J. Strife, J. Schwartz, M. Bier, and J. Zhou, Proceedings of the 43rd ASMS Con-

ference on Mass Spectrometry and Allied Topics, American Society for Mass Spectrometry, Atlanta, 1995, 1113.

87. S. D. Scullion, M. R. Clench, and M. Cooke, Proceedings of the 43rd ASMS Conference on Mass Spectrometry and Allied Topics, American Society for Mass Spectrometry, Atlanta, 1995, p. 841.

88. S. W. Morrall, D. L. Lawrence, and T. W. Federle, Proceedings of the 43rd ASMS Conference on Mass Spectrometry and Allied Topics, American Society for Mass Spectrometry, Atlanta, 1995, p. 842.

89. A. Raffaelli, S. Pucci, L. Campolmi, and P. Salvadori, Proceedings of the 43rd ASMS Conference on Mass Spectrometry and Allied Topics, American Society for Mass Spectrometry, Atlanta, 1995, p. 840.

90. G. A. Schultz and J. N. Alexander IV, Proceedings of the 42nd ASMS Conference on Mass Spectrometry and Allied Topics, American Society for Mass Spectrometry, Chicago, 1994, p. 174.

91. G. J. Harvey and J. C. Dunphy, Proceedings of the 40th ASMS Conference on Mass Spectrometry and Allied Topics, American Society for Mass Spectrometry, Washington, DC, 1992, p. 592.

92. K. Hiraoka and I. Kudaka, Rapid Commun. Mass Spectrom. 6:265 (1992).

93. D. D. Popenoe, S. J. Morris III, P. S. Horn, and K. T. Norwood, Anal. Chem. 66:1620 (1994).

94. P. Rovellini, N. Cortesi, and E. Fedeli, Riv. Ital. Sostanze Grasse 72:381 (1995).

95. S. D. Scullion, M. R. Clench, M. Cooke, and A. E. Ashcroft, J. Chromatogr., A 733:207 (1996).

96. I. Ogura, D. L. DuVal, S. Kawakami, and K. J. Miyajima, J. Am. Oil Chem. Soc. 73:137 (1996).

97. R. M. Facino, M. Carini, G. Depta, P. Bernardi, and B. Casetta, J. Am. Oil Chem. Soc. 72:1 (1995).

98. L. O. Hargiss, personal communication.

99. R. Weber, K. Levsen, G. J. Louter, A. J. H. Boerboom, and J. Haverkamp, Anal. Chem. 54:1458 (1982).

100. R. Large and H. Knof, J. Chem. Soc., Chem. Commun. 935 (1974).

101. H.-R. Schulten and D. Kümmler, Z. Anal. Chem. 278:13 (1976).

102. K. Levsen and H. Schwarz, Angew. Chem., Int. Ed. Engl. 15:509 (1976).

103. F. W. McLafferty, ACS Symp. Ser. 70: (1978).

104. E. Schneider, K. Levsen, A. J. H. Boerboom, P. Kistemaker, S. A. McLuckey, and M. Przybylski, Anal. Chem. 56:1987 (1984).

105. P. Daehling, F. W. Roellgen, J. J. Zwinselman, R. H. Fokkens, and N. M. M. Nibbering, Fresenius' Z. Anal. Chem. 312:335 (1982).

106. E. Schneider and K. Levsen, Comm. Eur. Communities, EUR 10388, Org. Micropollut. Aquat. Environ.: 14 (1986).

107. K. Levsen, E. Schneider, F. W. Roellgen, P. Daehling, A. J. H. Boerboom, P. G. Kistemaker, and S. A. McLuckey, Comm. Eur. Communities, EUR 8518, Anal. Org. Micropollut. Water: 132 (1984).

108. E. Schneider, K. Levsen, P. Daehling, and F. W. Roellgen, Fresenius' Z. Anal. Chem. 316:488 (1983).

109. B. Schueler and F. R. Krueger, Org. Mass Spectrom. *14*:439 (1979).
110. R. Stoll and F. W. Röllgen, Org. Mass Spectrom. *14*:642 (1979).
111. K. Balasanmugam and D. M. Hercules, Anal. Chem. *55*:145 (1983).
112. D. Weil and P. Lyon, Proceedings of the 39th ASMS Conference on Mass Spectrometry and Allied Topics, American Society for Mass Spectrometry, Nashville, 1991, p. 819.
113. D. Weil, P. Lyon, R. Pachuta and C. Moore, Proceedings of the 40th ASMS Conference on Mass Spectrometry and Allied Topics, American Society for Mass Spectrometry, Washington, DC, 1992, p. 842.
114. M. P. Chiarelli, M. L. Gross, and D. A. Peake, Anal. Chim. Acta *228*:169 (1990).
115. F. Hillenkamp, M. Karas, R. C. Beavis, and B. T. Chait, Anal. Chem. *63*:1193A (1991).
116. A. McIntosh, T. Donovan, and J. Brodbelt, Anal. Chem. *64*:2079 (1992).
117. B. Thomson, Z. Wang, A. Paine, A. Rudin, and G. Lajoie, J. Am. Oil Chem. Soc. *72*:11 (1995).
118. F. W. Karasek and R. E. Clement, *Basic Gas Chromatography/Mass Spectrometry: Principles and Techniques*, Elsevier, Amsterdam, 1988.
119. W. McFadden, *Techniques of Combined Gas Chromatography/Mass Spectrometry: Applications in Organic Analysis*, John Wiley and Sons, New York, 1973.
120 B. J. Guizinowicz, M. J. Guzinowicz, and H. F. Martin, *Fundamentals of Integrated GC/MS, Part III: The Integrated GC/MS Analytical System* Marcel Dekker, New York, 1977.
121. B. S. Middleditch, ed., *Practical Mass Spectrometry, A Contemporary Introduction*, Plenum Press, New York, 1979.
122. P. MacCarthy, R. W. Klusman, S. W. Cowling, and J. A. Rice, Anal. Chem. *67*:525 (1995).
123. P. MacCarthy, R. W. Klusman, S. W. Cowling, and J. A. Rice, Anal. Chem. *65*:244R (1993).
124. P. MacCarthy, R. W. Klusman, S. W. Cowling, and J. A. Rice, Anal. Chem. *63*:301R (1991).
125. T. Reemtsma, J. Chromatogr., A *733*:473 (1996).
126. M. L. Trehy, W. E. Gledhill, J. P. Mieure, J. E. Adamove, A. M. Nielsen, H. O. Perkins, and W. S. Sckhoff, Environ. Toxicol. Chem. *15*:233 (1996).
127. M. J.-F. Suter, R. Reiser, and W. Giger, J. Mass Spectrom. *31*:357 (1996).
128. J. A. Field, D. J. Miller, T. M. Field, S. B. Hawthorne, and W. Giger, Anal. Chem. *64*:3161 (1992).
129. J. A. Field, T. M. Field, T. Poiger, and W. Giger, Environ. Sci. Technol. *28*:497 (1994).
130. H. Brueschweiler, H. Felber, and F. Schwager, Tenside, Surfactants, Deterg. *28*:348 (1991).
131. C. F. Tabor and L. B. Barber II, Environ. Sci. Technol. *30*:161 (1995).
132. P. Romano and M. Ranzani, Water Sci. Technol. *26*:2547 (1992).
133. A. M. Nielsen and R. L. Huddleston, Dev. Ind. Microbiol. *22*:415 (1981).
134. H. Honnami and T. Hanya, Water Res. *14*:1251 (1980).
135. J. A. Field, J. A. Leenheer, K. A. Thorn, L. B. Barber II, C. Rostad, D. L. Macalady, and S. R. Daniel, Contam. Hydrol. *9*:55 (1992).

136. M. L. Trehy, W. E. Gledhill, and R. G. Orth, Anal. Chem. *62*:2581 (1990).

137. L. Cavalli, A. Gellera, and A. Landone, Environ. Toxicol. Chem. *12*:1777 (1993).

138. J. McEvoy and W. Giger, Environ. Sci. Technol. *20*:376 (1986).

139. I. Otvos, B. Bartha, Z. Balthazar, and G. Palyi, J. Chromatogr. *94*:330 (1974).

140. R. D. Smith, H. T. Kalinoski, and H. R. Udseth, Mass Spectrom. Rev. *6*:445 (1987).

141. J. D. Pinkston, T. E. Delaney, K. L. Morand, and R. G. Cooks, Anal. Chem. *64*:1571 (1992).

7

Chromatographic Processes for the Analysis of Anionic Surfactants

HENRIK T. RASMUSSEN and BRUCE P. McPHERSON   Department of
Analytical and Microbiological Sciences, Colgate-Palmolive Company,
Piscataway, New Jersey

I. INTRODUCTION

This chapter provides a review of the applicability of chromatographic and related techniques for the analyses of anionic surfactants in raw materials and finished products. The use of chromatography for the analysis of surfactants in
environmental samples and the use of mass spectrometry (MS) coupled to chromatography are reviewed in Chapters 3 and 6, respectively. To avoid overlap,
these topics will only be discussed when the chromatographic procedures described in these applications have broad-based utility and are potentially applicable to the analysis of raw materials and finished products.

The content of this chapter is organized by major compound type, followed
by chromatographic technique in an effort to make the information readily ac-
cessible. The use of emerging technologies such as supercritical fluid chromatography (SFC) and capillary electrophoresis (CE) has been included, and
approaches that are clearly outdated due to technological advances have been
omitted or are discussed only briefly.

II. SULFURIC ACID ESTERS

A. Sulfated Alcohols

1. Gas Chromatography

As is the case with all ionic compounds, the inherent nonvolatility of sulfated alcohols does not allow for direct analysis by gas chromatography (GC). As a result, derivatization or pyrolysis is necessary prior to analysis. Due to this
requirement and because sulfated alcohols are relatively simple mixtures and can
easily be separated directly by high-performance liquid chromatography
(HPLC), GC is being used less and less.

The carbon chain distribution of sulfated alcohols may be determined from
the hydrophobic oils produced by hydrolysis in dilute acid or alkali [1] or after treatment with hydriodic acid to form the iodo derivatives [2]. Using pyrolysis or acid pyrolysis (with phosphorus pentoxide) [3,4], C-S and C-O
bonds are broken at 400–550°C. However, since none of these approaches is
specific for the analysis of alkyl sulfates, the utility is limited to the determination of carbon chain distribution of raw materials or fractions isolated from
products.

2. High-Performance Liquid Chromatography

Due in part to the relative simplicity of their distributions, sulfated alcohols and alkanesulfonates may be separated by numerous HPLC methods. General approaches for the analysis of alkyl sulfates include the use of reversed-phase chromatography with mobile phase additives (including ion-pairing reagents) [5–16] and chromatography based on ion exchange or mixed mode ion-exchange/reversed-phase chromatography [17–21]. Numerous papers provide comprehensive discussions of the influence of operating variables on solute retention behavior [5,6,14,16–18,20,21] and potentially serve as useful guidelines for separation optimization. Detection has most frequently been achieved using refractive index detection [5–7,10,20,22], indirect spectophotometric detection [11,16–20], and (suppressed) conductivity or indirect conductivity detection [8,9,13–15,17,18,20,21].

Chromatographic parameters for several approaches are summarized in Table 1 and may serve as a useful starting point for method development. It is, however, also useful to examine the documented applicability and analytical performance of the methods. For example, Nakamura et al. [5] have shown their method to be applicable to the determination of carbon chain distribution. HPLC results compared favorably with those obtained by GC analysis of trimethylsilyl derivatives, and coefficients of variation of relative peak areas were less than 3% ($n = 5$). Steinbrech et al. [10] have demonstrated the use of their method for the determination of octylhydrogen sulfate and alkylsulfosuccinate in industrial mixtures with a coefficient of variation of 2–5% for the complete analytical procedure. Konig and Strobel [12] have developed a method suitable for the determination of numerous surfactants (including alkyl sulfates) in toothpastes. Using the latter approach, surfactants are identified via the capacity factors and quantitated using external calibration standards.

Additional work by Nakamura and Morikawa [6] has shown the viability of determining C_{10}–C_{18} alkyl sulfates in the presence of homologous mixtures of N-acylsarcosinates, N-acyl-L-glucamates, fatty alcohol diethanolamides, and alkyldimethylaminoacetic acid betaines. A chromatogram of alkyl sulfates obtained in the presence of N-acylsarcosinates and N-acyl-L-glucamates is shown in Fig. 1. Using similar conditions [7], sodium dodecyl sulfate (SDS) was separated from 14 other dodecyl surfactants. Optimized conditions allowed for the analysis of 50 commercial shampoos and household detergents for SDS and various other surfactants. Based on five replicate determinations, a recovery of $99 \pm 2.5\%$ was obtained for SDS.

In addition to varying chromatographic conditions to effect the desired selectivity, detection more specific than indirect photometric, conductivity, or refractive index may be utilized. Frei et al. [22,23] have used postcolumn–ion pair extraction and fluorimetric detection to selectively determine anionic surfactants including alkyl sulfates. Separation may be performed by either normal-phase

TABLE 1 Chromatographic Conditions for the Separation of Alkyl Sulfates

Column	Mobile phase	Detection method	Ref.
200 × 6 mm TSK-LS 410 (ODS/silica)	1.0 M NaClO$_4$ in 15/85 (v/v) water/methanol (pH 2.5 w. H$_3$PO$_4$); 1.5 mL/min	Refractive index	6
250 × 4.6mm MPIC-NS1 (neutral divinylbenzene resin)	0.01 M ammonium hydroxide-acetonitrile (65:35 v/v); 1 mL/min	Suppressed conductivity	8,9
250 × 4 mm RP/phenyl-anion	10 μm naphthalenetrisulfonate in 85% acetonitrile; 2.2 mL/min	Indirect UV; indirect fluorometric; conductivity	17
250 × 4 mm Nucleosil C8 (10 μm)	0.01 M tetrabutylammonium-hydrogen sulfate/methanol (23/77 v/v), pH 3.0 w. HCl; 2 mL/min	Refractive index	10
250 × 4 mm Nucleosil C18 (7 μm)	0.25M NaClO$_4$ in 80/20 (v/v) methanol/water; 1.5 mL/min	Refractive index	12

250 × 4 mm NS1 (neutral divinylbenzene resin)	Gradient: Eluent 1 = water, Eluent 2 = 30% acetonitrile/ 10 mM NH$_4$OH, Eluent 3 = 90% acetonitrile; 0 minutes = 33/67 (v/v) eluent 1/eluent 2. 1 minute = 33/67 (v/v) eluent 1/eluent 2. 4 minutes = 83/17 (v/v) eluent 2/eluent 3; hold for duration of run; 1 mL/min	Suppressed conductivity	15
250 × 4.6 mm IonPac AS11 (13 μm pellicular polystyrene-divinylbenzene)	40/60 (v/v) acetonitrile/water w. 1.0 mM LiOH; 1 mL/min	Suppressed conductivity	21
50 × 4.1 mm PRP-1 (polymeric reversed phase)	30/70 (v/v) acetonitrile/water with 0.1mM citrate buffer, 0.1 mM iron (II) 1,10-phenanthroline, pH 7.35; 1 mL/min	Indirect UV (510 nm)	16

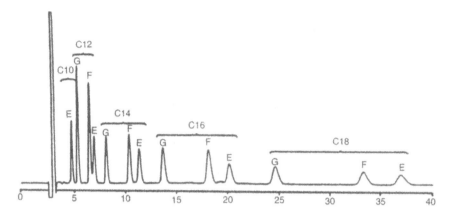

FIG. 1 HPLC Separation of (E) alkyl sulfates, (F) *N*-acylsarcosinates and (G) *N*-acyl-L-glucamates. Column: 200 × 6 mm TSK-LS410. Mobile phase: 1.0 M NaClO₄ in water/ethanol (pH 2.5 w. H₃PO₄) at 1.5 mL/min, refractive index detection. (Adapted from Ref. 6.)

HPLC [22] or reversed-phase HPLC [23], but the use of reversed-phase is generally preferred since separation by alkyl chain length is observed. Following reversed-phase separation using water–acetone–sodium dihydrogen phosphate as the mobile phase, the effluent is mixed with chloroform and a solution of acridinium chloride. The ion pairs formed between the anionic surfactant and the acridinium ion are extracted into the chloroform phase, which is then separated from the aqueous phase in an extraction coil. After phase separation the chloroform phase is monitored fluorimetrically (excitation 400 nm; emission 470 nm cut-off).

As an extension of previously cited work [6,7], Kanesato et al. [24], have also used postcolumn reaction detection. Using their approach, methylene blue in chloroform is added to the column effluent to pair with the anionic surfactants. For quantitation, the chloroform phase is monitored spectrophotometrically at 630 nm. For alkyl sulfates, the technique was shown to provide carbon chain distributions similar to those obtained by GC. Additionally, the method shows broad-base applicability to the determination of other surfactants. For example, by changing the postcolumn reagent to Orange II, cationic and amphoteric surfactants are selectively detected. Additional specific detection approaches have included the use of plasma atomic emission detection [25] and mass spectrometry [26,27].

3. Electrophoresis

Numerous authors have demonstrated the feasibility of separating alkyl sulfates by capillary electrophoresis [28–33] using indirect photometric detection. Key operating parameters for the various approaches are summarized in Table 2. The

TABLE 2 Capillary Electrophoresis Conditions for the Separation and Indirect Detection of Alkyl Sulfates

Buffer	Capillary[a]	Voltage	Detection	Ref.
12 mM veronal (pH 8.6)	70 cm × 50 μm	+25 kV	UV: 240 nm	28
1 mM potassium dichromate/ 1 mM sodium tetraborate/ 30 mM boric acid in 70/30 (v/v) water/acetonitrile (pH 8.0)	100 cm × 100 μm	+10 kV	UV: 265 nm	29
5.0 mM phosphate/5.0 mM) salicylate/1.0 mM Mg²⁺ (pH 7.0	60 cm × 50 μm	+20 kV	UV: 230 nm	30
5 mM naphthalenemonosulfonate, naphthalenedisulfonate, or naphthalenetrisulfonate in 100 mM boric acid/5 mM sodium tetraborate (pH 8.0)	75 cm × 50 μm	+30 kV	UV: 214 nm	31
5 mM naphthalenemonosulfonate/ 100 mM boric acid and 50% methanol (pH 6.0)	75 cm × 50 μm	+30 kV	UV: 206 nm	32
5 mM dihydroxybenzoic acid in 5% methanol (pH 8.1 w. NaOH)	100 cm × 75 μm	+30 kV	UV: 250 nm	33

[a]Underivatized fused silica capillaries of the dimension specified.

papers include numerous investigations of the influence of operating voltage, pH, addition of organic solvent, and buffer concentration and composition on resolution.

While most of the work cited provides only qualitative data, reproducibility studies were included by Nielen [28] and show the peak height and area reproducibility of 10^{-4} M standard solutions of C_8, C_9, C_{10}, and C_{12} alkyl sulfates to be less than 5% rsd based on six replicate injections. Furthermore, response linearity was excellent (R2 > 0.999 for each species) over the range of $0–2 \times 10^{-3}$ M. Analysis of a Teepol HB7 sample was performed, but no studies were conducted to relate the results to those obtained using other methodology.

Gibbons and Hoke [33] applied their method to the analysis of sodium dodecyl sulfate in a simulated stream water at levels ranging from 0.8 to 50 mg/L. Calibration was shown to be linear over the region of interest, and recoveries from placebos (clean water) which were spiked with known amounts of SDS were generally within 5% of the expected.

4. Thin Layer Chromatography

Thin layer chromatography (TLC) has long been a mainstay in the analysis of alcohol sulfates. Most of the recent publications refer to earlier studies, which will not be singled out in this work but are generally available when pursuing any of the references cited herein. Modes of separation offered by TLC are contrary to those of HPLC for the most part since, as previously illustrated, the HPLC provides carbon chain separations through the reversed-phase mechanism while common, normal-phase, TLC techniques produce "class" separations.

Qualitative, systematic separations of surfactants in shampoos, bubble bath formulations, and soaps, as reported by Mattisek [34], utilized silica gel 60 with a solvent mixture of 10:10:5:2 (v/v/v/v) n-propanol/chloroform/methanol/10 N ammonia [35] for long-chain primary alkylsulfates. The R_f value was 0.55 and an orange color developed with visualization using pinacryptol yellow in water under long-wavelength UV light (366 nm). An alternative system was also employed: silica gel G (impregnated with ammonium sulfate) [36] was eluted with a mixture of 80:19:1 (v/v/v) chloroform/methanol/0.1 N H_2SO_4. An R_f value of 0.30 resulted and pinacryptol yellow was used again. Several other materials coelute with the alkylsulfates under these conditions, so caution is advised when making unknown identifications based on R_f values and/or visualization alone. A third system offered freedom from interference with other surfactants: silica gel 60, activation with 30–50% relative humidity, developed with 9:1 (v/v) acetone/tetrahydrofuran. The reported R_f value was 0.28.

A thorough qualitative and quantitative study by Yonese et al. [37] employed nine different developing solvents to find optimum conditions to separate sodium dodecyl sulfate from dodecanesulfonate and dodecylbenzenesulfonate. R_f values for the alkylsulfate ranged from 0.25 to 0.95 on silica gel G impregnated with 10% ammonium sulfate. Best separations were reported with either 1:1:1 (volume ratio) methyl acetate/benzene/ethyl alcohol (R_f 0.84) or 20:6:1.6:1 (volume ratio) 2-methyl-4-pentanone/propyl alcohol/0.1 N acetic acid/acetonitrile (R_f 0.45). Using the latter elution solvent mixture, seven different adsorbents were compared, yielding R_f values for the dodecyl sulfate which varied from 0 to 0.88. Silica gel G impregnated with ammonium sulfate or silver nitrate provided the best separations. Variations in visualization/detection techniques included charring after spraying with 2–5% phosphomolybdic acid in ethanol or by charring unsprayed and using densitometer measurement with calculation by the triangle method. The preferred charring temperature was found to be 250°C (15 minutes) and sensitivity was improved without the phosphomolybdate spray. A calibration curve was linear from 4 to 15 µg, and variation coefficients/mean error statistical data indicated that the procedure could be applied to quantitative determinations.

A unique two-dimensional TLC separation of surfactants by Armstrong and Stine [38] uses both silica gel and C_{18} reversed-phase sorbents. Surfactants were spotted on the reversed-phase strip for class separation and developed with 75% ethanol. The anionics traveled to the solvent front, cationics remained near the origin, and nonionics separated between the anionics and cationics. The plate was then cut into three sections and developed in the second (perpendicular) direction, which incorporated the silica gel portion. Sodium dodecyl sulfate eluted to an R_f value of 0.15 after development with 8:1 (v/v) $MeCl_2$/MeOH in the second direction. Visualization was with I_2 vapor. Scanning densitometry was performed directly in the absorbance-reflectance mode at 215 nm, although detection limits were lower when using I_2 and scanning at 405 nm in the absorbance-transmittance mode. It was also reported that sensitivity and selectivity can be enhanced by using a variety of visualization or charring techniques [37,39].

Quantitative determination of sodium dodecyl sulfate in hydrophilic ointments was reported by Yamaha et al. [40] using a Chromarod flame ionization detector. A calibration curve was produced by applying 0.8, 1.6, 2.4, 3.2, and 4.0 μg of standard reference dodecyl sulfate onto silica gel–coated rods and developing in solvent containing 50:10:2 (v/v/v) ethyl acetate/methanol/28% NH_4OH in water. The rods were then dried and passed through a flame ionization detector (Iatroscan TH-10). Several ointment products were analyzed using the same technique and recovery studies and statistical data including CV were provided. The products were also analyzed using colorimetry [41] and HPLC, and it was confirmed that TLC-FID is better in terms of simplicity and accuracy.

Sodium lauryl sulfate was one of several surfactants separated by Hohm [42], who used primulin (CI 49000 Direct Yellow 59) and/or Thioflavin S for visualization under long-wave UV light in his studies. Fluorescence detection (365 or 435 nm excitation and 390 or 460 nm emission, respectively) was employed. Normal silica gel 60 plates can be used as well as NH_2 or CN-modified silica gel 60 HPTLC plates. Quantitative determinations were reported with relative error rates of 2–3% using fluorescence. Applications were illustrated for shampoos and shower/bath preparations. Preferred elution solvent mixtures for sodium lauryl sulfate were either 8:2:2 n-butanol/water/glacial acetic acid (v/v/v) or 35:30:25:5 chloroform/ethanol/dimethoxyethane/ammonia (v/v/v/v).

Sodium n-dodecyl sulfate was among 28 different surfactants separated on HPTLC silica plates by Simunic and Soljic [43]. Building upon the work of Hohm [42], Henrich [44], and Kruse et al. [45], they reported new detection techniques with water as well as known agents of detection such as UV light (254 and 366 nm) and iodine. The anionics show characteristic white spots with gas bubbles after development with ethanol and detection with water. Auxiliary methods in classification by charge type color reactions employ iodine and/or indicator dyes such as methylene blue, KI_3, and bromophenol blue. Separation using 95% ethanol and silica HPTLC plates produces an R_f of 0.88 for Na n-dodecyl sulfate, which exhibits elliptically shaped white gas bubbles upon dipping the plate in distilled water.

Alumina E–coated TLC plates can also be employed for the separation of sodium lauryl sulfate from most common surface active agents as follows (B. P.

McPherson, unpublished results): after applying the samples and reference standards to the plate, microwave for 3 minutes, allow to cool briefly, then develop with a 20:3 (v/v) mixture of methanol/NH$_4$OH in a filter-paper lined chromatography tank. Spraying lightly with 0.05% pinacryptol yellow in 3A alcohol (an anhydrous mixture of 95:5 ethanol/methanol) and viewing under long-wave UV light visualizes the lauryl sulfate at an R$_f$ value of 0.55. Densitometric quantitation has been performed on products such as liquid and powdered laundry detergents, raw materials, dishwashing liquids, and shampoos. Limits of detection are 4 µg for the pure sodium lauryl sulfate.

B.   Sulfated Ethoxylated Alcohols

1.   Gas Chromatography

Like alcohol sulfates, ethoxylated alcohol sulfates cannot be analyzed directly by GC. However, due to the complexity of the mixtures, full characterization (separation by both alkyl chain length and degree of oligomerization) is difficult to achieve by other means. As a result, acid hydrolysis to the parent ethoxylated alcohols followed by GC remains a popular method of analysis.

GC analyses have been performed on both underivatized and derivatized ethoxylated alcohols [46–49] The primary advantage of derivatization is to increase the volatility of the species, thereby allowing for compounds with higher degrees of ethoxylation to be analyzed. For this reason it is also generally useful to use gas chromatographic columns, which may be used at high temperatures. Columns that can be operated at temperatures of up to 425°C have been developed [50,51].

Using a 10 m × 0.32 mm i.d. SimDist-CB capillary column with a 0.1 µm stabilized bonded polysiloxane film, Silver and Kalinosky [48] compared the analyses of underivatized ethoxylated alcohols with trimethylsilyl derivatives. The column utilized allowed for temperature programming of up to 375°C. For underivatized species the analysis is limited to oligomers with degrees of ethoxylation of up to 13; for trimethyl silyl derivatives, oligomers with up to 22 ethoxy groups were separated. In addition to separating by degree of ethoxylation, this approach also separates the oligomers by carbon chain length, even for complex mixtures, which may be based on C$_{12}$, C$_{13}$, C$_{14}$, and C$_{15}$ alcohols (Fig. 2). A further advantage, resulting from the use of a FID, is that quantitative information about the sample may be obtained using effective carbon number (ECN) theory [52]. Using this approach, the simultaneous determination of alkyl chain distribution, oligomeric weight % distribution, and the mole average degree of ethoxylation were achieved. The experimentally determined degrees of ethoxylation were shown to be in excellent agreement with results obtained by NMR.

The method developed by Silver and Kalinosky [48] was repeated, with minor modification, by Rasmussen et al. [49] and extended to the analysis of

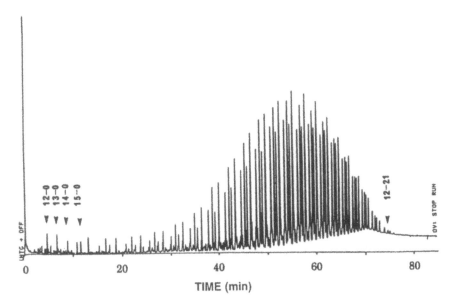

FIG. 2 High-temperature gas chromatogram of silylated C_{12}-C_{15} alcohol ethoxylate with an average of 11.3 moles of ethylene oxide. (From Ref. 48.)

ethoxylated alcohol sulfate raw materials and ethoxylated alcohol sulfates in commercially available dishwashing liquids. Notably in this work, the procedure for hydrolyzing ethoxylated alcohol sulfates and isolating the resulting ethoxylated alcohols was validated by comparing the EO distribution of a native ethoxylated alcohol with the corresponding sulfate.

The acid hydrolysis, isolation and trimethylsilyl derivatization were performed as follows: about 300 mg of ethoxylated alcohol sulfate and 25 mL of 25% aqueous HCl were refluxed in a small round bottom flask for 90 minutes. (The flask was fitted with a Graham 41 × 500 mm condenser with a dry ice trap to eliminate loss of volatiles). After refluxing, the reaction was left to cool and transferred to an extraction flask. 3A alcohol was added through the top of the condenser to wash the apparatus and distilled water was added to rinse any remaining sample left in the round-bottom flask. Both washes were added to the extraction flask. The sample was extracted with two 5-mL aliquots of chloroform. The combined aliquots were treated with 1 mL of HPLC grade Pyridine and 1 mL of bis(trimethylsilyl)trifluoroacetamide (BSTFA) and heated at 60°C for 20 minutes to form the trimethylsilyl derivatives. For chromatography, samples were diluted with chloroform, as needed, to produce reasonably sized peaks.

As shown in Fig. 3, the weight % distributions of the sulfated and corresponding unsulfated materials are very similar and the EO number agrees closely with

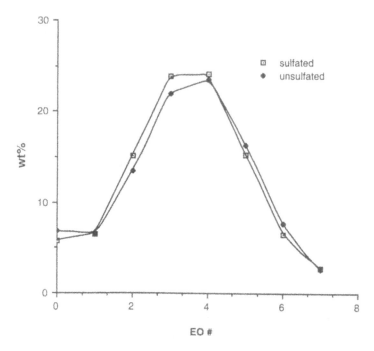

FIG. 3 Comparison of weight % distributions of sulfated and unsulfated C_{10}-3EO. (From Ref. 49.)

the manufacturer's specification of 3.0 EO. Specifically, the degrees of ethoxylation determined experimentally using the outlined procedure and ECN theory were 2.962 and 2.995, respectively. The method was also shown to be viable for the determination of EO sulfate distributions in products such as commercially available dishwashing liquids. However, several steps are required to isolate the ethoxylated alcohol sulfates from the product prior to analysis, and the isolation procedures may vary depending on the sample composition.

2. Supercritical Fluid Chromatography

As for GC, ethoxylated alcohol sulfates must be desulfated prior to analysis by supercritical fluid chromatography (SFC). Numerous qualitative chromatograms of ethoxylated alcohols, obtained using SFC with supercritical carbon dioxide and flame ionization detection, have appeared in the literature [47,53]. A primary advantage of SFC over GC is the potential of analyzing species with higher degrees of ethoxylation. As for GC, derivatization to increase volatility extends the range of applicability. Chester et al. [54] have analyzed ethoxylated alcohols as trimethylsilyl derivatives and have thereby been able to separate oligomers with

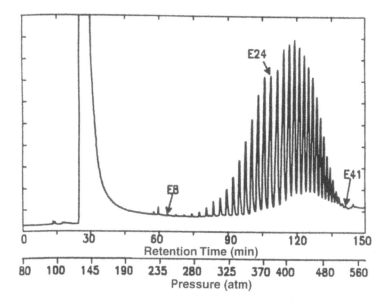

FIG. 4 SFC separation of the trimethylsilyl derivative of a high molecular weight ethoxylated alcohol mixture. (From Ref. 54.)

a degree of ethoxylation of up to 41 ethoxy groups (Fig. 4). Notably, however, these results were achieved using custom-made instrumentation, which allowed for the use of pressures up to 560 atm. Using commercially available instrumentation, where the upper pressure limit is 400 atm, the separation is limited to oligomers with 24 ethoxy groups. For acetate derivatives, Silver and Kalinosky [48] note similar limits and demonstrate that, while separation by degree of ethoxylation is observed, only partial resolution is observed between carbon chain oligomers.

As an alternative to quantitating oligomers via ECN theory [52], Geissler [55] has determined the response factors of EO standards as a function of oxygen-to-carbon ratio. Such an approach allows for calculation of response factors for oligomers for which no standards are available. Using this approach allowed for the determination of the EO distributions for simple mixtures (one carbon chain length) and provided EO numbers that were in excellent agreement with results obtained by NMR. For the analyses of mixed carbon alcohol ethoxylates, some peak overlap was observed. For such situations results were obtained by mathematical deconvolution.

As an alternative to the use of higher pressures to extend the elution range in SFC, Brossard et al. [56] have used packed columns and supercritical carbon dioxide with polar modifiers. While the modifiers do not allow for flame ionization detection, evaporative light-scattering detection is shown to be a viable alternative. Although no quantitative studies of the types outlined previously [48,49,55] were conducted, it was noted that all of the ethoxylated products have similar response factors.

3. High-Performance Liquid Chromatography

Due to the complexity of EO sulfates and the inherently smaller peak capacity of HPLC, complete separations by both alkyl chain length and degree of ethoxylation, as shown by GC (see above), have not been realized. However, since HPLC does not have the volatility constraint of GC, it is possible to separate higher molecular weight distributions. Studies (see Refs. 17,22,23,27) include ethoxylated alcohol sulfates analyses using methodologies similar to those specified in Sec. II.A. However, the resultant chromatograms may be complex and difficult to interpret. In this respect, the advent of MS detection is promising for the identification of individual oligomers.

Numerous methods have been developed to examine aspects of the composition of both alkyl and alkyl aromatic ethoxylated alcohol sulfates. Jandera et al., for example, have demonstrated the use of both reversed-phase [57] and normal-phase [58,59] HPLC for the separation of nonionic nonylphenol ethoxylates from sulfated nonylphenol ethoxylates.

A rather unique methodology by Bear [60] employs HPLC with an evaporative light-scattering detector for rapid separation and detection of sulfated ethoxylated alcohols without separating by carbon chain length or moles of ethoxylation. Components of the raw materials were separated into inorganic salt, sulfates, and unreacted alcohol using a rapid reversed-phase chromatography (Fig. 5). The column used was a 2.5 cm × 0.2 cm i.d. column packed with 10 μm C_{18}. The solvent system was a 4-minute gradient program of water and THF running in several steps from 10% THF to 100% in 2.6 minutes, then back to 10% in 4 minutes, all at a flow of 1 mL/min with a column compartment at 40°C. Calibration curves were presented which compared peak area responses for the sulfated material as well as the unreacted, unsulfated materials. Quantitative data comparing the HPLC-ELS and the standard mixed-indicator two-phase titration methods indicated that the concentrations calculated from the ELS are believed to be more accurate.

Benning et al. [61] use HPLC to determine the carbon chain distributions of sulfated ethoxylated alcohols by cleavage with HI and UV detection at 252 nm. A 30 cm × 3.9 mm Waters μBondapak column was used with a gradient elution starting at 75:25 methanol/water under a first-order linear gradient to 50:50 methanol/isopropanol in 30 minutes with a subsequent hold for 5 minutes, all at 1.5 mL/min. Quantitative data obtained by HPLC compared favorably to GC determinations of the iodides. The HPLC was advantageous in that its UV detection offered specificity not afforded by the GC-FID and in that the external

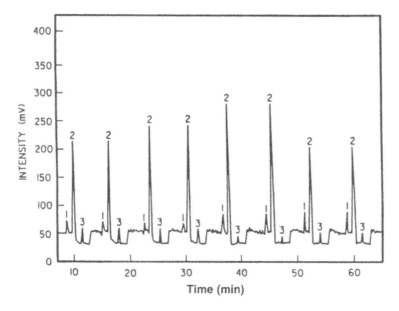

FIG. 5 HPLC analysis of sulfated alkyl ether surfactants (2), inorganic salt (1), and un-reacted alcohol (3). (From Ref. 60.)

standard method used by HPLC is more stable than the internal standard procedure used in GC.

Conductivity detection was used by Stemp et al. [62] in the analyses of ethoxylated alcohol sulfates for number of moles of ethoxylation and/or carbon chain length information. Ammonium lauryl sulfate was also analyzed for carbon chain length in this study. A 5 μm C_8 column (250 × 4.6 mm) was used with a mobile phase of 45:55 methanol/water containing 17.5 mg/L ammonium acetate. Distributions were obtained for carbon chain lengths from C_{10} to C_{16} and number of moles of ethoxylation up to eight. Several raw materials were analyzed, as were shampoos. Using UV detection in series, additional information was obtained about components aromatic in character such as ammonium xylenesulfonate, dodecylbenzenesulfonate, and sodium lauryl sarcosinate.

4. Electrophoresis

Goebel et al. [29] have attempted to analyze ethoxylated alcohol sulfates directly (underivatized) by capillary electrophoresis with indirect UV detection using the conditions specified in Table 2. Separations of simple mixtures such as a C_{10} 3EO ethoxylated alcohol sulfate were achieved, but analyses of more complex distributions led to overlap between oligomers of different alkyl chain lengths.

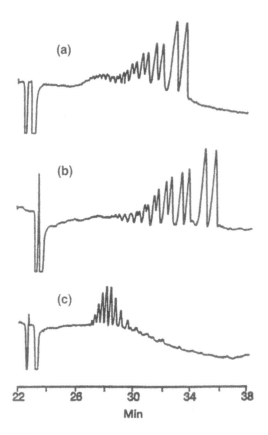

FIG. 6 CE separation of (a) a commercial dishwashing formulation, (b) Neodol 23-2A, and (c) a 6.5 EO sulfate. (From Ref. 29.)

Nevertheless, feasibility was shown for identifying the nature of the AEOS distribution in a commercially available dishwashing liquid (Fig. 6). A primary advantage of the method is the simplicity of the sample preparation. Both AEOS mixtures and dishwashing liquid required only dilution prior to analysis.

Analyses of ethoxylated alcohol sulfates by capillary electrophoresis (CE) have also been performed by Shamsi and Danielson [31] using the system described in Table 2. While fingerprints of the mixtures are observed, separation using this approach is also incomplete.

5. Thin Layer Chromatography

For the most part, any of the techniques or chromatographic conditions described above for thin layer chromatography of sulfated alcohols will be suitable for eluting and detecting their ethoxylated counterparts. Again, the information obtained will be the result of "class" separations rather than the carbon chain length–based separations of the HPLC methodologies in most instances.

References [34–36,42–45, and (McPherson, unpublished results)] include alcohol ether sulfates in their determinations, and with few minor exceptions they are handled exactly as described above under sulfated alcohols. The deviations involved are the characteristic "broader/streaked" spot shape when the ethoxylate is present and subtle differences in detection capabilities based upon the changes in chemistry afforded by ethoxylation. As is also the case with GC/HPLC subjects (see above), desulfation to the parent ethoxylated alcohol is the preferred alternative for determining the number of moles of ethoxylation with any certainty.

A study by Hohm [42] is unique among the TLC references cited in that the conditions used for the determination affords separation by homologs of moles of ethoxylation for additional characterization. Densitometric scans are illustrated, and quantitative determinations of ether sulfates in shampoos are covered.

III. SULFONIC ACID SALTS

A. Alkane- and Olefinsulfonates

1. Gas Chromatography

Sulfonates may separated by GC after conversion to volatile derivatives via acid desulfonation [63,64], alkali fusion [65,66], sulfonyl chlorination [67–71], sulfonyl fluorination [72], sulfonyl esterification [67,73–77], silylation [78–81], or conversion to sulfonamide derivatives [83]. As for other surfactants, it is, however, desirable to perform analyses without derivatization, and as a result more recent efforts have been directed towards the use of HPLC or CE.

GC remains useful (especially in combination with MS) in the analyses of raw materials where it may be desirable to fully characterize impurities, reactants, isomers, etc. For the analyses of α-olefinsulfonates, for example, characterization with respect to hydroxyalkanesulfonate [69] or isomer [73] content has been performed.

2. High-Performance Liquid Chromatography

Several of the methods for the determination of alkyl sulfates (Sec. II.A) are also applicable (as outlined or with minor modifications) to the analyses of alkanesulfonates [7–9,13–23,25]. Indeed, the two surfactant types behave very similarly. As a result, it is possible not only to use the same conditions for analysis, but, as shown in Fig. 7, to separate homologous series of both surfactant types simultaneously [14]. In addition to separations of linear alkanesulfonates, methods have been reported for the separation of branched alkanesulfonates [23,83,84], hydroxyalkanesulfonates [8–9,85,86], and α-olefinsulfonates [85–87].

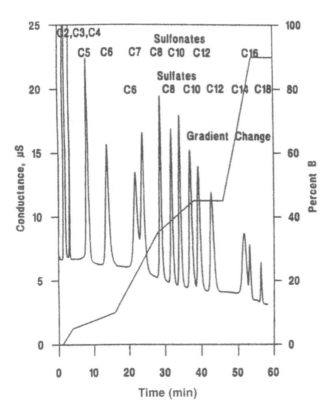

FIG. 7 Separation of a mixture of alkyl sulfonates and alkyl sulfates. Solvent A = water, 10 mM LiOH. Solvent B = 4:1 acetonitrile: water, 1 mM LiOH. Column = PRP-1 (polymeric reversed phase). Flow rate at 1 mL/min. Suppressed conductivity detection. Gradient conditions are given on the inset. (From Ref. 14.)

Smedes et al. [23] have used the postcolumn–ion pair extraction and fluorimetric detection technique described in Sec. II.A to separate a commercial mixture of C_{13}–C_{17} secondary alkyl sulfonates. Separation is achieved using a hypersil ODS column with a water/acetone/sodium dihydrogen phosphate gradient. Liebscher et al. [83] have separated C_{12}–C_{18} monosulfonates into the constituent positional isomers (Fig. 8). For each alkyl chain length, separation between individual isomers was observed for all but the centrally located pairs of isomers. Isomer distributions were obtained with good precision, allowing for the method to be used to determine the relative reactivity of each carbon site during sulfochlorination. In subsequent work [84], the influence of column chemistry, mobile phase composition, and concentration and nature of the ion-interaction reagent was investigated.

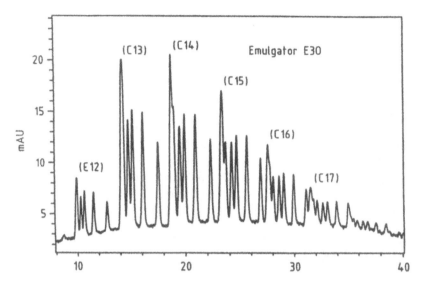

FIG. 8   Separation of branched alkyl sulfonates. Column: 2 × 200 mm × 4.6 mm Hypersil ODS, 5 μm. Mobile phase: (A) 3/8 (v/v) methanol/water with 0.25 mM N-methylpyridinium chloride, (B) 8/2 (v/v) methanol/water with 0.25 mM N-methylpyridinium chloride. Gradient: 40/100% B in 40 min at a flow rate of 1.3 mL/min. Temperature 40°C. Detection: indirect UV at 260 nm. (From Ref. 83.)

Characterization of commercial blends of α-olefinsulfonates are complex in the sense that full characterization involves not only separation by alkyl chain length and position of unsaturation, but also the separation from positional isomers of hydroxyalkanesulfonates and disulfonates. Using a Zorbax TMS column with a mobile phase consisting of 75/25 (v/v) methanol/water with 0.4 M sodium nitrate, Johannessen et al. [88] have demonstrated the separation of C_{14} and C_{16} alkenesulfonates, achieving separation by carbon chain length and partial resolution by double bond position. Analysis of C_{14} α-olefinsulfonate shows separation from C_{14} 4-hydroxyalkanesulfonate, C_{14} 3-hydroxyalkanesulfonate, and 3-alkanesulfonate. For mixtures of C_{14} and C_{16} blends, however, coelution is observed between C_{16} 3-hydroxyalkanesulfonate and C_{14} 2-alkenesulfonate. Using this system, disulfonates elute early in the chromatogram due to the increased polarity and as such do not interfere.

Castro and Canselier [86] have also examined the problem of separating alkenesulfonates, hydroxyalkanesulfonates, and disulfonates. Using their approach, separation is accomplished on an Altex Ultrasphere 5 μm RP 18 column (250 × 10 mm) using a mobile phases consisting of methanol/water with 1.1 mM nitric acid. A moving wire flame ionization detector was used for quantitation. The method is shown to provide (partial) resolution between disulfonates, 4-hydroxysulfonate, 3-hydroxysulfonate, and alkenesulfonates for both C_{14} and C_{16} mixtures. Separation of a commercial $C_{14}-C_{16}$ sample was demonstrated.

Beranger and Holt [85] have studied a broad range ($C_{14}-C_{30}$) of α-olefinsulfonate raw materials using reversed-phase HPLC with both refractive index and UV (210 nm) detection. The use of dual detection allows for identification of saturated (hydroxy) and unsaturated (alkene) sulfonates due to differences in the relative responses with the two detectors. Using Supelcosil LC-18 (4.6 × 250 mm) and Lichrospher 100 CH-18 super (4 × 250 mm) reversed-phase columns and column temperatures of 55°C, capacity factors of 1–10 are achieved with 85:15 methanol/water with 0.2 M NaNO$_3$ (RI detection) or 0.2 M NaCl (UV detection) for $C_{14}-C_{18}$ raw materials. Optimum separation is achieved by reducing the water content to 10% for $C_{16}-C_{24}$ raw materials and to 5% water for $C_{20}-C_{30}$ raw materials. Use of an efficient column is noted to partially avoid the coelution between C_{16} 3-hydroxyalkanesulfonate and C_{14} 2-alkenesulfonate observed by Johannessen et al. [88]. However, complete resolution between all the different chemical entities was not observed for complex samples.

Weiss [8,9] has used ion chromatography with suppressed conductivity detection to separate C_{14}-2-hydroxy-, C_{16}-3-hydroxy-, and C_{16}-2-hydroxyalkanesulfonates and C_{16} olefinsulfonate (*trans*-3-ene) from C_{16}-alkanesulfonate using the conditions described in Table 1. Use of the method, as well as the influence of different ion-pairing reagents, was also investigated for technical olefinsulfonate. To provide a means for using UV detection to monitor the hydroxyl and alkenyl forms of C_{16} α-olefinsulfonate, Kudoh and Tsuji [87] have performed derivatization with 4-diazomethyl-*N*, *N*-dimethylbenzenesulfonamide. Derivatized components are separated by reversed-phase HPLC and monitored at 240 nm.

3. Electrophoresis

Several of the methods detailed in Sec. II.A [30–32] for the determination of alkyl sulfates by CE have also shown applicability for the analysis of alkanesulfonates. The method of Chen and Pietrzyk [30] outlined in Table 2 is applicable to C_4-C_{12} alkanesulfonates and may be used for the simultaneous separation of alkanesulfonates and alkyl sulfates with minimal peak overlap. Separation is complete within 6 minutes. The conditions may also be used for the separation of sulfosuccinates.

Using the approach of Shamsi and Danielson [31] (Table 2, with the exception that the separation was monitored at 274 nm), surfactant alkanesulfonates may be separated in less than 5 minutes. In contrast, typical run times for HPLC are approximately 12 minutes. Utilizing the conditions described in a subsequent paper [32] (Table 2), improved separation was observed especially for homologs with longer alkyl chain lengths, albeit at the expense of analysis time. Separation between the alkanesulfonates and alkyl sulfates was observed. Minor modification of the buffer to incorporate an aromatic cationic species for indirect detection allowed for the simultaneous separation of sulfonates and quarternary ammonium salts.

Romano et al. [89] have separated C_4–C_{12} alkanesulfonates using 10 mM naphthalenesulfonate with 30% acetonitrile (pH 10.0) as the run buffer and have demonstrated the feasibility of using this approach for the analysis of alkanesulfonates in shampoo base. An advantage of this method is that sample preparation was limited to 200:1 dilution in water. Some interferences were observed for some of the peaks, but several others were used for quantitative analysis.

4. Thin Layer Chromatography

The alkane- and olefinsulfonates are other types of the anionic surfactants that are very amenable to TLC separations as described above for alcohol sulfates and/or ethoxylated alcohol sulfates. Studies described in those sections that specifically include alkanesulfonates include those by Yonese et al. [37], Hohm [42], and Simunic and Soljic [43].

Specific details indicated by Yonese et al. [37] included separations of sodium dodecanesulfonate with either 1:1:1 (volume ratio) methyl acetate/benzene/ethyl alcohol (R_f 0.74) or 20:6:1:6:1 (volume ratio) 2-methyl-4-pentanone/propyl alcohol/0.1 N acetic acid/acetonitrile (R_f 0.23). Silica gel G impregnated with ammonium sulfate or silver nitrate provided the best separations. Visualization by charring unsprayed at 250°C (15 minutes) provided best sensitivity.

Reported R_f values for alkanesulfonates in the Simunic and Soljic [43] study are just below those of sodium n-dodecyl sulfate, and the visualization technique yields gas bubbles of the same form and color.

A thorough investigation of polyamide TLC adsorbents by Takeshita et al. [90] for separating alkanesulfonates and alkylbenzenesulfonates produced high-sensitivity detection with pinacryptol yellow visualization. Separations of alkanesulfonates with carbon chain lengths from 4 to 18 were accomplished with elution solvent mixtures of 0.1–1.0 N aqueous ammonia/pyridine (15:1). Addition of methanol to the developing solvent increased R_f values of the highly retained, long-chain sulfonates. Detection limits as low as 0.05 µg were reported with the pinacryptol yellow under UV light (254 nm).

Alkene- and hydroxyalkanesulfonates are covered in the work by Allen and Martin [91] using ammonium sulfate–impregnated silica gel G TLC plates. The elution solvent mixture was 70:32:6 (v/v/v) chloroform/methanol/0.1 N sulfuric acid. Fuming SO_3 charring [92] on a block plate for 20 minutes at 150°C produced detection limits of less than 5 µg, and densitometry was used for quantitation. R_f values of 0.55, 0.45, and 0.20 were reported for alkenemonosulfonate, hydroxyalkanemonosulfonate, and the associated disulfonates, respectively. Statistical data is illustrated for compositional analyses of several lots of olefinsulfonate raw materials.

B. Linear Alkylbenzenesulfonates, Petroleumsulfonates, and Naphthalenesulfonates

1. Gas Chromatography

The goal of the separation of linear alkylbenzenesulfonates (LAS) is frequently to characterize mixtures with respect to positional isomerism and alkyl chain distribution. Full separation with respect to this criteria mandates separation of LAS into the (typically) 20 or more individual species present. Resolution of this number of structurally similar compounds is difficult to achieve by any method other than GC, and as a result GC continues to be widely used.

The most widely used approach for desulfonation of LAS to the parent hydrocarbon is hydrolysis with phosphoric acid [63,69,93–97]. Our current procedure for the analysis of LAS uses this approach combined with high-resolution capillary GC (M. DeMouth, unpublished results). Specifically, 100 mg of LAS and 2 mL of phosphoric acid are placed in a Carius tube, the tube is sealed, and the contents are heated at 300°C for 15 minutes in a Carius oven. After desulfonation, the tube is cooled and opened and the sample is transferred to a separatory funnel containing 25 mL of pentane and 50 mL of water. Following swirling, to extract the alkylbenzenes into the pentane, the aqueous layer is discarded. The pentane layer is subsequently washed with 50 mL of water. An additional 50 mL of water is then added, the mixture is neutralized with 10% NaOH (phenolphthalein endpoint), and the aqueous layer is discarded. After one additional wash with 50 mL of water, the pentane layer is isolated and used for chromatographic analysis.

Samples prepared as outlined are separated using the conditions of Table 3. A chromatogram of a sample obtained under these conditions is shown in Fig. 9. Carbon chain and phenyl isomer distributions are calculated based on normalized area percent.

Other approaches to the formation of volatile derivatives of LAS have involved the formation of sulfonate esters with diazomethane [47] and formation of sulfonyl chlorides [98,99]. The sulfonyl chloride can be further reacted to form the methylsulfonate derivative by reaction with methanol [100]. Trifluoroethyl derivatives have been formed via conversion to sulfonyl chlorides with phosphorus pentoxide followed by reaction with trifluoroethanol [101].

Direct approaches that do not rely on sample derivatization prior to analysis have also been developed. In the approach by Field et al. [102], LAS was extracted from the sample matrix (in this case, sewage sludge) as tetrabutyl ammonium ion pairs using supercritical carbon dioxide as the extraction fluid and chloroform as the collection solvent. Injection of the chloroform isolate into a GC injection port (fitted with a glass inlet liner with a plug of silanized glass wool and maintained at 300°C) led to the quantitative formation of butyl esters. The products were separated on a HP-5 column and detected using mass spectrometry.

As an extension of this work, Krueger and Field [103] isolated LAS (spiked

TABLE 3 GC Conditions for the Analysis of Desulfonated LAS

Column: 60 m × 0.25 mm i.d. capillary column with a 0.25 µm DB-1 film
Injection: 0.4 µL (cool on-column)
Flow rate: 0.5 mL/min helium
Temperature program: Temperature 1 = 60°C
                       Hold time 1 = 1 min
                       Rate 1 = 25°C/min
                       Temperature 2 = 150°C
                       Hold time 2 = 1 min
                       Rate 2 = 5°C/min
                       Temperature 3 = 200°C
                       Hold time 3 = 30 min
FID at 250°C

Source: M. DeMouth, unpublished.

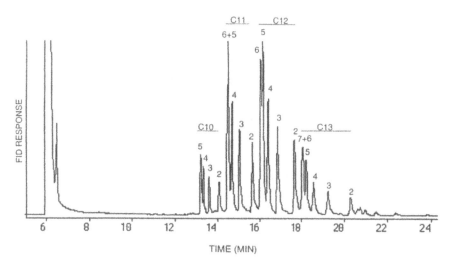

FIG. 9 GC chromatogram of desulfonated LAS. Peaks are labeled by alkyl chain length and position of substitution. For conditions, see Table 3. (Used by permission of Colgate-Palmolive Co.)

with C_8 LAS as surrogate standard) onto C_{18} Empore disks. The disks were subsequently placed in GC autosampler vials containing 5 mM tetrabutylammonium hydrogen sulfate and tetradecanesulfonic acid (internal standard) in chloroform and left for 40 minutes for equilibration. Upon injection into the GC (quartz wool lined glass liner, 300°C), LAS was converted to the butyl esters and separated and detected using a 30 m × 0.25 mm i.d. × 0.25 µm film thickness SE 54

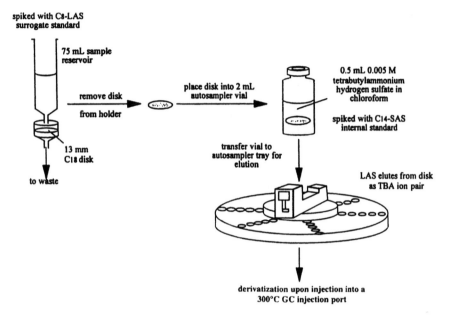

FIG. 10 Schematic diagram indicating steps involved in LAS isolation, in-vial elution, and derivatization. (From Ref. 103.)

capillary column and flame ionization detection. The process is shown schematically in Fig. 10.

The applicability of the Krueger and Field method [103] was shown for the analysis of aqueous LAS samples and samples of sewage effluent and commercial detergents. For several detergents studied, the precision of the analysis was 0.9–5%, and no interferences were observed in any of the chromatograms. Furthermore, more than 400 injections were performed using the same liner, indicating that the method is robust. To assess method accuracy, samples of deionized water and sewage effluent were spiked and the recoveries determined to be 99 ± 2% and 95 ± 5% LAS, respectively.

2. High-Performance Liquid Chromatography

HPLC conditions described in prior sections [13,15,21,22,60] also included applications for ABS separations. Bear's [60] rapid separation and evaporative light-scattering (ELS) detection techniques for sulfated ethoxylated alcohols is used as described for C_{12} and C_{16} alkylbenzenesulfonates. No inorganic salts or unreacted oils were found in the raw materials analyzed through their studies. Petroleum sulfonates as well as linear sodium alkylsulfonates are also analyzed by Bear with the ELS technique. No inorganic salts or unreacted oils were found in the linear alkylsulfonates, but the petroleum sulfonate raw materials contained relatively high levels of inorganic salt and wide ranges of unreacted oils. Quantitative data agreed favorably with two-phase titration methods for all materials studied.

Normal-phase HPLC with ion-pair extraction detection [22] resulted in "class" separations of sulfates from the sulfonates, and C_{10}–C_{14} alkylbenzenesulfonates were well retained after sulfated surfactants. Pan and Pietrzyk's [21] ion-exchange methodology, on the other hand, offers chain length separations of both short- and long-chain LAS derivatives. Conditions necessary for separations were very similar to those used for sulfated alcohols: Dionex Ion Pac AS11 columns with mobile phases of 30–40% acetonitrile in water (containing 1.5 or 10 mM LiOH or 50 mM Mg_2Cl). UV detection (220 or 225 nm) was used for the alkylbenzenesulfonate applications.

Nakae et al. [104] used fluorescence detection for determining trace amounts of alkylbenzenesulfonates. Separations were achieved on 5 μm LiChrosorb RP-18 columns (15 × 0.46 cm) with a mobile phase of 8:2 (v/v) methanol/water containing 0.1 M sodium perchlorate. 225 nm excitation and 295 nm emission wavelengths were used and the separations yielded alkyl chain distribution and partial phenyl isomer compositional information. Relative standard deviation for river water containing 0.097 μg/mL of ABS was 2%.

Similar studies by Marcomini and Giger [105] described the separation of linear 4-alkylbenzenesulfonates from alkylphenol polyethoxylates and nonyl phenol with applications including laundry detergent powders, sewage sludges, sludge-amended soils, and river sediments. Recoveries of 85–100% were obtained by Soxhlet extraction, relative standard deviations did not exceed 6%, and the detection limit was 80 ng for LAS. Excitation wavelengths of either 225 or 230 nm were used with emission wavelengths of 295 nm. The columns were 10 μm LiChrosorb RP-8 (100 × 4 mm) and 3 μm ODS II (250 × 4 mm). Gradient elution with the LiChrosorb column at 1.2 mL/min was as follows: first 6.5 minutes of 5% 2-propanol, 40% water, 55% acetonitrile/water (ACN/water, 45:55) + 0.02 M $NaClO_4$ were applied, followed by a 0.4 convex elution gradient for 10 minutes leading to 5% 2-propanol, 80% ACN/water (45:55) + 0.02 M $NaClO_4$, 15% ACN. An 8-minute linear gradient reestablished the initial conditions. With the ODS II column the following profile was used at 0.8 mL/min: 4 minutes of 3% 2-propanol, 12% water, 85% ACN/water (45:55) + 0.02 M $NaClO_4$ were followed by a 0.3 convex gradient elution of 23 minutes leading to 25% ACN and 75% ACN/water (45:55) + 0.02 M $NaClO_4$. Carbon chain lengths of 8–15 were included, and each of the isomers were studied.

Alkylarysulfonates, when used as additives for lubricating oils, require chain lengths of more than 16 carbon atoms to ensure solubility in the oil. Menez and Perez [106] describe their separation along with those of shorter chain lengths. Two RP-8 (200 × 4.6 mm, 10μm) analytical columns were connected in series, and UV detection was at 225 nm. To accomplish the gradient elution at 1.5 mL/min, the following solvents were prepared: Solvent A—water/acetonitrile (35:65 containing 0.1 M NaClO$_4$); Solvent B—methanol containing 0.1 M NaClO$_4$. The elution solvent profile was a convex curve from 50 to 100% B in 15 minutes followed by isocratic elution with pure solvent B. Various commercial ABS additives were illustrated as well as pure monoalkylbenzene sulfonate salts and a petroleum sulfonate, sodium salt.

Liquid pesticide formulations containing linear alkylbenzenesulfonates were studied using reversed-phase HPLC by Ban et al. [107]. Retention behavior of C$_8$–C$_{14}$ ABS homologs was studied as a function of the methanol concentration (65–85%) and pH of the eluent (2.7–6.5). With a Nucleosil C8 column (150 × 4 mm, 5μm), complete separation of ABS homologs and partial separation of positional isomers was achieved using 70% methanol/25 mM NaH$_2$PO$_4$/40 mM H$_3$PO$_4$, pH 3.5 (Fig. 11). A flow rate of 0.5 mL/min was used, and detection was with UV at 224 nm.

A study of differences in selectivity using either methanol or acetonitrile in reversed-phase separations with ion-suppression and ion-pair techniques by Marco-

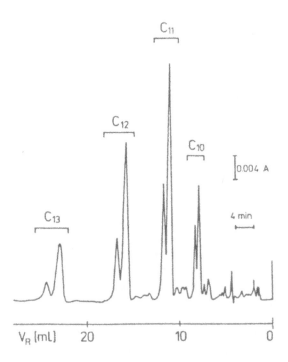

FIG. 11 HPLC separation of ABS homologs and partial separation of positional isomers of homologs. 20 μl of 115 mg/L ABSs. (From Ref. 107.)

mini et al. [108] resulted in complete separations of LAS from nonylphenol polyethoxylates and their carboxylic biotransformation products. Fluorescence detection
(225 nm excitation, 295 nm emission) was used with a LiChrospher RP-18 column
(250 × 4 mm, 5 μm). Mobile phase modifiers employed were trifluoroacetic acid,
tetrabutylammonium dihydrogen phosphate, and sodium perchlorate.

The effects of mobile phase ionic strength on reversed-phase separations of
LAS was illustrated by Chen and Pietrzyk [109]. Hamilton RP-1, 10 μm (150 ×
4.1 mm) or Zorbax ODS, 6 μm (150 or 250 × 4.6 mm) columns were used, and
detection was by UV at 225 nm or with conductivity after suppression. Various
acetonitrile/water mixtures were studied and enhancement in retention due to the
cation follows the order: $Al^{3+} > Ba^{2+} > Mg^{2+} (CH_3)_4N^+ > NH_4^+ > Na^+ > Li^+$. Major
improvements in efficiency and resolution are ionic strength and cation dependent, while selectivity is only modestly increased. The optimum mobile phase
cations are Na^+, Mg^{2+}, and Ba^{2+}, with Na^+ being the most convenient to use. LAS
homologs were resolved conveniently by isocratic elution, while resolution of
positional isomers required a mobile phase gradient. The most favorable gradient
was found to be one where the acetonitrile concentration increases and the NaCl
concentration decreases simultaneously in an acetonitrile/water solvent mixture.

As illustrated in Fig. 12, C_{10}–C_{14} linear alkylbenzenesulfonates are separated
on Zorbax octadecyl and octyl columns (4.6 × 250 mm) after extractions using
supercritical fluids in a study by Ashraf-Khorassani et al. [110]. CO_2 and CO_2
modified with methanol were used to extract and inject directly into the HPLC
system with a mobile phase consisting of 80:20 methanol/0.1 M $NaClO_4$. UV detection at 225 nm was employed. It was reported that under those conditions, the
direct injection of supercritical CO_2 onto the HPLC system does not cause baseline disruption if flow and mobile phase composition are controlled.

Reemtsma's review [111] highlights several HPLC applications in linear
alkylbenzenesulfonate separations including that of Marcomini et al. [112],
which achieves efficient separation of commercial LAS mixtures by both alkyl
chain length and the position of the benzene ring at the alkyl chain. Spherisorb
S3 ODS columns (3 μm, 250 × 4 mm) were used with fluorescence detection at
225 nm excitation and 295 nm emission (see Fig. 13). Altenbach and Giger [113]
report the separation of 25 sulfonic acids and 8 carboxylic acids using ion-pair
reversed-phase HPLC with detection limits in the range of 0.1–1 μg/L on Hypersil ODS (5 μm, 250 × 4 mm) with UV detection at 220 nm.

Petroleum sulfonates were also analyzed by Shamsi and Danielson [17] using
an Alltech silica mixed-mode RP/phenyl-anion exchange column (250 × 4.6 mm)
with indirect photometric, indirect fluorometric, and direct or indirect conductivity detection. The mobile phase consisted of 15:85 water/acetonitrile containing
10 μM 1,3,6-naphthalenetrisulfonate, and a flow rate of 2.2 mL/min was used.

Many naphthalenesulfonates are covered in the Reemtsma review [111], and
petroleumsulfonates including several naphthalene derivatives are studied by

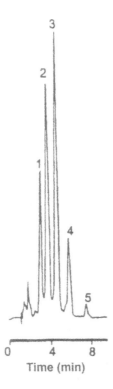

FIG. 12 On-line SFE/HPLC of linear alkylbenzenesulfonates. Extraction conditions: 10% MeOH modified supercritical CO_2, 4000 psi, 40°C. Range 0.1 aufs. 1, C_{10}; 2, C_{11}; 3, C_{12}; 4, C_{13}; 5, C_{14}. (From Ref. 110.)

Bear et al. [114]. Bear used a DuPont Zorbax SAX column (250 × 4.6 mm) with UV detection at 254 nm. A gradient was run with THF/water (50:50), THF/0.1 M KH_2PO_4 (pH 4.5) (50:50) and THF/0.2 M KH_2PO_4 (pH 6.5).

Diisobutyl- and diisopropylnaphthalenesulfonates were determined by Schreuder et al. [115] in pesticide formulations using reversed-phase HPLC. An isocratic separation on a 250 × 4.6 mm Polygosil 60 DCN 10 column used an eluent of methanol/water (4:6, v/v) acidified to pH 2.75 with phosphoric acid and UV detection at 235 nm. An alternate system was described employing gradient separations as follows on a LiChrosorb 100 RP-8 column, also 250 × 4.6 mm:

A = methanol/water (35:65, v/v) acidified to pH 2.75 with phosphoric acid, containing a little acetone (100 μL/L) to compensate for the absorbance difference with eluent **B**.

B = methanol/water (80:20, v/v) acidified to pH 3.0 with phosphoric acid.

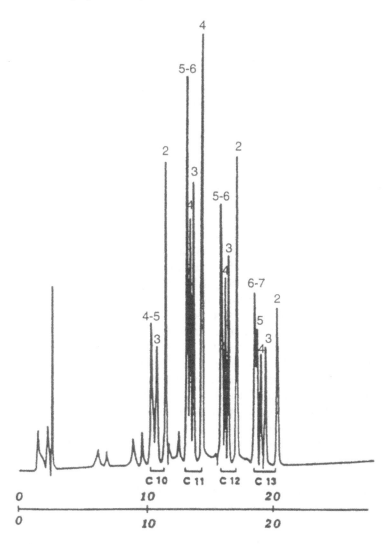

FIG. 13 RPLC chromatogram of a commercial LAS mixture. Numbers refer to the benzene isomers, i.e., the position of the benzene ring at the alkyl chain. (From Ref. 112.)

The gradient profile, at 1 mL/min, was as follows: 0 to 11 minutes, linear increase from 20 to 80% **B**; 11 to 16 minutes, linear increase from 80 to 90% **B**; 16 to 18 minutes, 90% **B**; 18 to 22 minutes, linear decrease from 90 to 20% **B**. The column was thermostated at 40°C, and UV detection was at 290 nm. Using either system produced the separation of mono, di, and tri isomers (mono eluting fastest, and tri exhibiting the most retention), and quantitative determinations agreed very closely between the two. Data are also presented for comparison to two-phase titrimetric determinations.

Twenty-three aromatic sulfonates, mostly naphthalene derivatives, were separated by Zerbinati et al. [116] using ion-pair mechanisms and fluorescence detection after solid-phase extraction from water. Detection limits reported were in the low µg/L range, and the column used was an Alltech Adsorbosphere C8 (5 µm, 250 × 4.6 mm). Effects on capacity factors were studied by varying pH, organic content, ion-pair reagent, and buffer concentration. The optimized mobile phase had a pH of 7.0, an acetonitrile/water ratio of 58:42, and contained 7 g/L of citric acid. Excitation and emission were set alternatively at 250/455 or 240/660 nm.

3. Electrophoresis

Altria et al. [117] have used capillary electrophoresis to quantitate trace levels of dodecylbenzenesulfonate (DBS) residues on pharmaceutical processing equipment after cleaning. As the purpose of the determination was to quantitate total DBS, no efforts were made to separate the DBS components, although it was noted that some peak splitting occurred in the presence of calcium and at reduced operating voltages. A typical electropherogram is shown in Fig. 14.

The chief advantages of this procedure versus HPLC are reduced sample pretreatment (samples were collected on cotton swabs and extracted into 10 mL of water), speed of analysis (2.5 min/sample), and sensitivity (4 µg based on the sample preparation used). The method shows no interference from dissolving water or from cotton wool extracts and was successfully validated with respect to sensitivity, recovery, precision, repeatability, robustness, detector linearity, and stability of standards, reagents, and samples.

Using a buffer consisting of 2:3 acetonitrile/water with 3.0 mM Mg^{2+} and 10 mM sodium acetate (pH 6.0), a 60 cm × 50 µm capillary (effective length = 40 cm), an operating voltage of 30 kV, and UV detection at 220 nm, Chen and Pietrzyk [30] have separated LAS by carbon chain length and obtain partial separation between positional isomers. In the absence of magnesium in the run buffer, C_{10}–C_{14} components are eluted as a single peak.

Desbene et al. [118,119] have investigated the use of CE and micellar electrokinetic chromatography (CE containing micellar SDS as a buffer component) for the analysis of sodium alkylarylsulfonates made by sulfonation of a refinery stream. Separations were achieved for numerous model compounds such as C_2 – C_{12} linear alkylbenzenesulfonates and for mixtures such as the alkylated compounds used as the bases for the preparation of industrial alkylbenzenesulfonates. Using a buffer consisting of 6.25 mM borate/boric acid (adjusted to pH 9) with 50 mM sodium dodecyl sulfate and 30% (v/v) acetonitrile numerous peaks, including dodecylbenzenesulfonate, were observed in commercial anionic surfactants. However, due to the complexity of the mixture and the speci-

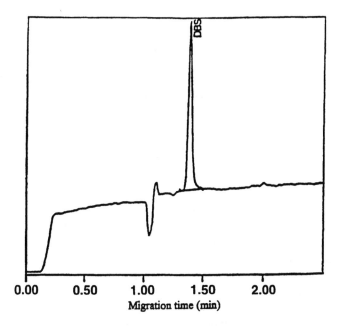

FIG. 14 Electropherogram of 10 μg/mL DBS. Capillary: 37 cm × 75 μm. Buffer: 2.5 mM borax/5.0 mM disodium hydrogen orthophosphate. Operating voltage: 25 KV; injection: 3 s (pressure); detection: UV at 200 nm. (From Ref. 117.)

ficity limitations of UV detection at 214 nm, full attribution of the identity of individual peaks was not possible.

Several naphthalenesulfonate derivatives are covered in the capillary electrophoresis section of the Reemtsma review [111]. Terabe and Isemura [120] separated positional isomers of naphthalenesulfonates and disulfonates using a polymeric cation [poly(diallyldimethylammonium)] in which sulfonate anions are transported through the capillary imbedded in the cationic polymer solely by electrophoresis.

Pfeffer and Yeung [121] separated isomeric naphthalenesulfonate derivaties by ion-pair electrochromatography using a wall-coated, open-tubular capillary. The capillary (50 cm × 10 μm ID, 40 cm separating distance) was coated with PS 264, and the buffer was 10 mM phosphate, pH 7.0, 1.25 mM TBA. A separation voltage of +21 kV produced efficient separations in less than 6 minutes, and the system was used with laser fluorescence for detection.

4. Thin Layer Chromatography

As described in the section above on TLC of sulfated alcohols, Mattisek's [34] separations of anionic surfactants in shampoos, bubble bath formulations, and soaps include alkylbenzenesulfonate. The preferred methodology was the system using silica gel 60, activation with 30–50% relative humidity, then development with 9:1 (v/v) acetone/tetrahydrofuran. The reported R_f value for LAS was 0.09.

The work of Yonese et al. [37] (see above) provided optimized conditions for the separation of dodecylbenzenesulfonate. R_f values ranged from 0.16 to 0.97 on silica gel G impregnated with 10% ammonium sulfate. Best separations for LAS were with either 8:2 (volume ratio) chloroform/methyl alcohol containing 5% 0.1 N sulfuric acid (R_f 0.68) or 20:6:1.6:1 (volume ratio) 2-methyl-1-4-pentanone/propyl alcohol/0.1 N acetic acid/acetonitrile (R_f 0.37). Again, the calibration curve was linear from 4 to 15 μg, and variation coefficients/mean error statistical data indicated that the procedure could be applied to quantitative determinations.

Armstrong and Stine's [38] two-dimensional TLC separation (see above) resulted in an R_f value of 0.09 for DBS after development in the second, normal-phase, direction. DBS was included in the Simunic and Soljic [43] study and, though complete separation from all other anionic surfactants was not achieved, the slight differences in R_f values along with visualizing characteristics were used to identify components.

Alumina E–coated TLC plates, when used as described in the section under alcohol sulfates (B.P. McPherson, unpublished results), produce an R_f value of 0.42 for DBS. Pinacryptol yellow (0.05%) in 3A alcohol spray and viewing under long-wave UV light is used for qualitative analyses of liquid and powdered laundry detergents, raw materials, dishwashing liquids, and shampoos.

The Takeshita et al. [90] investigation using polyamide adsorbents described above separated alkylbenzenesulfonates by carbon chain length. Pinacryptol yellow visualization yielded detection limits as low as 0.1 μg.

C. Lignosulfonates

Due to the high molecular weight of lignosulfonates, chromatographic analysis is limited to size exclusion chromatography (SEC). For example, Lewis and Yean [122] have separated various mixtures with weight average molecular weights of up to 2,000,000 using Waters I250/I125 or I125/I60 size exclusion columns, a mobile phase consisting of a 50 mM citric acid/disodium hydrogen phosphate buffer (pH 3.0) at a flow rate of 1.0 mL/min and UV detection at 280 nm. Majcherczyk and Huttermann [123] have noted that the addition of ion-pairing reagent (quaternary ammonium) to THF overcomes intermolecular association and adsorption of sample to the stationary phase. Such secondary equilibria lead to a reduction in the observed molecular weight and must be eliminated for accurate molecular weight distributions to be determined. In their approach, lignosulfonic acids are extracted from alkali solutions into a variety of organic solvents using methyltrioctylammonium chloride as an ion-pairing reagent.

Subsequent analyses by SEC using 20 mM quaternary ammonium in THF was applied to several technical lignosulfonic acids.

IV. MISCELLANEOUS SULFUR-CONTAINING ANIONIC SURFACTANTS

A. Gas Chromatography

Molever [124] has developed a procedure for the isolation, silylation, and gas chromatographic analysis of sodium lauroyl sarcosinate in personal care products such as liquid and bar soaps. Samples were treated with acidified dimethylformamide and the resulting extracts (filtered if necessary) reacted with BSTFA. Analyses were conducted using a 5 m × 0.53 mm capillary column with a 2.65 μm cross-linked methyl silicone film. The carrier gas was helium at 7 mL/min, and the oven was programmed from 150°C (2-min hold time) to 280°C (5-min hold time) at a rate of 5°C/min. Injector and FID temperatures were 275 and 290°C, respectively. Use of the method to measure lauroyl sarcosinate in precisely prepared liquid soap samples at the 5% level indicated recoveries of 104.1 ± 2.5%. Good precision was also observed for other products evaluated.

B. High-Performance Liquid Chromatography

The use of HPLC for the analyses of ethoxylated alkylphenolsulfonates has been described by several authors [125–127]. Hodgson and Stewart [125] have developed a rapid reversed-phase method for separating the surfactant sulfonate from nonionic precursor and anionic by products to monitor the degree of sulfonation. Escott and Chandler [127] have effected separations of phenylethoxysulfonate from the nonionic counterpart using reversed-phase HPLC with ammonium acetate as ion-pairing reagent and UV and MS detection.

The general HPLC conditions described by Nakamura and Morikawa [6,7] (Sec. II.A), have also shown applicability for the determination of N-acylsarcosinate, N-acyl-L-glucamate, N-acyl-N-methyl taurate, and di-2-ethylhexylsulfosuccinate in the presence of diverse ionogenic mixtures of surfactants. Use of the conditions reported by Konig and Strobel [12] to identify and quantitate various anionic surfactants in toothpaste allows for the determination of cocoyl sarcosinate, sulfated castor oil, α-sulfofatty acid methyl ester, lauryl sulfoacetate, fatty acid monoglyceride sulfonate, fatty acid isethionate, fatty acid taurate, and fatty acid methyl taurate. The method of Steinbrech et al. [10] has allowed for the determination of alkylsulfosuccinates in an industrial sample.

Ye et al. [128] have used HPLC for the separation of alkylated and sulfonated diphenyl oxides (Dowfax 8390). The mixture was separated into 11 components using a Waters Novapak C8 column (3.9 × 150 mm) and an acetonitrile/

methanol/water/NaHCO$_3$/Na$_2$CO$_3$ gradient. The compounds were detected by UV at 239 nm. Separation was also achieved on an IONpac NS1 column (4 × 250 mm) with a mobile phase consisting of 50% acetonitrile and 3 mM tetrabutylammonium hydroxide in water (pH 11.8). In the latter case both UV detection and suppressed conductivity detection is viable.

C. Thin Layer Chromatography

Mattisek [34] includes amidoether sulfate, alkylphenolethersulfonate, fatty acid methyl taurate, and fatty acid isethionate in his separation schemes. The three sets of TLC conditions described in that work under the prior section on alcohol sulfates were used as written to achieve varying degrees of separations among these surfactants, though the one employing the eluent of acetone/tetrahydrofuran was useful only for the amidethersulfate with an R$_f$ value of 0.18 and the others remained at the origin. A fourth scheme [129] was described as being advantageous for either alkylphenolethersulfonate, fatty acid methyl taurate or fatty acid isethionate: silica gel 60 with an eluent of 9.1 (v:v) ethanol/acetic acid. R$_f$ values of 0.90, 0.80 and 0.85, respectively, were obtained.

Armstrong and Stine [38] also reported the separation of sodium dioctylsulfosuccinate using their two-dimensional technique (see above), and an R$_f$ value of 0.28 was achieved.

Fatty acid taurate, fatty acid methyl taurate, alkylamidopolyglycolethersulfate, fatty acid sarcosinate, sodiumlaurylsarcosinate, and di-Na-laurethsulfosuccinate were among the surfactants separated by Hohm [42] using the conditions described above under alcohol sulfates. Although complete separations were not achieved for all test materials, quantitative determinations were illustrated for shampoos and shower/bath preparations.

The TLC methodologies described in the prior sections on alcohol sulfates, alkylsulfonates, and linear alkylbenzenesulfonates by Simunic and Soljic [43] can also be used for sodium diheptyl sulfosuccinate, sodium nonylphenolpentaethoxy sulfate, and disodium nonylphenolnonaethoxy sulfosuccinate. Again, complete separations are not achieved for all materials studied, but qualitative information is obtained via characteristic R$_f$ value and visualization comparisons.

V. CARBOXYLIC ACID SALTS

A. Gas Chromatography

Analyses of the salts of the fatty acids used in soap are typically performed following methylation with a suitable reagent. A comprehensive review of the advantages and disadvantages of various methylation procedures has appeared elsewhere [130] and will not be discussed at length here. The most prevalent means of preparing fatty acid methyl esters is via reaction with boron trifluoride/methanol. This reaction may be performed on soap that has been converted to the fatty acid via reaction with a strong acid, or directly from the salt [131]. Following the method of the Association of Official Analytical Chemists (AOAC) [132], derivatization of fatty acid is performed by mixing fatty acid with boron trifluoride/methanol and refluxing for 2 minutes. Subsequently, heptane is added and the mixture refluxed for an additional minute. After the mixture is cooled, saturated NaCl solution is added to float the heptane solution to the neck of the reaction vessel. An aliquot of the heptane solution is dried with Na_2SO_4 and diluted as necessary for GC analyses.

AOAC guidelines specify the use of a packed GC column for analysis. However, the determination may be performed faster or with greater resolution employing capillary columns. Capillary columns designed specifically for the analysis of fatty acids are available from most column manufacturers, and the choice of which column is most suitable is governed largely by the goals of the separation. A comprehensive review of the attributes of various stationary phases has been published [130]. To illustrate a typical separation, the chromatogram of a soap sample prepared by the AOAC method (with minor modification) is shown in Fig. 15 (M. DeMouth, unpublished results). Chromatographic conditions are given in Table 4. Following separation, AOAC guidelines specify quan-

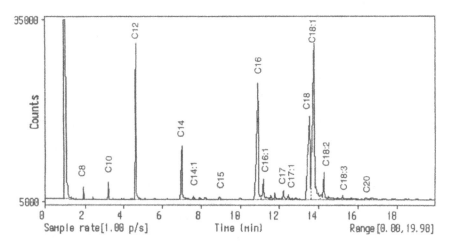

FIG. 15 GC chromatogram of fatty acids used in commercially available soap. Peaks are labeled as alkyl chain length: degree of unsaturation. For conditions see Table 4. (Used by permission of Colgate-Palmolive Co.)

TABLE 4   GC Conditions for the Analysis of Fatty Acid Methyl Esters

---

Column: 25 m × 0.32 mm i.d. capillary column with a 0.2 μm CP-Wax 57CB film
Injection: 0.5 μL (on-column)
Flow rate: 0.5 mL/min helium
Temperature program: Temperature 1 = 100°C
                        Hold time 1 = 2 min
                        Rate 1 = 25°C/min
                        Temperature 2 = 140°C
                        Hold time 2 = 5 min
                        Rate 2 = 10°C/min
                        Temperature 3 = 185°C
                        Hold time 3 = 10 min
FID at 250°C

Source: M. DeMouth, unpublished.

titation of individual fatty acids by simple area normalization or by corrected area normalization if component response factors can be determined.

In addition to the direct conversion of soap to fatty acid methyl esters described by Rotzche [131], soap has been converted directly to trimethylsilyl carboxylic acid derivatives [133]. Utilizing this approach, the soap is simply heated in the presence of Trisil (hexamethyldisilazane and trimethylchlorosilane in pyridine) and an aliquot of the reaction mixture is injected directly into the GC. No significant differences were observed between the direct analysis of soap and the analyses of soaps first converted to their fatty acids. Nimz and Morgan [134] have shown that trimethylsilyl derivatives of fatty acids can also be produced via coinjection of sample and bis(trimethylsilyl)trifluoroacetamide (BSTFA).

B.   High-Performance Liquid Chromatography

The analysis of fatty acids by HPLC has been reviewed previously [135]. As for GC, a vast amount of the HPLC literature is concerned with fatty acids not used in surfactant applications. Applications are noted for both derivatized and underivatized fatty acids. The chief advantage of derivatization is that it allows for the analysis of fatty acids at trace levels or in complex matrices. As an example, Moreno et al. [136] derivatized soaps with either bromo-methyl-methoxy-coumarin or *p*-bromophenacyl bromide to determine soap levels in sludge samples. For relatively simple separations an inherent advantage of HPLC is the ability to perform analyses on underivatized samples. For the separation of hexadecanoic and octadecanoic acid in industrial stearin, for example, Lian et al. [137] demonstrated an extremely simple procedure. Samples are diluted in ethanol and chromatographed isocratically using a Micropak

MCH-5 column (150×4 mm), 90:10 (v/v) acetonitrile/water (pH 2.5 with per-chloric acid) as the mobile phase, and UV detection at 210 nm. Using a flow rate of 1.5 mL/min and a temperature of 50°C, analyses are completed in less than 4 minutes.

Separations of more complex mixtures of underivatized fatty acids have also been reported. Tsuyama et al. [138] reported the separation of fatty acids in the C_{12}–C_{18} range using reversed-phase ion-pair chromatography with con-ductivity detection. $C_{12:0}$, $C_{14:0}$, $C_{18:3}$, $C_{18:2}$, $C_{16:0}$, $C_{18:1}$, and $C_{18:0}$ fatty acids are separated within 20 minutes. George [139] has developed methodology specifically for fatty acids in soap-related fats and oils, focusing on the analy-ses of beef tallow and coconut oil and their soaps. Analyses were conducted using two separate reversed-phase HPLC columns with a mobile phase con-sisting of 45:20:34.5:0.5 (by volume) acetonitrile/THF/water/glacial acetic acid. Refractive index was used for detection. Use of the specified mobile phase allows for soaps (dissolved in methanol) to be acid-hydrolyzed directly on the column. While coelution is observed for some species, coelution dif-fers for the two columns utilized. In combination,, the two systems allow for the determination of $C_{8:0}$, $C_{10:0}$, $C_{12:0}$, $C_{14:0}$, $C_{14:1}$, $C_{15:0}$, $C_{16:0}$, $C_{16:1}$, $C_{18:0}$ $C_{18:1}$ cis, $C_{18:1}$ *trans*, $C_{18:2}$, and $C_{18:3}$ fatty acids. A comparison of the fatty acid dis-tributions obtained using this HPLC method and the AOCS GC method [132] shows excellent agreement for six commercially available soaps. Further-more, the HPLC method has the advantage of differentiating between the $C_{18:1}$ *cis* and *trans* fatty acids.

An additional approach that has received considerable attention is the use of silver ion (or argentation) chromatography. This topic has been reviewed [140] previously and in-depth discussion is beyond the scope of this text. Separation is effected by impregnating a silica column with silver (nitrate) and performing the separation in the normal-phase mode. Retention is due (in part) to interaction be-tween fatty acid double bonds and the silver ion, and accordingly the fatty acids elute in reverse order of the degree of unsaturation.

C. Thin Layer Chromatography

The Hohm [42] work also includes carboxylic acid salts, which can be deter-mined using the conditions described above. Again, complete separations were not achieved for all components tested, but valuable qualitative information is obtainable.

Alvarez et al. [141] used a two-dimensional separation technique for the de-termination of free fatty acids. Fluorescence-labeled esters were made with 4-bromomethyl-6,7 -dimethoxy-coumarin. The plates consisted of a reversed-phase C_{18} layer (2×10 cm) interfaced with a $AgNO_3$-modified silica gel layer (10×10 cm). The reaction mixture was separated on the C_{18} layer first using acetonitrile/acetone/methanol/water (60:20:10:10, v/v/v/v), which separated the fatty acids by carbon chain lengths. Upon drying, the silica gel layer was impregnated with a saturated solution of AgNO$_3$ in methanol. The plates were then developed in the second dimension in chloroform/ethyl acetate/acetonitrile (90:8:2, v/v/v). Development in the AgNO$_3$-modified silica gel allowed separation based on the number of double bonds. Fluorescence scanning was done at 352 nm excitation and a cut-off filter at 400 nm. Fifteen fatty acids were tested and the lower limit of detection was 300 fmol.

Urea- and silver nitrate–impregnated silica gel plates were used by Rezanka [142] two-dimensionally to separate methyl esters of fatty acids. The 10% urea in silica gel layer (0.5 mm) was applied to a 20 × 5 cm portion of the plate, and the free surface (20 × 15 cm) was overlayered with silica gel with 10% AgNO$_3$. The methyl ester derivatives of the fatty acids were separated first on the urea layer using butyl acetate, then dried and turned 90 degrees for development with a mixture of hexane/diethyl ether/methanol (90:10:1). Resultant separations were by both the structure of the chain (branched) and the number of double bonds. Visualization was by spraying with a 0.1% ethanol solution of 2′,7′-dichlorofluorescein. Subsequent preparative isolations were also performed for confirmation of components by GC-MS.

Cocoa soap (sodium laurate) is developed and visualized at an R$_f$ value of 0.88 using the conditions previously described by Simunic and Soljic [43]. Soap elutes to an R$_f$ value of 0.30 with the alumina TLC plates using the conditions previously described under the alcohol sulfate section (B.P. McPherson, unpublished results).

VI. MISCELLANEOUS NON–SULFUR-CONTAINING ANIONIC SURFACTANTS

A. Supercritical Fluid Chromatography

Among the numerous surfactants separated by Sandra and David [47] were methyl esters of ethoxylated decanol phosphoric acid produced via reaction with diazomethane. Supercritical fluid chromatography analyses were conducted using a 20 m × 100 µm i.d. × 0.2 µm SE-54 column at a temperature of 100°C. The pressure was programmed from 10 to 20 MPa at 0.5 MPa/min and then at 0.25 MPa/min to 30 MPa.

B. High-Performance Liquid Chromatography

Derivatization with 4-diazomethyl-N,N-dimethylbenzenesulfonamide [87] followed by reversed-phase HPLC has shown applicability for the separation of monolauryl and dilauryl phosphate.

C. Electrophoresis

Additional applications of CE for the analysis of surfactants include separations of ethoxylated phosphates and phenol ethoxylated phosphate esters [143]. C_8-C_{10} ethoxylated phosphates were separated in a buffer consisting of 5 mM adenosine monophosphate, 100 mM H_3BO_3, 1 mM diethylenetriamine (DETA) (adjusted to a pH of 7.2 with NaOH) and detected (indirectly) using UV detection at 259 nm. Phenol ethoxylated phosphate esters were separated using a buffer consisting of 100 mM H_3BO_3 with 0.5 mM DETA (adjusted to a pH of 7.2 with NaOH) and detected using UV detection at 220 nm. Separations for both mixtures were performed in 75 cm (45–50 cm effective length) × 50 µm capillaries at +30 kV. A key to these separations is the addition of DETA to the run buffer as an electroosmotic flow modifier. While its addition does increase analysis times, significantly improved resolution is observed.

D. Thin Layer Chromatography

Sodium nonylphenolnonaethoxy phosphate is another of the anionic surfactants that give white spots with evolution of gas bubbles when separated and visualized according to the methodology previously described in the work of Simunic and Soljic [43].

REFERENCES

1.   J. D. Knight and R. House, J. Am. Oil Chem. Soc. *36*:195 (1959).
2.   S. Lee and N. A. Puttnam, J. Am. Oil Chem. Soc. *43*:690 (1966).
3.   H. Y. Lew, J. Am. Oil Chem. Soc. *44*:359 (1967).
4.   H. Y. Lew, J. Am. Oil Chem. Soc. *49*:665 (1972).
5.   K. Nakamura, Y. Morikawa, and I. Matsumoto, J. Am. Oil Chem. Soc. *58*:72 (1981).
6.   K. Nakamura and Y. Morikawa, J. Am. Oil Chem. Soc. *59*:64 (1982).
7.   K. Nakamura and Y. Morikawa, J. Am. Oil Chem. Soc. *61*:1130 (1984).
8.   J. Weiss, J. Chromatogr. *353*:303 (1986).
9.   J. Weiss, Tenside Deterg. *23*:5 (1986) (in German).
10. B. Steinbrech, N. Neugebauer, and G. Zulauf, Fresenius Z. Anal. Chem. *324*:154 (1986).
11. J. A. Boiani, Anal. Chem. *59*:2583 (1987).
12. H. Konig and W. Strobel, Fresenius Z. Anal. Chem. *331*:435 (1988).
13. J. B. Li and P. Jandik, J. Chromatogr. *546*:395 (1991).
14. D. Zhou and D. J. Pietrzyk, Anal. Chem. *64*:1003 (1992).
15. C. E. Hoeft and R. L. Zollars, J. Liq. Chromatogr. *17*:2691 (1994).
16. D. J. Pietrzyk, P. G. Rigas and D. Yuan, J. Chromatogr. Sci. *27*:485 (1989).
17. S. A. Shamsi and N. D. Danielson, Chromatographia *40*:237 (1995).
18. S. Maki, J. Wangsa and N. D. Danielson, Anal. Chem. *64*:583 (1992).

19. N. D. Danielson, S. A. Shamsi and S. A. Maki, J. High Resolut. Chromatogr. *15*:343 (1992).
20. S. A. Shamsi and N. D. Danielson, J. Chromatogr. Sci. *33*:505 (1995).
21. N. Pang and D. J. Pietrzyk, J. Chromatogr. *706*:327 (1995).
22. C. P. Terweij-Groen, J. C. Kraak, W. M. A. Niessens, J. F. Lawrence, C. E. Werkhoven-Goewie, U. A. T. Brinkman, and R. W. Frei, Int. J. Environ. Anal. Chem. *9*:45 (1981).
23. F. Smedes, J. C. Kraak, C. E. Werkhoven-Goewie, U. A. T. Brinkman, and R. W. Frei, J. Chromatogr. *247*:123 (1982).
24. M. Kanesato, K. Nakamura, O. Nakata, and Y. Morikawa, J. Am. Oil Chem. Soc. *64*:434 (1987).
25. K. J. Irgolic and J. E. Hobill, Spectrochim. Acta *42B*:269 (1987).
26. J. J. Conboy, J. D. Henion, M. W. Martin, and J. A. Zweigenbaum, Anal. Chem. *62*:800 (1990).
27. D. D. Popenoe, S. J. Morris III, P. S. Horn, and K. T. Norwood, Anal. Chem. *66*:1620 (1994).
28. M. W. F. Nielen, J. Chromatogr. *588*:321 (1991).
29. L. K. Goebel, H. M. McNair, H. T. Rasmussen, and B. P. McPherson, J. Microcol. Sep. *5*:47 (1993).
30. S. Chen and D. J. Pietrzyk, Anal. Chem. *65*:2770 (1993).
31. S. A. Shamsi and N. D. Danielson, Anal. Chem. *66*:3757 (1994).
32. S. A. Shamsi and N. D. Danielson, Anal. Chem. *67*:4210 (1995).
33. J. M. Gibbons and S. H. Hoke, J. High Resolut. Chromatogr. *17*:665 (1994).
34. R. Mattisek, Tenside Deterg. *19*:57 (1982).
35. K. Bey, Fette-Seifen-Anstrichmittel *67*:217 (1965).
36. H. K. Mangold and R. Kammereck, J. Am. Oil Chem. Soc. *39*:201 (1962).
37. C. Yonese, T. Shishido, T. Kaneko and K. Maryuyama, J. Am. Oil Chem. Soc. *59*:112 (1982).
38. D. W. Armstrong and G. Y. Stine, J. Liq. Chromatogr. *6*:23 (1983).
39. G. Zweig and J. Sherma, Handbook of Chromatography, Vol. II, CRC Press, Cleveland, Ohio, 1972.
40. K. Yamaoka, K. Nakajima, H. Moriyama, Y. Saito and T. Sato, J. Pharm Sci. *75*:606 (1986).
41. D. C. Abbott, Analyst *87*:286 (1962).
42. G. Hohm, Seife Oele Fette Wachse *116*:273 (1990).
43. S. Simunic and Z. Soljic, J. Liq. Chromatogr. Rel. Technol. *19*:1139 (1996).
44. L. H. Henrich, J. Planar Chromatogr. Mod. TLC *5*:103 (1992).
45. A. Kruse, N. Buschmann, K. Cammann, J. Planar Chromatogr. *7*:22 (1994).
46. L. Gildenberg and J. R. Trowbridge, J. Am. Oil Chem. Soc. *42*:70 (1965).
47. P. Sandra and F. David, J. High Resolut. Chromatogr. *13*:414 (1990).
48. A. H. Silver and H. T. Kalinosky, J. Am. Oil Chem. Soc. *69*:599 (1992).
49. H. T. Rasmussen, A. M. Pinto, M. W. DeMouth, P. Touretzky, and B. P. McPherson, J. High Resolut. Chromatogr. *17*:593 (1994).
50. S. R. Lipsky and M. L. Duffy, J. High Resolut. Chromatogr. Chromatogr. Com. *9*:376 (1986).

51. S. R. Lipsky and M. L. Duffy, J. High Resolut. Chromatogr. Chromatogr. Com. *9*:725 (1986).

52. J. C. Sternberg, W. S. Gallaway, and D. T. L. Jones, in *Gas Chromatography*, N. Brenner, J. E. Callen and M. D. Weiss (eds.), Academic Press, New York, 1962.

53. D. E. Knowles, L. Nixon, E. Campbell, D. W. Later, and B. E. Richter, Z. Anal. Chem. *330*:225 (1988).

54. T. L. Chester, D. J. Bowling, D. P. Innis, and J. D. Pinkston, Anal. Chem. *62*:1299 (1990).

55. P. R. Geissler, J. Am. Oil Chem. Soc. *66*:685 (1989).

56. S. Brossard, M. Lafosse and M. Dreux, J. Chromatogr. *591*:149 (1992).

57. P. Jandera and J. Urbanek, J. Chromatogr. *689*:255 (1995).

58. P. Jandera, J. Urbanek, B. Prokes, and H. Blazkova-Brunova, J. Chromatogr. *736*:131 (1996).

59. P. Jandera and B. Prokes, Chromatographia *42*:539 (1996).

60. G. R. Bear, J. Chromatogr. *459*:91 (1988).

61. M. Benning, H. Locke, and R. Ianniello, J. Liquid Chromatogr. *12*:757 (1989).

62. A. Stemp, V. A. Boriraj, P. Walling, and P. Neill, J. Am. Oil Chem. Soc. *72*:17 (1995).

63. J. D. Knight and R. House, J. Am. Oil Chem. Soc. *36*:195 (1959).

64. E. A. Setzkorn and A. B. Carel, J. Am. Oil Chem. Soc. *40*:57 (1963).

65. T. Nakagawa, M. Miyajima, and T. Uno, J. Gas Chromatogr. *6*:292 (1968).

66. S. Siggia, L. R. Whitlock, and J. C. Tao, Anal. Chem. *41*:1387 (1969).

67. J. J. Kirkland, Anal. Chem. *32*:1388 (1960).

68. J. B. Himes and I. J. Dowbak, J. Gas Chromatogr. *3*:194 (1965).

69. T. Nagai, S. Hashimoto, I. Yamane, and A. Mori, J. Am. Oil Chem. Soc. *47*:505 (1970).

70. J. S. Parson, J. Chromatogr. Sci. *11*:659 (1973).

71. A. Amer, E. G. Alley, and C. U. Pittman, Jr., J. Chromatogr. *362*:413 (1986).

72. J. S. Parson, J. Gas Chromatogr. *5*:254 (1967).

73. T. A. Taulli, J. Chromatogr. Sci. *7*:671 (1969).

74. A. Heyward, A. Mathias, and E. Williams, Anal. Chem. *42*:1272 (1970).

75. K. M. Baker and G. E. Boyce, J. Chromatogr. *117*:471 (1976).

76. A. El Homsi, B. Gilot, and J. P. Canselier, J. Chromatogr. *151*:413 (1978) (in French).

77. A. M. Nardillo, R. C. Castells, E. L. Arancibia, and M. L. Casella, Chromatographia *25*:618 (1988).

78. K. A. Caldwell and A. L. Tappel, J. Chromatogr. *32*:635 (1968).

79. J. Eagles, Anal. Chem. *43*:1697 (1971).

80. O. Stokke and P. Helland, J. Chromatogr. *146*:132 (1978).

81. L.-K. Ng and M. Hupe, J. Chromatogr. *513*:61 (1990).

82. H. Kataoka, T. Okazaki, and M. Makita, J. Chromatogr. *473*:276 (1989).

83. G. Liebscher, G. Eppert, H. Oberender, H. Berthold and H. G. Hautal, Tenside Surf. *26*:195 (1989) (in German).

84. G. Eppert and G. Liebscher, J. Chromatogr. Sci. *29*:21 (1991).

85. A. Beranger and I. T. Holt, Tenside Surf. *23*:247 (1986).

86. V. Castro and J.-P. Canselier, J. Chromatogr. *325*:43 (1986).
87. M. Kudoh and K. Tsuji, J. Chromatogr. *294*:456 (1984).
88. R. O. Johannessen, W. J. DeWitt, R. S. Smith, and M. E. Tuvell, J. Am. Oil Chem. Soc. *60*:858 (1983).
89. J. Romano, P. Jandik, W. R. Jones, and P. E. Jackson, J. Chromatogr. *546*:411 (1991).
90. R. Takeshita, N. Jinnai, and H. Yoshida, J. Chromatogr. *123*:301 (1976).
91. M. C. Allen and T. T. Martin, J. Am. Oil Chem. Soc. *48*:790 (1971).
92. T. T. Martin and M. C. Allen, J. Am. Oil Chem. Soc. *48*:752 (1971).
93. R. D. Swisher, J. Am. Oil Chem. Soc. *43*:137 (1966).
94. I. Otvos, B. Bartha, Z. Balthazar, and G. Palyi, J. Chromatogr. *94*:330 (1974).
95. V. H. Leidner, R. Gloor, and K. Wuhrmann, Tenside *13*:122 (1976) (in German).
96. I. Zeman, Tenside *19*:6 (1982).
97. Q. W. Osburn, J. Am. Oil Chem. Soc. *63*:257 (1986).
98. J. McEvoy and W. Giger, Environ. Sci. Technol. *20*:376 (1986).
99. Y. Yamini and M. Ashraf-Khorassani, J. High Resolut. Chromatogr. *17*:634 (1994).
100. H. Hon-Nami and T. Hanya, J. Chromatogr. *161*:205 (1978).
101. M. L. Trehy, W. E. Gledhill, and R. G. Orth, Anal. Chem. *62*:2581 (1990).
102. J. A. Field, D. J. Miller, T. M. Field, S. B. Hawthorne, and W. Giger, Anal. Chem. *64*:3161 (1992).
103. C. J. Krueger and J. A. Field, Anal. Chem. *67*:3363 (1995).
104. A. Nakae, K. Tsuji, and M. Yamanaka, Anal. Chem. *52*:2275 (1980).
105. A. Marcomini and W. Giger, Anal. Chem. *59*:1709 (1987).
106. H. R . Menez and C. L. Perez, J. High Resolut. Chromatogr. *12*:562 (1989).
107. T. Ban, E. Papp, and Janos Inczedy, J. Chromatogr. *593*:227 (1992).
108. A. Marcomini, A. DiCorcia, R. Samperi, and S. Capri, J. Chromatogr. *644*:59 (1993).
109. S. Chen and D. J. Pietrzyk, J. Chromatogr. *671*:73 (1994).
110. M. Ashraf-Khorassani, M. Barzegar, and Y. Yamini, J. High Resolut. Chromatogr. *18*:472 (1995).
111. T. Reemtsma, J. Chromatogr. *733*:473 (1996).
112. A. Marcomini, S. Stelluto, and B. Pavoni, Int. J. Environ. Anal Chem. *35*:207 (1989).
113. B. Altenbach and W. Giger, Anal. Chem. *67*:2325 (1995).
114. G. R. Bear, C. W. Lawley, and R. M. Riddle, J. Chromatogr. *302*:65 (1984).
115. R. H. Schreuder, A. Martijn, and C. Van De Kraats, J. Chromatogr. *467*:177 (1989).
116. O. Zerbinati, G. Ostacoli, D. Gastaldi, and V. Zelano, J. Chromatogr. *640*:231 (1993).
117. K. D. Altria, I. Gill, J. S. Howells, C. N. Luscombe, and R. Z. Williams, Chromatographia *40*:527 (1995).
118. P. L. Desbene, C. Rony, B. Desmazieres, and J. C. Jacquir, J. Chromatogr. *608*:375 (1993).
119. P. L. Desbene and C. M. Rony, J. Chromatogr. *689*:107 (1995).

120. S. Terabe and T. Isemura, Anal. Chem. *62*:652 (1990).
121. W. D. Pfeffer and E. S. Yeung, J. Chromatogr. *557*:125 (1991).
122. N. G. Lewis and W. Q. Yean, J. Chromatogr. *331*:419 (1985).
123. A. Majcherczyk and A. Huttermann, J. Chromatogr. *764*:183 (1997).
124. K. Molever, J. Am. Oil Chem. Soc. *70*:101 (1993).
125. P. K. Hodgson and N. J. Stewart, J. Chromatogr. *387*:546 (1987).
126. J. A. Pilc and P. A. Sermon, J. Chromatogr. *398*:375 (1987).
127. R. E. A. Escott and D. W. Chandler, J. Chromatogr. Sci. *27*:134 (1989).
128. M. Ye, R. Walkup, and K. Hill, J. Liq. Chromatogr. Rel. Technol. *19*:1229 (1996).
129. D. Frahne, S. Schmidt, and H. G. Kuhn, Fette-Seifen-Anstrichmittel *79*:32, 122 (1977).
130. N. C. Shanta and G. E. Napolitano, J. Chromatogr. *624*:37 (1992).
131. H. Rotzche, J. Chromatogr. *552*:281 (1991).
132. *Official Methods of Analysis of the Association of Official Analytical Chemists*, 15th ed., Association of Official Analytical Chemists, Washington, DC, 1990.
133. L.-K. Ng and M. Hupe, J. Chromatogr. *637*:104 (1993).
134. E. L. Nimz and S. L. Morgan, J. Chromatogr. Sci. *31*:145 (1993).
135. P. R. Brown, J. M. Beebe, and J. Turcotte, Crit. Rev. Anal. Chem. *21*:193 (1989).
136. A. Moreno, J. Bravo, J. Ferrer, and C. Bengoechea, J. Am. Chem. Soc. *70*:667 (1993).
137. H.-Z. Lian, L. Mao, and J. Miao, J. Liq. Chromatogr. *17*:4081 (1994).
138. Y. Tsuyama, T. Uchida, and T. Goto, J. Chromatogr. *596*:181 (1992).
139. E. D. George, J. Am. Oil Chem. Soc. *71*:789 (1994).
140. B. Nikolova-Damyanova, *Advances in Lipid Methodology* (W. W. Christie, ed.), Oily Press, Ayr, 1992, pp. 181–237.
141. J. G. Alvarez, X. G. Fang, R. L. Grob, and J. C. Touchstone, J. Liquid Chromatogr. *13*:2727 (1990).
142. T. Rezanka, J. Chromatogr. *727*:147 (1995).
143. S. A. Shamsi, R. M. Weathers, and N. D. Danielson, J. Chromatogr. *737*:315 (1996).

8

Analysis of Anionic Fluorinated Surfactants

ERIK KISSA Consultant, Wilmington, Delaware

I. DETERMINATION AND CHARACTERIZATION

Fluorinated surfactants are the supersurfactants [1]. The fluorocarbon hydrophobe can lower the surface tension of water below the lower limit reached by hydrocarbon-type surfactants. In addition to providing extremely high surface activity in water, the fluorinated hydrophobe is oleophobic as well.

Like all surfactants, fluorinated surfactants are either ionic or nonionic. Unlike nonionic surfactants, ionic surfactants can dissociate into ions in an aqueous medium. The hydrophobic part can belong to a negative or positive ion. Some surfactants have negative and positively charged functional groups on the same backbone. The fluorinated surfactants can therefore be classified into four types:

Anionic surfactants—the hydrophobic part is an anion, for example, $R_fCOO^-Na^+$, where R_f is a fluorine-containing hydrophobe.

Cationic surfactants—the hydrophobic part is a fluorine-containing cation, for example, $C_7F_{15}CONH(CH_2)_3N^+(CH_3)_3I^-$.

Amphoteric surfactants—have at least one anionic and one cationic group at their isoelectric point, for example, $C_9F_{19}OC_6H_4SO_2NH(CH_2)_3N^+(CH_2)_3CH_2COO^-$.

Nonionic surfactants—do not dissociate into ions, for example, $C_7F_{15}CH_2CH_2O(CH_2CH_2O)_nH$.

Anionic surfactants are the most important class of fluorinated surfactants. The hydrophobe can be a fully or partially fluorinated alkyl group having a linear or a branched alkyl chain. The hydrophobic part of some fluorinated surfactants may include nonfluorinated alkyl and aryl groups as well. The hydrophobe may be attached via a sulfide, carbonamido, or sulfonamido linkage to the rest of the surfactant molecule.

The hydrophobe of anionic fluorinated surfactants can contain other elements (O, N, Cl, S, and Si) as well, as shown in the following examples:

C, H, F

C_nF_{2n+1} –

$C_nF_{2n+1}CH_2CH_2$ –

C, H, F, O

$C_nF_{2n+1}OCF_2CF_2$ –

$C_nF_{2n+1}OC_6H_4$ –

C, H, F, O, N
$C_nF_{2n+1}CONH(CH_2)_3N<$

C, H, F, O, S
$C_nF_{2n+1}SO_2NH(CH_2)_3N<$

C, F, Cl
$CF_3Cl_2(CF_2CFCl)_{n-1}CF_2-$

C, F, Si
$C_8F_{17}CH_2CH_2Si\,(CH_3)_2-$

Based on the hydrophile structure, anionic fluorinated surfactants can be divided into four main categories:

Carboxylates: $R_fCOO^-\,M^+$
Sulfonates: $R_fSO_3^-\,M^+$
Sulfates: $R_fOSO_3^-\,M^+$
Phosphates: $R_fOP(O)O_2^{--}M^+_2$

where R_f is a fluorine-containing hydrophobe and M^+ an inorganic or organic cation. Some anionic surfactants contain nonionic hydrophilic polyoxyethylene segments, which increase their solubility and compatibility with cationic or amphoteric surfactants.

The extremely high chemical resistance of the carbon-fluorine bond permits the use of fluorinated surfactants where hydrocarbon surfactants do not survive. However, the high resistance to chemical attack complicates the elemental analysis of fluorinated surfactants.

Analytical techniques are employed to determine the purity or concentration of anionic fluorinated surfactants and to characterize a fluorinated surfactant and its solutions. Since most fluorinated surfactants are mixtures of homologs, the term purity has to be redefined for each particular case. In most cases the determination of purity begins with the analysis of intermediates used to synthesize the surfactant. Usually the intermediates can be readily analyzed by chromatography and the homolog distribution determined. Gas chromatography has only a limited value for the analysis of fluorinated surfactants proper since most perfluorinated anionic surfactants are not sufficiently volatile for gas chromatography. Derivatization techniques are needed to convert them to an adequately volatile form.

In general, the concentration of a fluorinated surfactant in solution can be determined by conventional volumetric or spectroscopic methods used for hydrocarbon-type surfactants [1–5]. In addition to the functional groups utilized for the analysis of hydrocarbon-type surfactants, the fluorine content is a unique feature useful for the determination of fluorinated surfactants. For example, if a fluorinated surfactant of a known fluorine content is the only fluorine-containing species in a solution, then the fluorine content of the solution indicates the concentration of the fluorinated surfactant.

The high surface activity and the unusual characteristics of the fluorinated hydrophobe have inspired numerous studies of the solution behavior of fluorinated surfactants. Sophisticated analytical procedures have been developed for physical-chemical investigation of the association of fluorinated surfactants in solution and the behavior of fluorocarbon-hydrocarbon surfactant mixtures.

II. ELEMENTAL ANALYSIS

Elemental analysis is more important for fluorinated surfactants than for hydrocarbon-type surfactants because the fluorine content can indicate the concentration of a fluorinated surfactant in admixture with hydrocarbon-type surfactants or nonfluorinated chemicals. Hence the concentration of a fluorinated surfactant can be determined by elemental analysis without having to resort to complicated separation schemes.

Fluorine in an organic substance can be determined by nondestructive methods or by destruction of the organic matter by combustion or fusion. Nondestructive methods for elemental fluorine analysis, such as neutron activation [6] and X-ray fluorescence, are rapid but require unusual equipment or are not adequately accurate, sensitive, or versatile.

Fluorine in organic compounds is usually determined by converting organic fluorine to an inorganic fluoride. Various combustion methods are routinely used for this purpose. However, the carbon-fluorine bond is exceptionally strong, and extremely vigorous conditions are needed for a quantitative mineralization. Conventional combustion conditions used for the determination of carbon and hydrogen in nonfluorinated organic compounds are not adequate for a quantitative analysis of fluorinated surfactants.

The most vigorous analytical technique for the determination of fluorine in organic fluorochemicals is combustion in an oxyhydrogen flame. The original torch designed by Wickbold [7] used an oxygen/hydrocarbon gas mixture. The Wickbold torch was modified in a Du Pont laboratory by Sweetser [8] (Fig. 1), who replaced the hydrocarbon with hydrogen.

Dobratz, in Du Pont's Jackson Laboratory, raised the temperature of the combustion chamber by cooling it with air instead of water and provided the Sweetser apparatus with a bypass system to allow a continuous operation during introduction of samples and collection of analyte without disrupting the operation of the oxyhydrogen flame (unpublished). A stopcock, installed between the joint J and the stopcock H, allows the flow from the condenser I to bypass the absorption tower D and enter a glass vessel, which is in turn connected to the vacuum line E. The vessel serves as a sump to collect the water produced by burning hydrogen while the absorption tower D is being serviced. The vessel is equipped

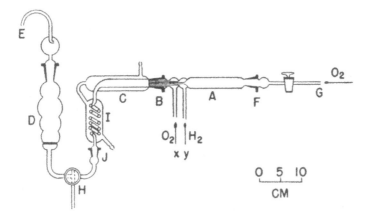

FIG. 1 The oxyhydrogen combustion apparatus-A, pyrolysis tube; B, oxyhydrogen torch; C, flame chamber; D, absorber; E, spray trap; F, removable joint; G, sweep oxygen inlet; H, 3-way stopcock; I, spiral condenser; J, joint; x, oxygen inlet; y, hydrogen inlet. (From Ref. 8. Reproduced with permission of American Chemical Society.)

with side arms connecting it to the vacuum system or venting it to the atmosphere. At the bottom is a stopcock for emptying the vessel.

In order to avoid sudden pressure changes when flows are switched, the system is equipped with a large surge tank. Every day before the torch is operated, the flows through the receiver, through the surge tank, and through the sump have to be balanced by adjusting appropriate needle valves in the vacuum lines.

The flow from the torch is directed into the sump, the joint F is opened, and the sample boat is placed into the pyrolysis chamber A. The joint F connecting the oxygen sweep line is then closed again and the flow is directed to the absorption tower D containing water or an alkaline solution. The sample is pyrolyzed slowly with a gas burner moving on a rack below the chamber. The vapors and pyrolysis products are swept into the oxygen/hydrogen flame burning at 2000°C. Organic fluorine is converted to hydrofluoric acid, which is collected in the absorption tower and determined with a fluoride ion–selective electrode [9].

Kissa (unpublished) made some modifications of the apparatus and developed procedures for the determination of fluorine in metal-containing samples and biological samples, such as blood and tissues. Metal-containing samples, for example, anionic surfactants, give low results due to retention of fluorine as a metal fluoride in the pyrolysis residue. The loss of fluorine can be prevented by an addition of 20% sulfuric acid to the sample prior to combustion. A dilute solution of sodium hydroxide is then placed into the absorption tower instead of water to neutralize carry-over acid.

The combustion apparatus has a memory, meaning that a minute fraction of fluoride is absorbed in the system and may be desorbed during the following run. If the fluorine content of samples is within the same order of magnitude, a steady state is achieved. However, when the analysis of a sample containing minute (ppm) amounts of fluorine follows the analysis of a sample with a high fluorine content, the results for the ppm fluorine analysis are likely to be high. For this reason, two combustion apparatuses are needed: one designated for regular samples and the other for trace analysis in the ppm range (>100 µm F). Alternatively, the quartz and glass parts exposed to the combustion products can be replaced with similar parts dedicated to samples of either a high or low fluorine content.

Two oxyhydrogen torch units are routinely operated in Du Pont's Jackson Laboratory, and during their 30-year history no serious incidents have occurred. Although the mixture of oxygen and hydrogen is potentially hazardous, elaborate safety devices, such as automatic electronically controlled shut-down valves built into the apparatus, assure safe operation.

Other versions of the oxyhydrogen torch have been developed [10]. A combustion apparatus is commercially available from Heraeus [11]. However, a two-stage combustion procedure, including pyrolysis of the sample prior to combustion in the torch, is essential for accurate results. Aspiration of a fluorinated surfactant solution directly into the torch results in low fluorine values.

Combustion in an oxygen Parr bomb [12,13], although less vigorous than combustion in an oxyhydrogen flame, has given quantitative results for perfluorooctanoic acid and its salts. However, the method is not suitable for volatile organic fluorine compounds. Aqueous samples, e.g., blood, have to be dried and pelletized.

Combustion in an oxygen flask [14,15] is convenient, but the results tend to be low. The combustion converts fluorine to the fluoride ion, which is determined titrimetrically [16–23] or with a fluoride ion–selective electrode [9,24–27]. Satisfactory results have been obtained for perfluorodecalin and perfluorotripropylamine [28]. However, the fluorine recovery of samples containing trifluorobenzoic acid and p-fluorobenzoic acid was found to be low (82–87%) by the oxygen flask combustion method [29]. While the method is simple and easy to use, it is unsuitable for samples of a low fluorine content since the sample size is limited to 50 mg. Like the oxygen Parr bomb, the oxygen flask is suitable for dry samples only.

Ashing [29] and fusion with metallic potassium [30] or with sodium biphenyl [31–33] may give low results caused by either losses of fluorine or incomplete mineralization.

The limitations of the various combustion and fusion methods leaves the oxyhydrogen flame as the most powerful technique for quantitative mineralization of a fluorinated surfactant.

III. VOLUMETRIC METHODS AND ION-PAIR SPECTROSCOPY

Volumetric methods used for hydrocarbon-type surfactants [2,3] are applicable to fluorinated surfactants, unless the solubility of the fluorinated surfactant imposes some limitations. Anionic fluorinated surfactants can be titrated potentiometrically with benzethonium chloride (Hyamine 1622), using a surfactant-selective [34] or a nitrate-selective electrode (Fig. 2). Alternatively, turbidity of the titration medium or conductivity can be used as an endpoint indication.

Two-phase titration methods [2,3] are less convenient and are applicable to a fluorinated surfactant only if a suitable water-immiscible solvent can be found for the ion pair formed by the fluorinated surfactant and the titrant.

Anionic hydrocarbon-type surfactants can be determined spectrophotomet-

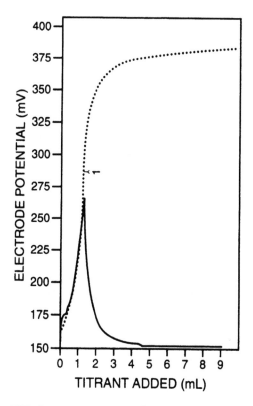

FIG. 2   Titration of Zonyl® FSA with 0.05 N Hyamine 1622. Metrohm Model 670 titrator, Orion Model 9342BN surfactant-selective electrode and Model 90-02 double junction reference electrode. (From Ref. 1.)

rically by forming an ion pair with a cationic dye [2,3]. The ion pair is extracted into a water-immiscible solvent, for example, chloroform or dichloroethane, and determined spectrophotometrically. Jones [35] suggested the use of methylene blue for the determination of anionic surfactants, but his original method gave occasionally erroneous results. Numerous modifications of the method have been published [1,36]. Sharma et al. [37] used a methylene blue method to determine perfluorinated carboxylic acids with a 7- to 10-carbon perfluorinated alkane chain. Mixtures of perfluorinated carboxylic acids and alkyl sulfates were analyzed by extracting the ion pairs from a citrate buffer and from 0.25 N H_2SO_4. In a medium of high pH, both surfactants—sodium dodecyl sulfate and sodium perfluorooctanoate—form a colored ion pair. At low pH (sulfuric acid) the sodium salt of perfluorooctanoic acid is present as the free acid, which is extracted as a colorless species. Hence the difference between the two absorbancies represent the perfluorooctanoic acid concentration.

IV. CHROMATOGRAPHY

Gas chromatography (GC) is the main tool for analyzing fluorinated surfactant intermediates, such as telomer iodides and telomer alcohols. However, most anionic fluorinated surfactants are not sufficiently volatile for gas chromatography and require a conversion to a volatile species [38–40]. Various derivatization reactions are used to increases the volatility of the surfactant [2,39]. Perfluorooctanoic acid has been derivatized with diazomethane and determined as its methyl ester by gas chromatography [40].

High-performance liquid chromatography (HPLC) is an important method for the quantitative analysis of anionic surfactants and is employed also for the analysis of fluorinated surfactants. For example, Chung-Li Lai et al. [41] determined C_6, C_7, and C_8 perfluorocarboxylic acids and their sodium salts by HPLC, using a Varian 4270 Integrator, a Tracor 951 LC pump, a Rheodyne injector valve, and a Wescan conductivity detector. The 150 mm long, 4.6 mm i.d. column was packed with 5 μm Nucleosil C18. The mobile phase was 65/35 (v/v) methanol-water, with stepwise gradient elution, using 25/75 (v/v) methanol-water for elution of sodium chloride and 100% methanol for eluting the surfactant.

Asakawa et al. [42] determined lithium perfluorooctanesulfonate using a Model BIP-1 HPLC chromatograph (Japan Spectroscopic Co., Ltd.) equipped with a differential refractometer and a data module. The chromatographic column, 150 × 4.6 mm i.d.) was packed with Finepak SIL $C_{18}S$ (5 μm, spherically shaped octadecylsilane). Aqueous surfactant solutions were injected into the HPLC column using a sample loop injector (Rheodyne). The surfactants were eluted with acetonitrile-water (5:4, v/v) containing 10 mM tetra-*n*-butylammonium bromide. Mixtures of anionic fluorinated surfactant and hydrocarbon-type surfactant monomers, separated from their micellar solutions by ultrafiltration (see Sec. XIV) and gel chromatography, have been analyzed by HPLC.

Gel chromatography [43–47] is a partition chromatography technique that separates surfactant monomers from their micellar solution according to their molecular size. Gel chromatography employs a column packed with a porous separating medium, usually a gel obtained by reacting dextran and epichloro-hydrin [43] and marketed under the trade name Sephadex. Nakagawa and Ji-zomoto [47] developed a gel chromatography method for binary surfactant solutions.

Asakawa et al. [42] analyzed mixtures of lithium perfluorooctanesulfonate and lithium tetradecyl sulfate by gel chromatography. The solutions contained monomers and micelles of the surfactants. Sephadex G-50, a cross-linked dex-tran gel with a particle size of 20–80 μm, was allowed to swell in distilled water and poured into the column (gel height of 34 or 46 cm). Two methods were used for the separation. In the tail analysis method, sample solutions were applied continuously to the top of the column and eluted with distilled water at a flow rate of 20 mL/h. In the sandwich method, the gel column was equilibrated with a monomer solution, followed by injection of the sample solution and elution, as a sandwich, with the same monomer solution. A conductance flow cell was used to detect surfactants and record conductances against the elution volume. Each eluted fraction (1 mL), collected by an automatic fraction collector, was ana-lyzed by HPLC or isotachophoresis.

Supercritical fluid chromatography (SFC) uses a fluid held above its critical point as the mobile phase [48–50]. The advantages of SFC are high resolution in a short analysis time and the high solvent power and low viscosity of the mobile phase. In some ways, SFC combines the advantages of both HPLC and GC. Like HPLC, SFC does not require the sample to be volatile. Like gas chromatography, SFC can use a flame ionization detector. A variety of com-pounds have been used as the SFC mobile phase, but CO_2 is the most common. Carbon dioxide is nonflammable, thermally stable, innocuous, and available in high purity. Supercritical fluid chromatography, with CO_2 as the mobile phase, can be used to determine the telomer and homolog distribution in fluorinated surfactants.

V. UV AND IR SPECTROSCOPY

The analysis of anionic fluorinated surfactants by ultraviolet (UV) spectroscopy is limited to surfactants featuring functional groups that absorb in the UV region, such as aromatic nuclei and long perfluorocarbon chains. However, UV spec-troscopy can be useful for the determination of UV-absorbing impurities in an-ionic fluorinated surfactants.

Aliphatic carboxylic acids and their anions are known to absorb in the ultraviolet region as a result of $n - \pi^*$ and $\pi - \pi^*$ transitions. Mukerjee at al. [51] found that long-chain perfluorocarboxylates, such as perfluorooctanoate ($\varepsilon = 344$ L mol^{-1} cm^{-1} at 25°C), have higher molar absorptivities in the 205–230 nm region than perfluoroacetate ($\varepsilon = 57$ L mol^{-1} cm^{-1} at 25°C). The absorptivity of perfluoroalkanoates is sufficient for a quantitative determination of the fluorinated surfactant down to the 10^{-5} M concentration range using a 10-cm cell. Mukerjee et al. [51] observed that below the critical micelle concentration (cmc), perfluoroheptanoate and perfluorooctanoate solutions obeyed Beer-Lambert's law within 1%. A somewhat better linear relationship was obtained by relating absorbance data to the fluorinated surfactant concentration by:

$$A = a_1 + b_1 c \tag{1}$$

where A is absorbance, a_1 and b_1 are constants, and c is the fluorinated surfactant concentration. At the cmc the absorptivity increases markedly. The increase (Fig. 3) is large enough to permit the determination of the cmc from UV absorption data.

Anionic fluorinated surfactants of the structure $R_fCH_2CH_2SCH_2CH_2COOLi$ do not absorb in UV, although they have a perfluoroalkyl chain and a carboxylate function.

Infrared (IR) spectroscopy [52] is used mainly for identification and charac-

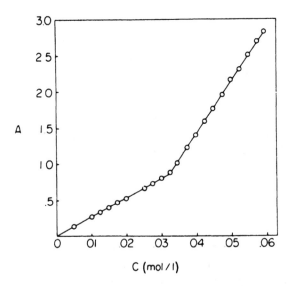

FIG. 3 Plot of absorbance (A) at 230 nm versus perfluoroheptanoic acid concentration. (From Ref. 51. Reproduced with permission of American Chemical Society.)

terization of fluorinated surfactants. A beam of infrared radiation is passed through the sample and focused at a monochromator, which disperses the radiation into a spectrum. An infrared spectrum is recorded by plotting the percent transmission of the sample as a function of frequency or wavelength. IR spectra can be used to identify functional groups of the sample or to identify a compound by comparing its spectra to reference spectra of a known pure compound. Conventional dispersive IR spectroscopy is not very sensitive, the detection limit of a component in a mixture is above 0.2–1.0%. Quantitative analysis of surfactants by IR spectroscopy was of minor importance until Fourier transform spectroscopy was developed [52]. In Fourier transform spectroscopy the entire frequency range of interest is transmitted through an interferometer. The output signal is recorded as a function of interference, and the resulting interferogram is converted to a spectrum using a Fourier transform and a computer. Fourier transform infrared (FTIR) spectroscopy increases the accuracy of absorption data by accumulating repetitive spectra and combining digitalized data electronically. The precision of IR spectroscopy is limited mainly by the signal-to-noise ratio. Since noise is random, cumulative collection of absorption data by Fourier transform spectroscopy increases the precision of quantitative IR spectroscopy.

Fluorinated surfactants exhibit absorption bands arising from CF stretching and CF_2 vibration modes.

Raman spectroscopy analyzes frequency changes in scattered monochromatic radiation. Light passing through a material medium is transmitted, absorbed, and/or scattered. When scattering involves only a direction change, the scattered light has the same frequency as the unscattered light. However, when light interacts with matter and various transitions are involved, the scattered light will have gained or lost energy. The resulting change in frequency is characteristic of the material studied. Since the intensity of Raman scattering is low, Raman spectroscopy was almost dormant until the development of the laser provided a high-density monochromatic light source.

Raman spectroscopy has been very useful for conformation studies of hydrocarbon-type surfactants in aqueous systems but has only a limited value for fluorinated surfactants. Unlike the strong IR absorption band arising from the CF stretching mode, the intensity of Raman bands is low for these vibrations. In contrast to the characteristic CH stretching mode, the CF and CF_2 modes are in a region where other molecular modes occur and complicate absorption patterns. Recent developments, such as Fourier Raman spectrometry and the charge-coupled device (CCD) detector, have enhanced the usefulness of Raman spectroscopy for fluorinated surfactants.

Amorim de Costa and Santos [54] have shown Raman spectroscopy to be useful for structural and conformation analysis of anionic fluorinated surfactants (Fig. 4).

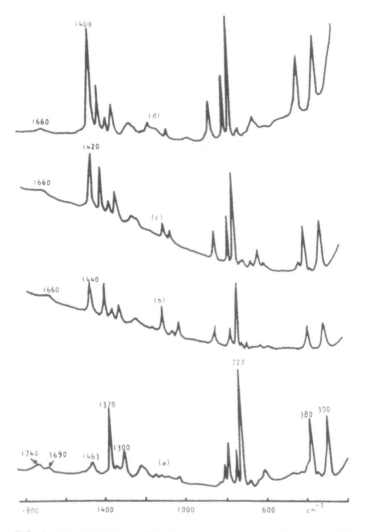

FIG. 4 The 200-1800 cm⁻¹ Raman spectra of perfluorodecanoic acid (a) and its lithium (b), sodium (c), and ammonium (d) salts at 22°C. (From Ref. 54. Reproduced with permission of Academic Press, Inc.)

Ito et al. [55] recorded Raman spectra for fluorocarbon-hydrocarbon hybrid anionic surfactants, using a spectrometer fitted with a multireflection cell, a fiber-optic light-collecting device, and an intensified photodiode-array detector [56]. The data were processed by a computer. Raman spectra were obtained in the concentration range well above the cmc, because the Raman intensity was too weak to be measured below the cmc.

VI. MASS SPECTROMETRY

Mass spectrometry (ms) [57–65] involves four steps: (1) isolation of the component of interest, (2) ionization, (3) separation of the ions in a combination of electric and magnetic fields according to their mass/charge ratio, and (4) detection. The molecular ions and ionic fragments are detected by an electrometer, and their relative abundances are recorded in the mass spectra. The sensitivity of detection can be increased with an electron multiplier.

The first step in mass spectrometry isolates a component of the sample by (1) vaporization using a direct insertion probe to heat the sample to about 200–300°C, (2) flash desorption at a very rapid heating rate to minimize thermal degradation, or (3) chromatography. Mass spectrometry is made more powerful by adding a chromatographic "front end" to separate the components of the sample before each enters the mass spectrometer. Most fluorinated surfactants, even when derivatized, are not sufficiently volatile to be analyzed by GC-MS, the most commonly used "hyphenated mass spectrometer." HPLC is more useful for fluorinated surfactants. Several techniques have been developed for interfacing liquid chromatography (LC) with a mass spectrometer [65]. Over a dozen LC-MS interfaces are commercially available, including a transport interface using a belt to transport the eluent through a desolvation chamber to the ionization source, direct liquid introduction into the ion source, particle beam, thermospray, electrospray, and others [65].

The thermospray technique [66–69] uses a heated vaporizer from which the HPLC eluent containing the dissolved electrolyte is sprayed as a jet into a heated chamber. The ions and molecules are pumped into the mass spectrometer via a sampling orifice positioned normal to the axis of the vaporizer probe. Electron impact or collision-activated ionization, although optional, provides structural information.

Schröder [45,70,71] analyzed fluorinated surfactants in water and wastewater using HPLC coupled by a thermospray interface to a tandem mass spectrometer (MS/MS). Alternatively, the chromatographic column was bypassed and the analyte was injected into the mass spectrometer (FIA, flow injection analysis).

Supercritical fluid chromatography using CO_2 as the mobile phase eliminates the problems associated with the evaporation of a liquid eluent and is therefore more compatible than liquid chromatography with MS.

The second step in mass spectrometry, ionization of the sample, is accomplished by one of several techniques, some of which include sample fractionation as well. Usually the sample is bombarded with a beam of electrons or energetic particles. Electron impact ionization employs electrons from a heated filament to ionize a gas phase sample. The energy of the commonly used 70 eV electrons is in excess of the energy required for removing an electron from a molecule to produce a molecular ion. Hence the electron impact ionization causes fragmentation of the sample and provides a fragmentation pattern that gives useful structural information. In field ionization a volatile sample passes through a strong electric field (10^7–10^8 V/cm), which generates molecular ions with little fragmentation. Both ionization techniques, electron impact ionization and field ionization, require a vaporizable sample, limiting their application to volatile surfactants or degradation products of nonvolatile surfactants.

Modern "soft" ionization techniques have overcome the sample volatility requirement by combining the first two steps in mass spectrometry, sampling and ionization. The "soft" ionization techniques used for the analysis of surfactants include fast atom bombardment (FAB), field desorption (FD), desorption chemical ionization (DCI, also called direct chemical ionization), secondary-ion mass spectrometry (SIMS), and laser desorption methods.

FAB directs a beam of energetic inert gas atoms onto a sample in a viscous liquid matrix, usually in glycerol or triethanolamine. A flow of the sample in the solution replenishes the sample on the surface and limits degradation by the particle beam. The surfactant suppresses ionization of glycerol and, by providing a cleaner background, increases the sensitivity of the FAB method [72,73]. The FAB techniques usually provide quasimolecular ions with only a few fragment ions. Electrolytes added to the sample solution facilitate the formation of quasi-molecular ions, for example, adduct ions with alkali metal cations.

Heller et al. [74] observed large cluster ions of perfluoroalkanesulfonates desorbed under FAB conditions. Cesium perfluorohexanesulfonate formed clusters containing as many as 29 anions and 30 cations. The abundance of these high mass ions produced by perfluoroalkanesulfonates was much higher than the mass ions formed by CsI. For example, the abundance of the cluster $Cs(C_6F_{13}SO_3Cs)^+_{19}$ at m/z 10240.4 was 8 times greater than the abundance of the $Cs_{40}I_{39}^+$ at 10265.5. Discontinuities in surface tension of cesium perfluorohexanesulfonate and in the slope of cluster abundance occurred at the same surfactant concentration range in tetraglyme solutions, suggesting that the cluster formation and aggregation in solution are related.

Lyon et al. [75] characterized fluoroalkanesulfonates by FAB ionization combined with tandem MS/MS spectrometry. The samples were dissolved in glycerol or triethylamine, placed on the copper target of the FAB probe and bombarded with 8 keV xenon atoms. The ions formed were accelerated into the analyzer of the mass spectrometer. Normal spectra were recorded by scanning the first mass spectrometer, MS-I, and leaving the second mass spectrometer, MS-II, fixed to pass all ions. Tandem mass spectrometry was used to enhance the FAB technique by collision-activated dissociation (CAD). An appropriate ion selected with MS-I was subjected to collisions with helium atoms in the collision cell and the CAD spectra recorded by the MS-II unit. Examples of the FAB and CAD spectra are shown in Figs. 5, 6, and 7.

The fragmentation of perfluoroalkanesulfonates involved the cleavage of the

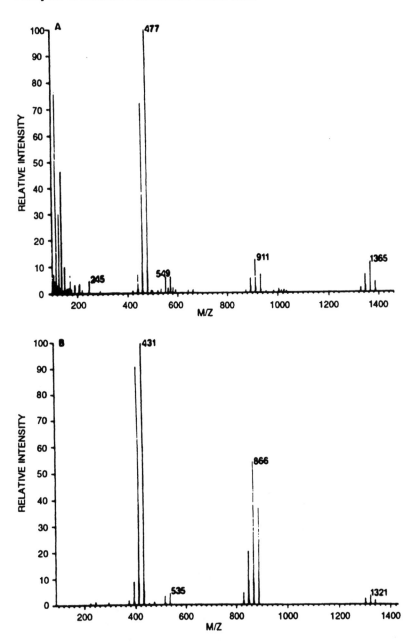

FIG. 5   FAB spectrum from Na 2-hydroperfluoroheptanesulfonate: (A) positive ions, (B) negative ions. (From Ref. 75. Reproduced with permission of American Chemical Society.)

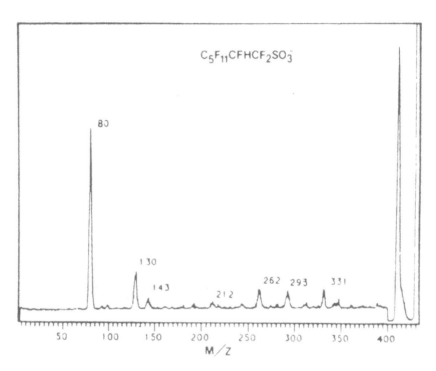

FIG. 6 CAD spectrum of negative ions from m/z 431, the 2-hydroperfluoroheptanesulfonate anion. (From Ref. 75. Reproduced with permission of American Chemical Society.)

C-C bond with the loss of a $C_nF_{2n}+1\cdot$, followed by the detachment of tetrafluoroethylene:

$$CF_3(CF_2)_n - CF_2CF_2CF_2(CF_2)_mSO_3^- \rightarrow CF_3(CF_2)_n\cdot + \cdot CF_2CF_2CF_2(CF_2)_mSO_3^- \rightarrow$$
$$\rightarrow CF_2 = CF_2 + \cdot CF_2(CF_2)_mSO_3^-$$

The fragmentation reaction sequence is analogous to the thermal decomposition mechanism of poly(tetrafluoroethylene).

Substitution of hydrogen for one of the terminal fluorine atoms changes the fragmentation mechanism. Formation of HF becomes then the main reaction in the fragmentation process.

An amphoteric fluorinated surfactant, Du Pont's Zonyl®FSB, has been used as a calibration standard for high-resolution FAB-MS measurements [76].

In field desorption the sample is deposited directly onto carbon dendrites serving on the anode as activated emitters. Salts of perfluoroalkanesulfonaic acids are desorbed as high mass clusters under FD conditions [74,77–79].

Desorption chemical ionization requires the sample to be placed onto a direct-

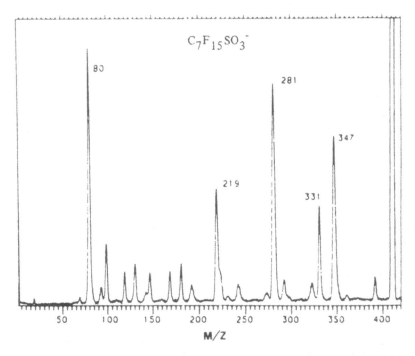

$$C_7F_{15}SO_3^-$$

FIG. 7 CAD spectrum of negative ions from m/z 411, the perfluoroheptanesulfonate anion. (From Ref. 75. Reproduced with permission of American Chemical Society.)

insertion probe located within the chemical ionization plasma [80–82]. Cationic surfactants produce molecular ions and decomposition ions useful for quantitative analysis [73,83], but the DCI technique is less informative for anionic surfactants.

VII. NUCLEAR MAGNETIC RESONANCE SPECTROSCOPY

Nuclear magnetic resonance (NMR) is a very powerful tool for investigating surfactant systems. The theory of NMR spectroscopy has been described in several books [84–93] and will not be discussed here in detail. The application of NMR to surfactant systems have been reviewed by Lindman et al. [94].

NMR spectroscopy is based on the allowed orientation (Zeeman energy levels) of nuclei with nonzero angular momentum when the sample is placed into a magnetic field. The nuclei can be realigned by varying the external magnetic field or by radio-frequency irradiation. When the applied energy matches the energy required for the transition between Zeeman levels, resonance results.

All nuclei do not have the same resonance frequency because their chemical environment can change the applied magnetic field. As a result of differences in shielding, nuclei in functional groups have characteristic resonance frequencies. The difference in the resonance frequencies of two chemically and/or magnetically unequal nuclei indicates the chemical shift, expressed in ppm. To calculate the chemical shift, the difference between the resonance frequency of the sample peak and the resonance frequency of the reference peak is divided by the resonance frequency of the reference peak or by the "observed frequency" given by the instrument manufacturer [92]. The chemical shifts observed by NMR depend on the concentration of the species and the solvent [95]. The solvent effect has been used to investigate the environment of atoms within the fluorinated surfactant micelle [96–99].

NMR spectroscopy yields structural information on surfactants and their micelles, values of the free energy of micellization, ΔG_m°, and the corresponding enthalpy and entropy changes, ΔH_m° and ΔS_m°. For the analyses of fluorinated surfactants, ^{2}H-, ^{2}H-, ^{13}C-, ^{19}F-, ^{14}N-, and ^{133}Cs-NMR spectroscopy has been employed.

^{1}H-NMR spectroscopy can provide information on the environment of the fluorinated surfactant in the micelle. The high sensitivity of the ^{1}H nucleus is a definite advantage. Monduzzi et al. [100] utilized the Fourier transform pulsed gradient spin-echo (FTPGSE) ^{1}H-NMR technique to determine the self-diffusion coefficients of water in water-in-oil (W/O) microemulsions containing perfluoropolyether (PFPE) oils and an anionic surfactant with a PFPE hydrophobe. The self-diffusion data provided quantitative information on the amount of water in the composition range where continuous water coexists with water in droplets. Yoshino et al. (101) synthesized anionic hybrid surfactants with a fluorocarbon hydrophobe, a hydrocarbon hydrophobe, and a -SO$_3$Na hydrophile. Pulsed Fourier transform 300 MHz ^{1}H-NMR spectra of the anionic surfactants were recorded in DCCl$_3$ or trifluoroacetic acid with trimethylsilane (TMS) as the internal reference.

^{13}C-NMR spectroscopy, because of its high resolution and wide chemical shift range [102–105], can give qualitative information on molecular conformation of fluorinated surfactants in solution and quantitative information on cmc values. However, the ^{13}C signal is relatively weak, for two reasons. The abundance of the ^{13}C isotope in carbon-containing substances is about 1.1% of carbon atoms. Furthermore, the ^{13}C has a lower magnetogyric ratio (lower magnetic strength) than ^{1}H, which reduces the sensitivity further to a total factor of about 1:5800. Several techniques have been developed to overcome this loss of sensitivity. Instead of sweeping the resonance frequencies successively, the Fourier transform method uses a radio-frequency pulse to excite all resonance frequencies at once, and the signal is enhanced by repeated pulses and signal averaging.

For NMR studies of fluorinated surfactants the most useful nucleus is ^{19}F, in addition to ^{13}C and ^{1}H nuclei. Changes in the ^{19}F chemical shift at cmc are larger

than changes in the proton chemical shifts and therefore provide more information on fluorinated surfactants and their micellar structures. ^{19}F-NMR spectra
have been recorded for structural characterization of perfluorononanoic acid
[106] and perfluoropolyether carboxylic acids [107]. Micelle formation in solutions of anionic fluorinated surfactants has been studied by measuring the ^{19}F
chemical shift [96–99,108–111] (Fig. 8). Muller and coworkers studied the effect of the environment on fluorine atoms in an anionic surfactant micelle by
^{19}F-NMR [96–99]. Carlfors and Stilbs [112] used the Fourier transform NMR
pulsed-gradient spin-echo (FT-PGSE) method [113–115] for the determination
of multicomponent self-diffusion coefficients in micellar solutions of sodium
perfluorooctanoate and sodium perfluorooctanoate/sodium decanoate. Partition
coefficients were calculated from the self-diffusion data for a homologous series
of *n*-alkanols, benzene, and benzyl alcohol.

Palepu et al. [116] measured the ^{19}F chemical shift changes for 1:1 mixtures
of sodium perfluorooctanoate with α- and β-cyclodextrins. In α-cyclodextrin
mixtures the shifts for terminal fluorine atoms changed more than those for fluo-

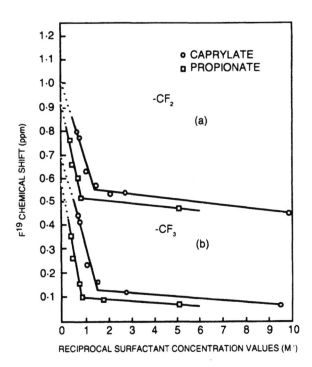

FIG. 8   Plot of ^{19}F chemical shift against the inverse concentration of sodium perfluoro-
caprylate and sodium perfluoropropionate: (a) CF_2; (b) CF_3. (From Ref. 108. Reproduced
with permission of American Chemical Society.)

rine atoms in the middle of the chain. By contrast, in β-cyclodextrin mixtures, the fluorine atoms in the middle of the chain were affected more than the terminal ones. Guo et al. [117] employed ^{19}F-NMR to investigate the association of α-, β-, and γ-cyclodextrins with sodium perfluorobutanoate, sodium perfluoroheptanoate, sodium perfluorooctanoate, and sodium perfluorononanoate. Trifluoroacetic acid was used as the external reference, and the difference between the ^{19}F chemical shift for the mixed system and that for the solution containing only the fluorinated surfactant was measured. The results of this systematic study showed a weak association between α-cyclodextrin and the fluorinated surfactants. Fluorinated surfactants with a short chain formed a 1:1 complex with β-cyclodextrin, whereas those having a longer chain formed a 2:1 complex, especially at higher β-cyclodextrin concentrations. γ-Cyclodextrin formed a 1:1 complex with the fluorinated surfactants. The association constants for the 1:1 complexes were calculated from the ^{19}F chemical shifts measured for various cyclodextrin concentrations. The results were explained in terms of the cavity size of the host cyclodextrins. The cavity of α-cyclodextrin is apparently too small to accommodate fluorinated surfactants and form an inclusion complex.

Boden et al. [118] used ^{133}Cs-NMR to map the phase diagram of the cesium pentadecafluorooctanoate-water system, to characterize phase transitions, and to probe the micelle size and orientational ordering. ^{133}Cs-NMR spectra were measured at 35.44 MHz, using a pulse width of 9 μs (45° pulse). Monduzzi et al. [119] employed ^{2}H- and ^{14}N-NMR techniques to identify lyotropic crystalline phases of ammonium salts of perfluoropolyether carboxylic acids.

The dynamic parameters of fluorinated surfactant solutions have been studied by NMR relaxation methods. The theory of NMR relaxation has been discussed in detail by Henriksson and Ödberg [120] and reviewed by Lindman et al. [94]. Spin-lattice relaxation transfers energy from the higher energy level to the lattice as thermal energy (The term "lattice" is used here to denote molecules other than the fluorinated surfactant in the sample). Since the resulting temperature change is too small to be detected, the relaxation time of recovery of the absorption signal from saturation is measured. Fluorine relaxation times can be measured both in water and in deuterium oxide, and the different magnetic properties of protons and deuterons can provide information about the environment of the fluorocarbon segments.

Henriksson and Ödberg [120] determined ^{19}F spin-lattice relaxation time for heptafluorobutyric acid by the π/2, π/2 pulse method and concluded that fluorocarbon chains in the heptafluorobutyric acid micelles are to some extent exposed to water. Ulmius and Lindman measured ^{19}F spin-lattice relaxation time for various perfluorinated or partially fluorinated carboxylic acids and concluded that the fluorocarbon chains come in contact with water only at the micellar surface [121]. Since the deuterium relaxation time is affected by the state of water (bound or free), Burger-Guerisi et al. [122] studied phase transitions in fluorinated microemulsions by measuring ^{2}H-NMR relaxation times. Tiddy [123] measured ^{19}F spin-lattice and spin-spin relaxation times for the lamellar phase of the ammonium perfluorooctanoate-water system. The ^{19}F relaxation rates were found to be qualitatively similar to the relaxation rates of protons in analogous hydrocarbon surfactant systems. The spin-lattice relaxation times indicated that the CF_2 groups at 298°K rotate about the long axis of the hydrophobic chain more slowly than do the CH_2 groups in hydrocarbon systems. The activation energy of rotation is similar to or smaller than the activation energy for analogous hydrocarbon systems.

VIII. CHEMICAL RELAXATION METHODS

Chemical relaxation methods (CRM) rely upon establishing a thermodynamic equilibrium between a mixture of reactants and reaction products and then perturbing this equilibrium by generating a rapid but very small change in one of the parameters affecting that equilibrium, such as pressure or temperature [124]. As a result of the perturbance, the system shifts to another equilibrium position commensurate to the change in a parameter. This shift of equilibrium is characterized by one or several time constants, the relaxation times, which are related to the rate constants of the chemical reactions studied.

CRM has been very useful in studies of micellization kinetics, based on the theory of Aniansson and Wall [125–127], modified by Kahlweit and coworkers [128–132]. Chemical relaxation techniques have been described in several articles and books [133–138] and reviewed by Lang and Zana [124].

Dilute micellar solutions of surfactants are characterized by two well-separated relaxation times [139–142]. The theory of Aniansson and Wall [125,143] assumes a stepwise aggregation of surfactant monomers to form micelles [144]. The fast relaxation time is attributed to the exchange of monomeric surfactants between the micelles and the intermicellar solution. The slow relaxation time is attributed to micelle formation breakdown. The theory and its modifications by Kahlweit [128–132] has been the basis for most interpretations of chemical relaxation times and has provided valuable information on kinetics of micellization.

The chemical relaxation methods usually employed are the temperature-jump, pressure-jump, shock-tube, ultrasonic absorption, and stopped flow methods.

The temperature-jump method utilizes rapid heating techniques, such as heating by a microwave pulse, discharge of a charged coaxial cable in the solution, discharge of a capacitor, or heating by a pulse of laser light. The temperature-jump relaxation technique [133] has been used by Hoffmann and Ulbricht [146] with optical detection, utilizing a pH indicator (Thymol Blue) to observe relaxation processes of a 1:1 mixture of perfluorooctanoic acid and its sodium salt.

The pressure-jump method utilizes an autoclave with a thin metal diaphragm, which bursts and allows the pressure of the autoclave to drop very rapidly to atmospheric pressure. A pressure-jump apparatus with conductivity detection and twin cell arrangement is shown in Fig. 9. One cell contains the sample investigated, and the other cell contains an electrolyte of similar conductivity but no relaxation.

The shock tube technique is somewhat similar. The bursting of the diaphragm generates a pressure drop, which propagates through a tube half filled with water or ethanol. Reflections of the pressure jump at the bottom and the top of the tube cause an addition and subtraction of the incident and reflected pressure waves. As a result, the equilibrium is shifted by a rectangular change in pressure.

Pressure-jump and a shock-wave methods with conductivity detection have been used by Hoffmann and coworkers in their studies of micelles formed by perfluorinated surfactants [145–151], including salts of perfluorooctanesulfonic acid and perfluorooctanic acid.

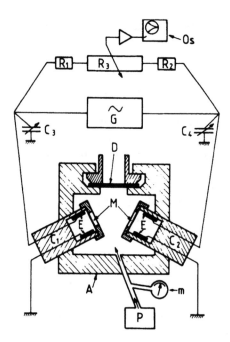

FIG. 9 Pressure-jump apparatus: A, autoclave; C_1, C_2, conductivity cells; E, electrodes; M, elastic membrane; D, metal diaphragm; P, pressure pump; m, manometer; G, 40 kHz generator for the conductivity bridge; C_3, C_4, tunable capacitors; R_1 and R_2, helipot resistors; R_3, potentiometer; Os, oscilloscope. (From Ref. 124.)

In the ultrasonic absorption method the equilibrium is shifted periodically by harmonic changes of pressure and temperature caused by the propagation of ultrasonic waves in fluids. Ultrasonic relaxation techniques have been used to study the kinetics of micelle formation in surfactant solutions [152–163], including those of alkali metal salts of perfluorooctanoic acid.

The periodic fluctuations in temperature and pressure caused by the acoustic wave are several orders of magnitude less than the temperature or pressure perturbations of jump techniques. Rassing et al. [161] suggested that the ultrasonic and jump methods measure different modes of micelle formation whose relaxation times differ by several orders of magnitude. Ultrasonic absorption techniques [162] have also been used to measure relaxation spectra of cesium perfluorooctanoate in water and in deuterium oxide [163].

The stopped flow method involves a rapid mixing of two solutions in less than a millisecond. Since the mixture observed is not in equilibrium, the stopped flow method is not truly a chemical relaxation method. The stopped flow method is useful, nevertheless, for observing perturbations by composition jumps.

IX. SMALL-ANGLE SCATTERING METHODS

Small-angle scattering allows measurement of distances in the range of 0.5–50 nm. Small-angle scattering methods differ in principle from imaging methods, such as microscopy, which collect and focus radiation scattered by the objects being studied and reconstruct their image. In contrast, small-angle scattering methods produce an interference pattern of the radiation scattered by the objects under study. The interference pattern can be converted to reconstruct an average image and interpreted to obtain basic information on surfactant micelles. It is important to keep in mind that small-angle scattering techniques provide only an average image in the space of correlation functions. Averaging severely limits the amount of information obtainable because thermal agitation in surfactant solutions produces large fluctuations. However, this limitation also has an advantage. All structures are described by several averaged parameters, and, in that sense, averaging facilitates interpretation of data [164].

Small-angle and wide-angle X-ray diffraction techniques have been reviewed in several articles and books [164–168].

Fontell and Lindman [169] investigated phase equilibria of two-component systems consisting of water and perfluorononanoic acid or its salts. Small-angle X-ray scattering (SAXS) showed liquid crystalline regions besides regions of micellar solutions. The thickness of fluorocarbon layers in the liquid crystalline region and the area per polar head group were estimated.

Holmes et al. [170] studied nematic and disrupted lamellar phases in cesium pentadecafluorooctanoate, using small angle X-ray and neutron scattering.

Small-angle neutron scattering (SANS) [164,171–173] is a very useful tool for studying perfluorinated surfactants. The contrast between water and the fluo-rinated surfactant micelles is strong, due to the large difference in scattering length between ^{19}F and ^{1}H [174].

Hoffman et al. [175,176] measured small-angle neutron scattering (SANS) of lithium perfluorooctanoate, diethylammonium perfluorononanoate, and tetraethylammonium perfluorooctanesulfonate micelles in D_2O or in mixtures of D_2O and H_2O. The radii, micelle concentrations, and aggregation numbers were calculated.

Burkitt et al. [177,178] used SANS to examine the size and shape of micelles in solutions containing ammonium perfluorooctanoate or mixtures of ammonium perfluorooctanoate with ammonium decanoate. The SANS measurements were made using a neutron diffractometer and a computer to process the basic data to give the intensity of scattering $I(Q)$, as a function of Q relative to water. The scattering vector Q for elastic scattering is defined by

$$Q = (4\pi/\gamma) \sin (\Theta/2) \tag{2}$$

where λ is the wavelength of the radiation (neutrons) in the medium and θ is the scattering angle.

Burkitt et al. [177,178] concluded that SANS is an excellent method for the determination of micellar weights of ammonium octanoate, ammonium de-canoate, and ammonium perfluorooctanoate. The scattering data suggested that the ammonium perfluorooctanoate micelles are cylindrical.

X. LIGHT SCATTERING

Light scattering has provided a wealth of information on the shape, size, polydis-persity, and micellization of surfactant solutions and microemulsions. A large number of papers has been published and the theory of light scattering has been reviewed in several books [179–182].

The theoretical aspects of light scattering are in several ways similar to those of small-angle scattering. However, important differences exist in the status of experimental techniques. Dynamic light scattering is now in routine use, dy-namic neutron scattering is a new technique recently developed, but the practical feasibility of dynamic x-ray scattering is uncertain [182].

The use of light-scattering methods for studying micellar structures of fluori-nated surfactants is limited to partially fluorinated surfactants. Micelles of per-fluorinated surfactants are very weak light scatterers. The refractive index – concentration slope, dn/dc, is 10–100 times smaller than that for hydrocarbon chain surfactants [150]. However, when the surfactant aggregates are large, the turbidity of the system can be measured. Tomašić et al. [183] measured light scattering of the anionic surfactant $CF_3O\ (CF_2CFCF_3)O)_n CF_2COOH$ ($n \approx 1.6$) as a function of its concentration in water.

XI. LUMINESCENCE PROBING METHODS

Luminescence probes are molecules or ions that, upon photoexcitation, emit light having characteristics dependent on the immediate environment of the probe [184]. The light emitted can therefore serve to characterize the environment of the luminescence probe.

Luminescent probes can be divided into two groups: fluorescence probes and phosphorescence probes. Fluorescence is an emission of light associated with the transition of excited single states to the ground state. Phosphorescence is emission of light associated with the transition from the lowest triplet state to the ground state (Fig. 10) [185]. Radiative lifetimes of fluorescence generally range from 10^{-10} to 10^{-7} s, and radiative lifetimes of phosphorescence range from 10^{-5} to several seconds [184, 186].

Some molecules or ions can function as quenchers and inhibit luminescence. Quenching, caused by interactions between the luminescent probe and a quencher, may be reversible or irreversible. Excimer formation is a case of reversible quenching. Some luminescent probes react, in the excited state, with an identical molecule in the ground state and form an excimer:

$$^1P^* + P \underset{k_{-E}}{\overset{k_E}{\rightleftharpoons}} {}^1P_2^* \tag{3}$$

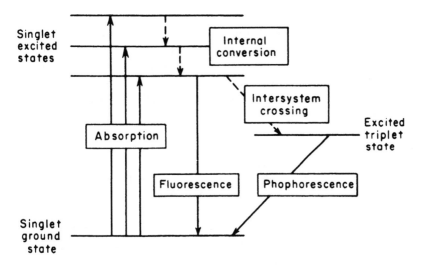

FIG. 10 Molecular energy levels involved in photochemical processes. (From Ref. 185.)

where k_E and k_{-E} are the rate constants for excimer formation and dissociation [184]. A probe in the excited state may associate in an analogous manner with a dissimilar molecule in the ground state and form an exciplex. The fluorescence characteristics of excimers and exciplexes differ from those of the monomeric probe. The excimer or exciplex formation is sensitive to the viscosity of the environment around the probe and therefore provides useful information on the structure of surfactant solutions.

Another reversible quenching technique used in micellar studies is energy transfer from the probe in the excited state to an energy acceptor molecule.

The luminescence methods involve solubilization of the probe in micelles and the determination of fluorescence spectra and fluorescence polarization. Various steady-state and transient-state fluorescence methods have been employed, experimental details of which can be found in the literature [187–191]. Careful selection of the probe is an essential requirement for obtaining meaningful results. Luminescence methods are based on the assumption that the probe does not affect the fundamental nature of the solution and the micelles. This assumption must be validated for each system being studied.

The fluorescence intensity of ammonium 1-anilinonaphthaline-8-sulfonate (ANS) in a solution of a hydrocarbon-type surfactant is constant below the cmc of the surfactant, but increases linearly with increasing surfactant concentration above the cmc. The concentration dependence of fluorescence intensity indicates that the ANS probe is solubilized in the micelles of the hydrocarbon-type surfactant. In contrast, fluorinated surfactants do not solubilize ANS [192]. The ANS probe is therefore useful for investigating fluorinated surfactant and hydrocarbon-type surfactant mixtures. Asakawa et al. [102] studied the micellar environment of mixed fluorinated surfactants and hydrocarbon-type surfactants by fluorescence intensities of ANS, auramine, and pyrene. The ANS fluorescence intensity increased with increase in hydrocarbon-type surfactant concentration (LiDS = lithium dodecyl sulfate). Because the ANS probe was not incorporated in LiFOS (lithium perfluorooctane sulfonate) micelles, the fluorescence intensity increased very little with increasing fluorinated surfactant concentration.

Muto et al. [193] measured pyrene fluorescence lifetime, τ_o, and the ratio I_1/I_3 of the intensities of the first vibronic and the third vibronic band of the monomeric pyrene. The pyrene fluorescence data revealed the existence of a single type of mixed micelle in solutions of LiDS-LiFOS, LiFOS–hexaoxyethylene glycol dodecyl ether, or LiFOS–octaoxyethylene glycol dodecyl ether mixtures. The lifetime and the intensity ratio of vibronic peaks have been used to determine the cmc of fluorinated surfactant micelles [194]. However, the solubility of pyrene in micelles of fluorinated surfactants is not adequate for determining the micelle aggregation number [194,195].

The I_1/I_3 ratio is very sensitive to the polarity of the medium sensed by the pyrene probe. The pyrene fluorescence technique has therefore been utilized for the characterization of adsorbed layers of hydrocarbon surfactants and fluorinated anionic surfactants on alumina [196,197].

Asakawa et al. [198] prepared a new quencher, 1,1,2,2,-tetrahydroheptadecafluorodecylpyridinium chloride, which quenches pyrene fluorescence by interaction with its pyridinium group. The fluorescence quenching was measured in solutions of anionic fluorinated and hydrocarbon surfactant mixtures.

XII. X-RAY PHOTOELECTRON SPECTROSCOPY

X-ray photoelectron spectroscopy (XPS; also electron spectroscopy for chemical analysis, or ESCA) is eminently suited to the study of surfactant adsorption. The XPS method is highly sensitive to the surface composition and can characterize adsorbed surfactant layers without elaborate sample preparation.

The theory and practice of XPS has been reviewed in several monographs and journal articles [199–205]. The sample is placed into a chamber and positioned for analysis. The chamber is evacuated to a high vacuum of $<10^{-7}$ torr, and the sample is irradiated with soft X-rays, usually from a MgK_α (1253.6 eV) or AlK_α (1486.6 eV) source. The X-ray irradiation generates photoelectrons, which are emitted with kinetic energy, E_K, governed by the energy of the exciting radiation, $h\nu$, and the binding energy, E_B, of the electron:

$$E_K = h\nu - E_B - \phi \tag{4}$$

The "work function," ϕ, depends on the sample and the spectrometer used for measuring photoelectron emission.

The binding energies of the electrons are characteristic of the element and the environment of the atom in the molecule. Hence XPS can characterize the composition and the chemical state of the near-surface region.

XPS spectra are strongly affected by the orientation of the sample, the source, and the spectrometer. Almost all (about 95%) of the signal emerges from the distance 3λ within the solid, where λ is the inelastic mean free path of the electron, also called the attenuation length of the emerging electron. The sampling depth, d, of the subsurface analyzed by XPS is given by

$$d = 3\,\gamma \sin \alpha \tag{5}$$

where α is the exit angle of the emitted electron relative to the sample surface. The mean free path depends on the kinetic energy of the photoelectron which is in turn affected by the energy of the radiation source. The sampling depth has a maximum when $\alpha = 90°$ and is usually less than 50 Å (5 μm).

Although the method is considered to be nondestructive, sample damage and evaporative losses have been of concern [206]. The fluorine-to-carbon photoelectron peak intensity ratios, F(1s)/C(1s), have been found to decrease during an X-ray exposure of several minutes, depending on experimental conditions [207].

XPS can yield qualitative and quantitative information on adsorbed surfactant layers. The overlayer on the substrate decreases the intensity, I_e, of a photoelectron peak, originating from a component in the substrate, by:

$$\frac{I_c}{I_e} = \exp\left(\frac{-\delta}{\Lambda_e \sin \alpha}\right) \tag{6}$$

where δ is the thickness of the overlayer sampled, I_c is the intensity of the photoelectron peak originating from the covered substrate, Λ_e is the inelastic free path of the electron (IMFP) in the overlayer, and α is the electron takeoff angle relative to the sample surface [208]. IMFP, the average distance a photoelectron travels before an inelastic collision, depends on the binding energy of the photoelectron and the composition of the sample.

Angle-dependent XPS (variable takeoff angle) can confirm the surfactant overlayer thickness and determine the continuity of the surfactant overlayer [208] (Fig. 11). The effective sampling depth of XPS analysis is a function of the electron IMFP and the takeoff angle. By decreasing the takeoff angle, the signal intensity contributions to the photoelectron spectrum from the top surface region can be selectively enhanced. This relationship serves as the basis for angle-dependent depth profiling. The angle-dependent ratio of overlayer to substrate, R, can be calculated using a simplified expression given by Fadley [209]:

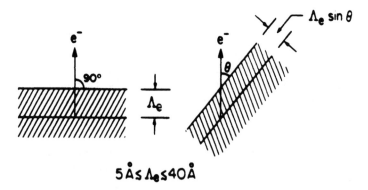

FIG. 11 The principle of angle-dependent XPS, where Λ_e is the IMPF of the electron being analyzed and Θ is the angle between the sample surface and the emitted electrons. (From Ref. 208. Reproduced with permission of American Chemical Society.)

$$R\left(\frac{\text{overlayer}}{\text{substrate}}\right) = K\{\exp(\tau/\sin\alpha) - 1\} \tag{7}$$

where K is a function of atom density, instrument response, the kinetic energies of the substrate and overlayer atoms within the measured levels, and the effective cross sections of the atoms. The effective overlayer thickness, τ, is given by

$$\tau = \frac{\delta}{\Lambda_e} \tag{8}$$

where δ is the actual overlayer thickness.

The orientation of an adsorbed surfactant can be determined by measuring the intensity of the peak for an atom on one end of the surfactant molecule relative to the intensity of a peak for an atom on the opposite end. Fluorine bound to carbon in $CF_3(CF_2)_n$- groups induce a chemical shift to a higher electron-binding energy. The resulting peak is readily distinguished in the C 1s spectrum from the peak for the carbon in the nonfluorinated portion of the molecule. If the surfactant molecule prefers a certain orientation, one peak is enhanced relative to the other for a given takeoff angle [208].

Gerenser et al. [208] found that the thickness of the adsorbed surfactant layer calculated by the angle-dependent method [Eq. (7)] is always greater than the thickness value calculated by the substrate-attenuation method [Eq. (8)]. The discrepancy was explained by orientation of the surfactant, which affects only the angle-dependent method and has no effect on the attenuation method.

Mitsuya [210] examined chemisorption of 11H-eicosafluoroundecanoic acid onto n-type (111) oriented silicon wafers by XPS. The samples were irradiated by X-rays from a magnesium target (MgK_α, 1253.6 eV). The irradiation angle was 60° with respect to the surface normal for optimum surface sensitivity and reproducibility of peak intensities.

XIII. ELECTROCHEMICAL METHODS

A. Electric Conductivity

Electric conductivity provides highly useful information on the purity and the association of surfactants in solution. The measurement of electric conductivity is experimentally straightforward, but requires certain precautions. The electric conductivity depends on the cell temperature, which therefore must be carefully controlled. Polarization has to be avoided by using AC or applying short pulses of opposing polarity.

Usually the specific conductivity is plotted against the surfactant concentration or the equivalent electric conductivity is plotted against the square root of surfactant concentration [211]. The increase in specific conductivity is rapid at

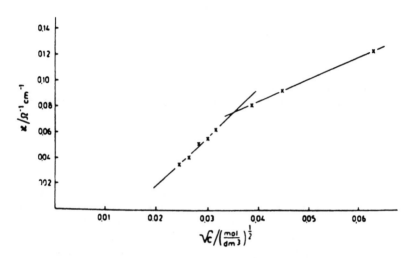

FIG. 12 The specific conductivity of tetraethylammonium perfluorooctanesulfonate plotted against the square root of its concentration. (From Ref. 213. Reproduced with permission of Carl Hanser Verlag.)

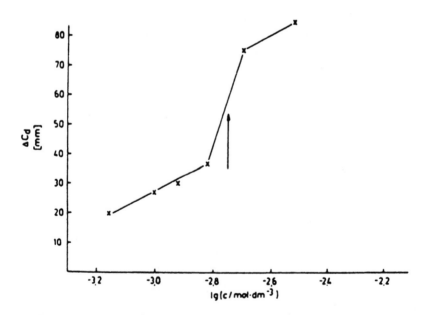

FIG. 13 The relative depression of the AC polarographic differential capacity, ΔC_a, plotted against the log concentration of tetraethylammonium perfluorooctanesulfonate. (From Ref. 213. Reproduced with permission of Carl Hanser Verlag.)

first, after which the slope decreases suddenly. The inflection point in the curve indicates the cmc. In contrast, the equivalent conductivity decreases slightly at first, then decreases sharply at the cmc.

Muzzalupo et al. [212] measured the equivalent conductivity of the alkali metal salts of perfluorooctanoic and perfluorononanoic acids in water.

Dörfler and Müller [213] determined the cmc value for tetraethylammonium perfluorooctanesulfonate from $\kappa/\sqrt{c}$ plots (cmc = $1.7 \cdot 10^{-3}$) (Fig. 12) and by AC polarography from the differential capacity/potential curves (cmc = $1.2 \cdot 10^{-3}$) (Fig. 13).

B. Transient Electric Birefringence

Electric birefringence can give useful information on the shape and size of micelles and fluorinated surfactant aggregates, especially when complemented by other physical methods [184,214–224]. Electric birefringence has been successfully used to determine the shape and size of polymers, polyelectrolytes, and surfactant micelles. Colloidal particles or molecular aggregates, which have a permanent dipole moment or are polarized anisotropically, orient in an electric field. The colloidal solution becomes optically anisotropic and exhibits electric birefringence, termed the Kerr effect [187,225,226].

An apparatus for electric birefringence measurements is shown schematically in Fig. 14 [222]. Rectangular high-voltage pulses of short duration are applied to the solution, and the build-up and decay of electric birefringence is measured. The beam of a He-Ne laser is polarized by a Glan prism set at 45° with respect to

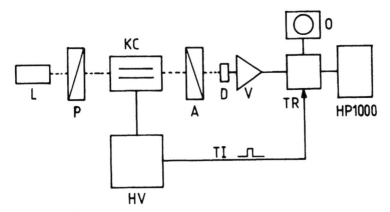

FIG. 14 Apparatus for electric birefringence measurements: L, He-Ne laser; P, polarizer; KC, Kerr cell; A, analyzer; D, photodiode detector; V, amplifier; TR, transient recorder; O, oscilloscope; HP, HP1000 computer; HV, high-voltage pulse generator; TI, trigger impulse to start the recording system. (From Ref. 222. Reproduced with permission of American Chemical Society.)

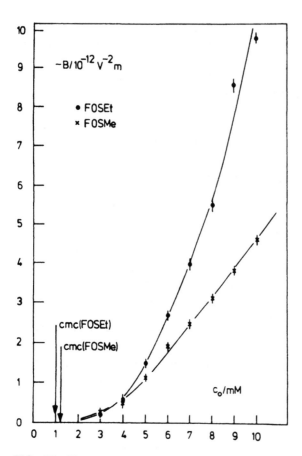

FIG. 15 Kerr constant B of tetraethylammonium perfluorooctanesulfonate (FOSET) and tetramethylammonium perfluorooctanesulfonate (FOSMe) as a function of total concentration C_o for $T = 20°C$. (From Ref. 217. Reproduced with permission of American Chemical Society.)

the electric field applied across the Kerr cell. The polarized light traverses the Kerr cell containing the sample and passes through the analyzer to the photodiode detector. The analyzer and the polarizer are in a crossed position. The signal of the detector is digitized by a recorder interfaced with a computer.

The electric birefringence, $\Delta n = n_\| - n_\perp$, is the difference in refractive indexes parallel and perpendicular to the direction of the applied electric field. Electric birefringence is related to the optical retardation or phase shift, δ, by the equation:

$$\Delta n = \frac{\delta \lambda}{(2\pi l)} \tag{9}$$

where l is the path length of the Kerr cell and λ is the vacuum wavelength [212]. The Kerr constant, B, is:

$$B = \lim_{E \to 0} \frac{\Delta n}{\left(\lambda E^2\right)} \tag{10}$$

where E is the field strength. The Kerr constant depends on the temperature and the surfactant concentration (Fig. 15) [217]. Below the cmc, the surfactant does not exhibit birefringence and only the solvent birefringence is observed. The Kerr constant of ionic surfactants can assume positive or negative values, depending on temperature and the counterion [217].

The rise and decay of the electric birefringence of a fluorinated surfactant is shown in Fig. 16 [217]. The birefringence relaxation time, τ_B, is related to the rotational diffusion constant, D_R,

$$\tau_B = (6D_R)^{-1} \tag{11}$$

The rotational diffusion constant, D_R, is proportional to L^{-3}, where L is the length of a rodlike micelle or the diameter of a disklike spheroid. Hence τ_B values cannot distinguish between disks or rods, unless complemented with Kerr constant values or data obtained by other methods.

Tamori et al. [224] used electric birefringence to estimate the micellar size and shape in mixed surfactant solutions containing hexaoxyethylene glycol dodecyl ether and lithium perfluorooctanesulfonate or lithium dodecyl sulfate.

The time dependence of birefringence is affected by intermicellar interaction, electrolytes, and polydispersity. If the aggregates are polydisperse, the time de-

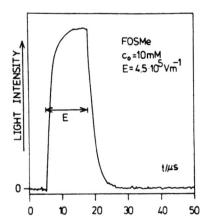

FIG. 16 The rise and decay of the electric birefringence of FOSMe. The duration of the applied electric field pulse is indicated with the arrow. (From Ref. 217. Reproduced with permission of American Chemical Society.)

pendence of birefringence deviates from a single exponential relationship. The size distribution function must be known, because the deviation depends on the width of the distribution function. [184]. In spite of these limitations, Shorr and Hoffmann [217,222] concluded that electric birefringence measurements are useful for the determination of the dimensions of anisotropic surfactant aggregates. While electric birefringence data alone are not sufficient to define the type of surfactant aggregates, the electric birefringence method complements other physical methods.

XIV. ULTRAFILTRATION

Ultrafiltration techniques [227–229] have been used to separate surfactant monomers from their micelles. Ultrafiltration is in principle similar to conventional filtration; the difference between the two processes is mainly the size of particles separated. Whereas conventional filtration separates particles larger than 1 μm, ultrafiltration can separate particles larger than 2 nm. The separation of very small particles requires a very small pore size and consequently a high hydrostatic pressure, up to 10 atm, to force the liquid through the filter.

Asakawa et al. [230] employed ultrafiltration to study mixtures of anionic hydrocarbon-type surfactants with anionic fluorinated surfactants, lithium perfluorooctane sulfonate, and sodium perfluorooctanoate. The separation method was based on the assumption that surfactant monomers can pass through the ultrafiltration membrane, but the pores of the membrane are sufficiently small to retain micelles. The membrane used (YC-05, Amicon Corp.) can exclude molecules having a molecular weight larger than 500. The filtration cell was equipped with a magnetic stirring bar suspended above the membrane. The surfactant solution was forced through the membrane by applying pressurized nitrogen gas at 3.0 kg/cm^2. The concentrations of the filtrand and the filtrate were determined by HPLC or isotachophoresis, a high-resolution electrophoretic method [231]. Ultrafiltration experiments were conducted as a function of initial surfactant concentration. Below the cmc the surfactant concentrations in the filtrand and filtrate were equal. Above the cmc the surfactant concentration in the filtrate became constant and was equal to the cmc, in accord with the postulate that the filtrate contains monomeric surfactant molecules in equilibrium with micelles that did not pass the filter.

XV. SURFACE TENSION

It is perhaps needless to state that surface tension is the most important physical property of a surfactant to be determined. This is especially true for fluorinated surfactants, which can lower the surface tension of water to exceptionally low values not attainable by surfactants with a hydrocarbon type hydrophobe.

Methods for surface and interfacial tension measurement have been the subject of numerous papers and review articles [232–247]. In spite of the apparent simplicity of surface tension measurement, correct and reproducible values are not always readily obtainable. In addition to the specific limitations of each technique, time dependence of surface tension of surfactant solutions can be a major complication. Surface tension depends on the adsorption and orientation of molecules at the liquid/air interface. Whereas in pure liquids only microseconds are needed for equilibrium orientation, in surfactant solutions hours or even days may be needed to attain equilibrium surfactant adsorption at a freshly created surface. Adsorption and orientation kinetics are especially critical when measuring surface tension of surfactant mixtures.

Surface tension methods measure either static or dynamic surface tension. Static methods measure surface tension at equilibrium, if sufficient time is allowed for the measurement, and characterize the system. Dynamic surface tension methods provide information on adsorption kinetics of surfactants at the air/liquid interface or at a liquid/liquid interface. Dynamic surface tension can be measured in a time scale ranging from a few milliseconds to several minutes [248]. However, a demarcation line between static and dynamic methods is not very sharp because surfactant adsorption kinetics can also affect the results obtained by static methods. It has been argued [249] that in many industrial processes, sufficient time is not available for the surfactant molecules to attain equilibrium. In such situations, dynamic surface tension, dependent on the rate of interface formation, is more meaningful than the equilibrium surface tension. For example, peaked alcohol ethoxylates, because they are more water soluble, do not lower surface tension under static conditions as much as conventional alcohol ethoxylates. Under dynamic conditions, however, peaked ethoxylates are equally or more effective than conventional ethoxylates in lowering surface tension [250].

Most techniques stretch the liquid/air surface at the moment of measurement. For example, the drop weight method [251] and the ring method [252–255] stretch the surface during detachment. However, instruments are now available that measure surface tension without detaching the ring from the liquid, for example, the Krüss Tensiometer K12.

The surface tension methods measure a force, pressure, or drop size (volume, weight, or dimensions). Examples of methods that measure a force are the ring method [252–255] and the plate method [256,257]. The capillary height [258–261,269] and maximum-bubble-pressure methods [262–269] measure pressure. Pendant drop [261,270–272], sessile drop [261,273], drop volume [274–276], drop weight [251,269,277–280], and spinning drop methods [281–283] measure the size or the dimensions of a drop. Special techniques [248] for measuring dynamic surface tension include the oscillating jet [284], the dynamic drop volume [248,285], the inclined plate [286], the strip [287], the free falling method [288], the pulsed drop [289], the dynamic maximum-bubble-pressure [249,264–267,290–292], and the dynamic capillary methods [248].

Modern tensiometers are interfaced with computers to increase the accuracy of the measurement and obtain dynamic surface tension readings within short but accurately measured time intervals.

The *ring method* [252–255] is one of the most frequently used techniques for surface tension measurement. A platinum ring, attached to a vertical wire, is placed horizontally into the liquid (Fig. 17). The force, P, needed to pull the ring through the interface is measured. If one assumes that the ring supports a cylinder of liquid:

$$P = 4\pi\gamma R \tag{12}$$

where R is the radius of the ring. Actually, the liquid column lifted by the ring is not a cylinder (Fig. 17), and a correction factor [252–255,293,294] is needed. The accuracy of surface tension values obtained by the ring method is limited by the accuracy of the correction factor. The ring method gives reproducible results only if certain precautions are taken [294–299].

It has been argued that the ring method is suitable only for measuring the surface tension of pure liquids. The applicability of the method for the measurement of surface tension of surfactants has been debated [257,300–302]. The wire loop method [303,304] is similar to the ring method.

The *Wilhelmy plate method* [256,257], the *sessile drop method* [261,273], and the *capillary height method* [258–261] measure equilibrium surface tension, if sufficient time is allowed for the adsorption of surfactant molecules at the surface to attain the state of equilibrium. The Wilhelmy plate method measures the force exerted on a vertical plate partially immersed in the liquid (Fig. 18). If wetting of the plate is complete the force, F, is proportional to the surface tension, γ, and the circumference, L, of the plate:

$$F = \gamma \times L \tag{13}$$

The Wilhelmy plate method has the advantages of measuring strictly static surface tension and being less sensitive to vibrations of the vessel or a slight deformation of the plate. The prerequisite is the complete wetting of the plate, indicated by a zero contact angle. Significant contact angles in the wetting of the plate by some liquid systems have been observed [305]. Wetting is facilitated by using a roughened plate or a platinized platinum plate from which the liquid does not recede. Tadros [306] used a glass plate for measuring the surface tension of anionic fluorinated surfactants by the Wilhelmy method. Hirt et al. [249] used a platinum wire instead of a plate for surface tension measurements of fluorinated surfactants. The Wilhelmy method is applicable to single fibers [307].

The *pendant drop method* has been described by Andreas at al. [308] and oth-

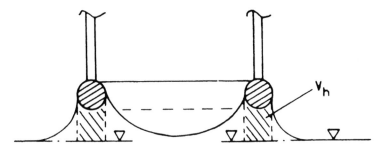

FIG. 17 The ring method. (Reproduced by permission of Krüss USA.)

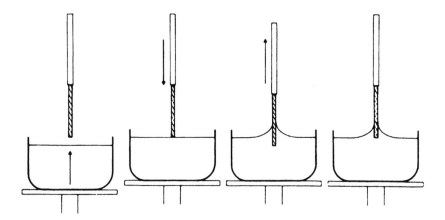

FIG. 18 The Wilhelmy plate method. (Reproduced by permission of Krüss USA.)

ers [261,270–272,309]. The apparatus is simple, but the technique requires skill for forming the drop and maintaining its size and shape during the measurement of its narrowest and widest diameters d_2 and d_1 (Fig. 19).

The *drop volume method* [274–276,310,311] requires only a burette or a syringe. Either the volume required to form the drop is measured or the number of drops formed by a measured volume of liquid is counted (Fig. 20).

The *drop weight method* measures the weight of a drop (or several drops) emerging from a capillary of known dimensions [251,269,277–280]. A slight vacuum is applied to the apparatus through a tubing until the drop, forming at the outlet of the capillary, assumes almost its full size. The drop is then allowed to detach itself from the capillary. Surface tension is calculated by the equation:

$$\gamma = (mg/r)F \tag{14}$$

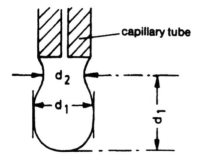

FIG. 19 The pendant drop method. (Reproduced by permission of Krüss USA.)

FIG. 20 The drop volume method. (Reproduced by permission of Krüss USA.)

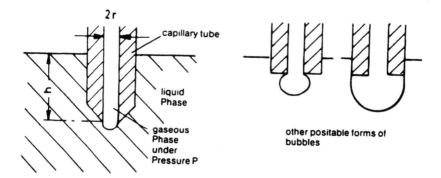

FIG. 21 Maximum-bubble-pressure method. (Reproduced by permission of Krüss USA.)

where m is the mass of the drop, g is the acceleration of gravity, r is the capillary radius, and F is a factor dependent on the drop volume and r [312]. The drop volume can be calculated from the drop weight and the density of the liquid. With proper corrections [313], the method is quite accurate [314,315].

The *maximum-bubble-pressure method* [247,249,262–269] (Fig. 21) measures pressure in a bubble formed at the end of a capillary when a gas, e.g., air, is blown through the capillary into the liquid. The pressure increases when the bubble grows and attains its maximum value when the bubble has obtained the shape of a hemisphere. The pressure decreases when the bubble grows further and finally bursts. Maximum-bubble-pressure methods have been compared [316], and equipment for automated surface tension determination by maximum-bubble-pressure measurement has been developed [317–320].

Dynamic surface tension measurements by Hirt et al. [249], based on the maximum-bubble-pressure method, revealed large differences between equilibrium and dynamic surface tension values of fluorinated surfactants. The surface tension transition from equilibrium values to dynamic diffusion-limited values depended on the surfactant type, concentration, and the bubble generation rate.

The *spinning drop method* [281–283,321,322] is used to determine interfacial tensions between two liquids. A capillary tube is mounted in a chamber, leaving the ends open. The chamber and the tube are filled with the heavier of the two liquids and the capillary rotated at a high speed (about 2000 rpm). A drop of the other liquid having a lower density is placed into the capillary. The drop moves into the center of the capillary tube and usually assumes the shape of a cylinder with curved edges. The radius of the drop is measured using a microscope or a camera interfaced with a computer.

Examples of methods used to measure surface tensions of anionic fluorinated surfactants are shown in Table 1.

XVI. ANIONIC FLUORINATED SURFACTANTS IN BIOLOGICAL SYSTEMS

The discovery of two types of fluorine, organic and inorganic, in human blood [337–341] intensified the interest in the absorption and retention of fluorochemicals in biological systems. The conjecture that organic fluorine in blood originated from exposure to fluorinated surfactants prompted analyses of blood and other biological samples for fluorinated surfactants. A study on exposure to fluorinated surfactants [342] found fluorine concentrations ranging from 1 to 71 ppm in the blood of workers handling ammonium perfluorooctanoate.

The methods used for the determination of fluorinated surfactants in biological samples can be divided into two groups: (1) the determination of organic fluorine, which represents the concentration of a fluorinated surfactant if other

TABLE 1 Methods Used for Measuring Surface Tension of Anionic Fluorinated
Surfactants

Method	Author	Ref.
Ring	Bernett and Zisman (1959)	323
	Caporiccio et al. (1984)	107
	La Mesa and Sesta (1987)	324
	Glöckner (1989)	325
	Monduzzi et al. (1994)	119
	Muzzalupo et al. (1995)	212
Wireloop	Thoai (1977)	304
Plate	Funasaki and Hada (1979)	326
	Tadros (1980)	306
	Gorodinsky and Efrima (1994)	329
	Esumi and Ogiri (1995)	330
	Yoshino et al. (1995)	101
Pendant drop	Jarvis and Zisman (1959)	331
	Matubayasi et al. (1977)	332
	Motomura (1989)	333
Drop volume	Zhao Guo-Xi and Zhu Bu-Yao (1983)	334
Drop weight	Shinoda and Nakayama (1963)	314
	Mukerjee and Honda (1981)	251
Maximum bubble pressure	Scholberg et al. (1959)	336
	Hirt et al. (1990)	249

fluorochemicals are absent, and (2) a specific method for the fluorinated surfactant of interest.

The determination of organic fluorine in biological samples such as whole blood, serum, and plasma involves destruction of organic matter by combustion or ashing to convert organic fluorine to inorganic fluoride and the determination of fluoride in the sample. The organic fluorine is calculated as the difference between fluoride found in the combusted sample minus inorganic fluoride present in the uncombusted sample. Ashing in the presence of magnesium carbonate [339], magnesium oxide [341], or calcium phosphate [343,344] leads to low results and has been superseded by combustion in a closed system with oxygen. Venkateswarlu [345] and Belisle and Hagen [13] have employed an oxygen Parr bomb for the determination of fluorine in whole blood or serum. Since liquid blood cannot be combusted in an oxygen Parr bomb, Belisle and Hagen [13] removed water from blood in vacuo at ambient temperature. The solids were pelletized after mixing with a preweighed amount of benzoic acid. The pellet was combusted in an oxygen Parr bomb which contained 0.025 N NaOH to absorb fluoride. The reaction mixture was acidified with perchloric acid and the fluoride was extracted into m-xylene containing triethylsilanol. The fluoride, converted to triethylfluorosilane, was determined by gas chromatography. The method gives accurate results for perfluorooctanoic acid in blood, but volatile fluorochemicals in the sample evaporate together with water when the blood sample is dried.

Kissa [346] studied the kinetics of postexposure elimination of a fluorinated anionic surfactant, $F(CF_2CF_2)_{3-8}CH_2CH_2SO_3X$, where $X = H$ and NH_4), from blood using rats as the test animal. The biological samples, including whole blood, serum, and various organic tissues, were combusted in an oxyhydrogen flame. A 0.5–1.0 g sample of whole blood, serum, or plasma was introduced into the combustion apparatus. Hydrofluoric acid formed during the combustion was collected in water and determined with a fluoride ion–selective electrode [9]. To calculate the organic fluorine concentration, the inorganic fluoride concentration of the sample was determined with an analyte addition method using a fluoride ion–selective electrode [347] and deducted from the total fluorine concentration. The method has the advantage of being applicable to liquid samples without having to convert them to a solid form. The oxyhydrogen combustion method can determine volatile fluorochemicals, which would be lost if a liquid sample has to be dried. A large number of samples of blood and organic tissues have been analyzed routinely by the oxyhydrogen combustion method in a Du Pont laboratory.

The direct determination of fluorinated surfactants is possible if the fluorinated surfactant is amenable to chromatography or spectroscopy. Belisle and Hagen [40] determined perfluorooctanoic acid in blood, urine, and liver tissue. Perfluorooctanoic acid was extracted from blood or other biological samples with hexane in the presence of hydrochloric acid and converted to its methyl ester with diazomethane. The recovery of known amounts of perfluorooctanoic acid added to human plasma was essentially quantitative. The precision of the method was inferior to that of the determination of perfluorooctanoic acid by elemental fluorine analysis, but could probably be improved by using a capillary chromatography column instead of the packed column used by the authors.

XVII. ANIONIC FLUORINATED SURFACTANTS IN THE ENVIRONMENT

The presence of fluorinated surfactants in air and in water or wastewater is of concern. Fluorinated surfactants can enter air as vapor if volatile or as a liquid or solid aerosol. Fluorinated surfactants have to be monitored in air at industrial sites where fluorinated surfactants are produced or used to protect the health of workers.

The determination of fluorochemicals in *air* usually involves two steps: collection of an air sample and determination of the fluorochemical in the sample collected. Sampling techniques used have included grab sampling, concentration by cryogenic techniques, trapping in a solvent, and adsorption on a solid adsorbent, with the last method being the most convenient. However, the common adsorbents hold fluorinated surfactants firmly, and desorption of the collected species for its determination is difficult and frequently not quantitative. Kissa [348] therefore introduced the concept of total organic fluorine concentration in air. The fluorinated surfactant in air is collected on a solid adsorbent, such as activated carbon, graphitized carbon, silica, Tenax, etc. The adsorbent is combusted in a oxyhydrogen torch. Hydrogen fluoride, formed by combustion of organic fluorine on the adsorbent, is collected in water and determined with a fluoride ion–selective electrode [9]. When only one fluorine-containing species is present in air, the fluorine content represents its concentration. When several fluorochemicals are present, the fluorine concentration indicates the maximum concentration a fluorinated surfactant may have in the mixture of fluorochemical air contaminants. If this maximum concentration value indicated by the total fluorine concentration is below the acceptable limit, a specific analytical method is not needed.

Fluorinated surfactants present in air as solid aerosols, such as dust, are collected on filters made of mixed cellulose esters and combusted together with the filter in the oxyhydrogen torch [348]. Volatile fluorochemicals can be collected on a solid adsorbent in an adsorption tube connected to the outlet of the filter.

The determination of fluorinated surfactants in *water* and *wastewater* is essential for (1) the detection of pollution by fluorinated surfactants, (2) study of biodegradation, and (3) determining the effect of fluorinated surfactants on aquatic life. If a specific method is not needed, the oxyhydrogen combustion method is the most effective. By introducing a 10-mL water sample into the oxyhydrogen torch in several portions, as little as 20–40 ppb fluorinated surfactant can be detected without the need to concentrate the sample before combustion.

When other fluorochemicals are present, the fluorinated surfactant has to be separated and determined by a specific analytical method. Some of the conventional methods for the analysis of hydrocarbon type surfactants [1] are applicable also for the determination of fluorinated surfactants (see Secs. III and IV). However, when the fluorinated surfactant has to be identified and structural information is needed, mass spectrometry [349] is the method of choice. Schröder [70,71] employed a tandem mass spectrometer (MS/MS) to analyze fluorinated surfactants in water and wastewater.

An anionic fluorinated surfactant (Fluowet PL 80) and a cationic fluorinated surfactant (Fluowet L 3658-1) were quantitatively analyzed in water. However, analyses of these fluorinated surfactants in wastewater containing sludge were complicated by strong adsorption of the surfactants on the sludge. Extraction of the surfactants with acidified methanol was incomplete. The determination of the anionic surfactant (Fluowet PL 80) by combustion in a oxyhydrogen flame gave a quantitative result.

REFERENCES

1. E. Kissa, *Fluorinated Surfactants*, Marcel Dekker, New York, 1994.
2. T. M. Schmitt, *Analysis of Surfactants*, Marcel Dekker, New York, 1992.
3. J. Cross, ed., *Anionic Surfactants—Chemical Analysis*, Marcel Dekker, New York, 1977.
4. M. J. Rosen and H. A. Goldsmith, *Systematic Analysis of Surface-Active Agents*, 2nd ed., Wiley-Interscience, New York, 1972.
5. M. R. Porter, ed. *Recent Developments in the Analysis of Surfactants*, Elsevier Applied Science, London, 1991.
6. E. A. M. England, J. B. Hornsby, W. T. Jones, and D. R. Terrey, Anal. Chim. Acta *40*:365 (1968).
7. R. Wickbold, Angew. Chem. *64*:133 (1952).
8. P. B. Sweetser, Anal. Chem. *28*:1766 (1956).
9. E. Kissa, Anal. Chem. *55*:1445 (1983).
10. H. J. Friese and A. Pukropski, Fr. Demande FR 2 592 162 (1987).
11. Heraeus PCL-B-40, Verbrennungsapparatur V5 (Heraeus Quartzschmelze GmbH, Hanau, Germany), Tenside Surf. Det. *23*:271 (1986).
12. J. J. Bailey and D. G. Gehring, Anal. Chem. *33*:1760 (1961).
13. J. Belisle and D. F. Hagen, Anal. Biochem. *87*:545 (1978).
14. W. Schöniger, Mikrochim. Acta 123 (1955).
15. T. S. Ma and R. C. Rittner, *Modern Organic Elemental Analysis*, Marcel Dekker, New York, 1979.
16. N. H. Gelman and L. M. Kiparenko, Zh. Anal. Khim. *20*:229 (1965).
17. F. H. Oliver, Analyst *91*:771(1966).
18. A. Steyermark, R. R. Kaup, D. A. Petras, and E. A. Pass, Microchem. J. *4*:55 (1960).
19. H. Trutnovsky, Mikrochim. Acta 498 (1963).
20. J. Horacek and V. Pechanec, Mikrochim. Acta 17 (1966).
21. T. S. Light and R. F. Mannion, Anal. Chem. *41*:107 (1969).
22. F. W. Cheng, Mikrochim. Acta 841 (1970).
23. W. Selig, Z. Anal. Chem. *249*:30 (1970).
24. H. J. Francis, Jr., J. H. Donarine, and D. D. Pesing, Microchem. J. *14*:580 (1969).
25. D. A. Shearer and G. F. Morris, Microchem. J. *15*:199 (1970).
26. J. Pavel, R. Kuebler, and H. Wagner, Microchem. J. *15*:192 (1970).
27. Y. Inoue and K. Yokoyama, Iyakuhin Kenkyu *14*:25 (1983).
28. Y. Asano and Y. Shimada, Koku Eisei Gakkai Zasshi *35*:526 (1985).
29. J. L. Stuart, Analyst *95*:1032 (1970).
30. T. S. Ma and J. Gwirtsman, Anal. Chem. *29*:140 (1957).
31. W. K. R. Musgrave, Analyst *86*:842 (1961).
32. P. P. Wheeler and M. J. Fauth, Anal. Chem. *38*:1970 (1966).
33. M. Noshiro, T. Yarita, and S. Yonemori, Bunseki Kagaku *32*(12):E403 (1983).
34. Model 93-42 Surfactant Electrode Instruction Manual, Orion Research, Cambridge, MA, 1988.
35. J. H. Jones, J. Assoc. Offic. Agric. Chemists *28*:398 (1945).
36. J. Waters and C. G. Taylor, in *Anionic Surfactants—Chemical Analysis* (John Cross, ed.), Marcel Dekker, New York, 1977, p. 193.

37. R. Sharma, R. Pyter, and P. Mukerjee, Anal. Lett. 22(4):999 (1989).
38. D. E. Elliott, J. Chromatogr. Sci. 15:475 (1977).
39. K. Blau and G. King, eds., *Handbook of Derivatives for Chromatography*, Heyde, London, 1977.
40. J. Belisle and D. F. Hagen, Anal. Biochem. 101:369 (1980).
41. C. Lai, J. H. Harwell, E. A. O'Rear, S. Komatsuzaki, J. Arai, T. Nakakawaji, and Y. Ito, Colloids Surfaces A104:231 (1995).
42. T. Asakawa, S. Miyagishi, and M. Nishida, Langmuir 3:821 (1987).
43. J. Porath and P. Flodin, Nature 183:1657 (1959).
44. F. Tokiwa, K. Ohti, I, Kokubo, Bull. Chem. Soc. Jpn. 41:2845 (1968).
45. T. Nakagawa and H. Jizomoto, Kolloid Z. Z. Polym. 23:79 (1969).
46. T. Nakagawa and H. Jizomoto, Kolloid Z. Z. Polym. 239:606 (1970).
47. T. Nakagawa and H. Jizomoto, J. Am. Oil. Chem. Soc. 48:571 (1971).
48. E. Klesper, A. H. Corwin, and D. A. Turner, J. Org. Chem. 27:700 (1962).
49. L. G. Randall, Sep. Sci. Tech. 17:1 (1982).
50. B. A. Charpentier and M. R. Sevenants, eds., *Supercritical Fluid Extraction and Chromatography*, ACS Symposium Series 366, American Chemical Society, Washington, DC, 1988.
51. P. Mukerjee, M. J. Gumkowski, C. C. Chan, and R. Sharma, J. Phys. Chem. 94:8832 (1990).
52. P. R. Griffits and J. A. de Haseth, *Fourier Transform Infrared Spectrometry, Chemical Analysis*, Vol. 83, J. Wiley & Sons, New York, 1986.
53. D. Lin-Vien, N. B. Colthup, W. G. Fateley, and J. G. Grasselli, *The Handbook of Infrared and Raman Characteristic Frequencies of Organic Molecules*, Academic Press, San Diego, CA, 1991.
54. A. M. Amorim da Costa and E. B. H. Santos, Colloid Polym. Sci. 261:58 (1983).
55. A. Ito, K. Kamogawa, H. Sakai, K. Hamano, Y. Kondo, N. Yoshino, and M. Abe, Langmuir 13:2935 (1997).
56. K. Kamogawa, M. Takamura, H. Matsuura, and T. Kitagawa, Spectrochim. Acta 50:1513 (1994).
57. K. D. Cook, in *Encyclopedia of Polymer Science and Engineering*, Vol. 9, John Wiley & Sons, New York, 1987, p. 319.
58. M. B. Bursey, in *McGraw-Hill Encyclopedia of Science and Technology*, Vol. 10, McGraw-Hill Book Co., New York, 1987, p. 492.
59. K. Levsen, *Fundamental Aspects of Organic Mass Spectrometry*, Verlag Chemie, Weinheim, 1978.
60. M. E. Rose and R. A. W. Johnstone, *Mass Spectrometry for Chemists and Biochemists*, Cambridge University Press, New York, 1982.
61. J. T. Watson, *Introduction to Mass Spectrometry*, Raven Press, New York, 1982.
62. F. W. McLafferty, ed., *Tandem Mass Spectroscopy*, Wiley, New York, 1983.
63. J. R. Chapman, *Practical Organic Mass Spectrometry*, Wiley, London, 1986.
64. J. B. Lambert, H. F. Shurvell, D. Lightner, and R. G. Cooks, *Introduction to Organic Spectroscopy*, MacMillan Publishing Co, New York, 1987.
65. W. M. A, Niessen and J. van der Greef, *Liquid Chromatography Mass Spectrometry*, Marcel Dekker, New York, 1992.

66. J. V. Iribarne and B. A. Thomson, J. Chem. Phys. *64*:2287 (1976).
67. B. A. Thomson and J. V. Iribarne, J. Chem. Phys. *71*:4451 (1979).
68. C. R. Blakely and M. L. Vestal, Anal. Chem. *55*:750 (1983).
69. M. L. Vestal and G. J. Fergusson, Anal. Chem. *57*:2373 (1985).
70. H. Fr. Schröder, Vom Wasser *73*:111 (1989).
71. H. Fr. Schröder, Vom Wasser *77*:227 (1991).
72. K. H. Ott, Fette Seifen Anstrichm. *87*:377 (1985).
73. E. Schneider, K. Levsen, P. Daehling, F. W. Roellgen, Fresenius' Z. Anal. Chem. *316*:488 (1983).
74. D. N. Heller, C. Fenselau, J. Yergey, R. J. Cotter, and D. Larkin, Anal. Chem. *56*:2274 (1984).
75. P. A. Lyon, K. B. Tomer, and M. L. Gross, Anal. Chem. *57*:2984 (1985).
76. J. M. Gilliam, P. W. Landis, and J. L. Occolowitz, Anal. Chem. *55*:1531 (1983).
77. P. Daehling, F. W. Roellgen, J. J. Zwinselman, R. H. Fokkens, and N. M. M. Nibbering, Fresenius' Z. Anal. Chem. *312*:335 (1982).
78. E. Schneider, K. Levsen, P. Daehling, and F. W. Roellgen, Fresenius' Z. Anal. Chem. *316*:488 (1983).
79. G. D. Roberts and E. V. White, Biomed. Mass Spectrom. *11*:273 (1984).
80. M. A. Baldwin and F. W. McLafferty, Org. Mass Spectrom. *7*:1353 (1973).
81. R. J. Cotter, Anal. Chem. *52*:1589A (1980).
82. C. D. Daves, Jr., Acc. Res. Chem. *12*:359 (1979).
83. R. J. Cotter, G. Hansen, and T. R. Jones, Anal. Chim. Acta *136*:135 (1982).
84. A. Abragam, *Principles of Nuclear Magnetism*, Oxford University Press, Oxford, 1961.
85. R. K. Harris, *Nuclear Magnetic Resonance Spectroscopy: A Physicochemical Approach*, Pitman, London, 1983.
86. M. L. Martin, J. J. Delpuech, and G. J. Martin, *Practical NMR Spectroscopy*, Heyden, London, 1983.
87. A. Rahman, *One and Two Dimensional NMR Spectroscopy*, Elsevier, New York, 1989.
88. C. P. Sichter, *Principles of Magnetic Resonance*, 3rd ed., Springer-Verlag, New York, 1989.
89. R. Kitamaru, *Nuclear Magnetic Resonance: Principles and Theory*, Elsevier, Amsterdam, 1990.
90. H.-O. Kalinowski, S. Berger, and S. Braun, *Carbon C-13 NMR Spectroscopy*, John Wiley & Sons, New York, 1991.
91. R. Ernst, G. Bodenhausen, and A. Wokaun, *Principles of Nuclear Magnetic Resonance in One or Two Dimensions*, Oxford Science Publications, Oxford University Press, New York, 1991.
92. H. Friebolin, *Basic One- and Two-Dimensional NMR Spectroscopy*, VCH Publishers, New York, 1991.
93. J. K. M. Saunders and B. K. Hunter, *Modern NMR Spectroscopy, A Guide for Chemists*, Oxford University Press, New York, 1991.
94. B. Lindman, O. Söderman, and H. Wennerström, in *Surfactant Solutions*, (R. Zana, ed.) Marcel Dekker, New York, 1987, p. 295.

95. P. Mukerjee and K. Mysels, Pap. Symp. 1974 ACS Symp. Series 9:239 (1975).
96. N. Muller and R. H. Birkhahn, J. Phys. Chem. 71:957 (1967).
97. N. Muller and R. H. Birkhahn, J. Phys. Chem. 72:583 (1968).
98. N. Muller and T. W. Johnson, J. Phys. Chem. 73:2042 (1971).
99. N. Muller and H. Simsohn, J. Phys. Chem. 75:942 (1971).
100. M. Monduzzi, A. Chittofrati, and M. Visca, Langmuir 8:1278 (1992).
101. N. Yoshino, K. Hamano, Y. Omiya, Y. Kondo, A. Ito, and M. Abe, Langmuir 11:466 (1995).
102. T. Asakawa, M. Muori, S. Miyagishi, and M. Nishida, Langmuir 5:343 (1989).
103. B.-O. Persson, T. Drakenberg, and B. Lindman, J. Phys. Chem. 80:2124 (1976).
104. R. J. E. M. de Weerd, J. W. de Haan, J. M. van de Ven, M. Achten, and H. M. Buck, J. Phys. Chem. 86:2523 (1982).
105. D. L. Chang, H. L. Rosano, and A. E. Woodward, Langmuir 1:669 (1985).
106. K. Fontell and B. Lindman, J. Phys. Chem. 87:3289 (1983).
107. G. Caporiccio, F. Burzio, G. Carniselli, and V. Biancardi, J. Colloid Interface Sci. 98:202 (1984).
108. R. Haque, J. Phys. Chem. 72:3056 (1968).
109. R. E. Bailey and G. H. Cady, J. Phys. Chem. 73:1612 (1969).
110. N. Muller and F. E. Platko, J. Phys. Chem. 75:547 (1971).
111. N. Muller, J. H. Pellerin, and W. W. Chen, J. Phys. Chem. 76:3012 (1972).
112. J. Carlfors and P. Stilbs, J. Colloid Interface Sci. 103:332 (1985).
113. E. L. Hahn, Phys. Rev. 80:580 (1950).
114. E. O. Stejskal, and J. E. Tanner, J. Chem. Phys. 42:288 (1965).
115. P. Stilbs, J. Colloid Interface Sci. 87:385 (1982).
116. R. Palepu and V. C. Rainsborough, Can. J. Chem. 67:1550 (1989).
117. W. Guo, B. M. Fung, and S. D. Christian, Langmuir 8:446 (1992).
118. N. Boden, K. W. Jolley, and M. H. Smith, J. Phys. Chem. 97:7678 (1993).
119. M. Monduzzi, A. Chittofrati, and V. Boselli, J. Phys. Chem. 98:7591 (1994).
120. U. Henriksson and L. Ödberg, J. Colloid Interface Sci. 46:212 (1974).
121. J. Ulmius and B. Lindman, J. Phys. Chem. 85:4131 (1981).
122. C. Burger-Guerisi, C. Tondre, and D. Canet, J. Phys. Chem. 92:4974 (1988).
123. G. J. T. Tiddy, Symp. Faraday Soc. No. 5:150 (1971).
124. J. Lang and R. Zana, in Surfactant Solutions (R. Zana, ed.), Marcel Dekker, New York, 1987, p. 405.
125. E. A. G. Aniansson and S. N. Wall, J. Phys. Chem. 78:1024 (1974).
126. E. A. G. Aniansson and S. N. Wall, J. Phys. Chem. 79:857 (1975).
127. E. A. G. Aniansson, S. N. Wall, J. Lang, C. Tondre, R. Zana, H. Hoffmann, I. Kielmann, and W. Ulbricht, J. Phys. Chem. 80:905 (1976).
128. M. Kahlweit and M. Teubner, Adv. Colloid Interface Sci. 13:1 (1980).
129. E. Lessner, M. Teubner, and M. Kahlweit, J. Phys. Chem. 85:1529. (1981).
130. E. Lessner, M. Teubner, and M. Kahlweit, J. Phys. Chem. 85:3167 (1981).
131. M. Kahlweit, J. Colloid Interface Sci. 90:92 (1982).
132. M. Kahlweit, Pure Appl. Chem. 53:2069 (1981).
133. M. Eigen and L. de Maeyer, in Technique of Organic Chemistry, Vol. VIII, (S. L. Friess, E. S. Lewis, and A. Weissberger, eds.), Wiley-Interscience, New York, 1963, p. 895.

134. D. Hague, *Fast Reactions*, Wiley-Interscience, New York, 1971.
135. G. C. Hammes, ed., *Technique of Organic Chemistry*, Vol. VI, Part II: *Investigation of Elementary Reaction Steps in Solution and Very Fast Reactions*, Wiley-Interscience, New York, 1974.
136. E. Wyn-Jones, ed., *Chemical and Biological Applications of Relaxation Spectrometry*, D. Reidel, Holland, 1975.
137. C. F. Bernasconi, *Relaxation Kinetics*, Academic Press, New York, 1976.
138. W. J. Gettins and E. Wyn-Jones, eds., *Techniques and Applications of Fast Reactions in Solution*, D. Reidel, Dordrecht, Netherland, 1979.
139. E. Graber, J. Lang, and R. Zana, Kolloid-Z. *238*:470 (1970).
140. E. Graber, J. Lang, and R. Zana, Kolloid-Z. *238*:479 (1970).
141. P. J. Sams, J. E. Rassing, and E. Wyn-Jones, Chem. Phys. Lett. *13*:233 (1972).
142. P. J. Sams, J. E. Rassing, and E. Wyn-Jones, J. C. S. Faraday II *70*:1247 (1974).
143. E. A. G. Aniansson and S. N. Wall, J. Phys. Chem. *79*:857 (1975).
144. G. E. A. Aniansson, in *Aggregation Processes in Solution* (E. Wyn-Jones and J. Gormally, eds.), Elsevier, Amsterdam, 1983.
145. G. Platz and H. Hoffmann, Ber. Bunsenges. Physik. Chemie *76*:491 (1972).
146. H. Hoffmann and W. Ulbricht, Z. Physik. Chem. Neue Folge *106*:167 (1977).
147. H. Hoffmann and B. Tagesson, Z. Physik. Chem. Neue Folge *110*:113 (1978).
148. H. Hoffmann, W. Ulbricht, and B. Tagesson, Z. Physik. Chem. Neue Folge *113*:17 (1978).
149. H. Hoffmann, B. Tagesson, and W. Ulbricht, Ber. Bunsenges. Physik. Chem. *83*:148 (1979).
150. H. Hoffmann, G. Platz, H. Rehage, K. Reizlein, and W. Ulbricht, Makromol. Chem. *182*:451 (1981).
151. H. Hoffmann, G. Platz, and W. Ulbricht, J. Phys. Chem. *85*:1418 (1981).
152. T. Yasunaga, H. Oguri and M. Miura, J. Colloid Interface Sci. *23*:352 (1967).
153. T. Yasunaga, S. Fujii and M. Miura, J. Colloid Interface Sci. *30*:399 (1969).
154. R. Zana and J. Lang, Compt. Rend. C *266*:893 (1968).
155. R. Zana and J. Lang, Compt. Rend. C *266*:1377 (1968).
156. E. Graber, J. Lang, and R. Zana, Kolloid. Z., Z. Polymere *238*:470 (1970).
157. E. Graber and R. Zana, Kolloid Z., Z. Polymere *238*:479 (1970).
158. P. J. Sams, E. Wyn-Jones, and J. Rassing, Chem. Phys. Lett. *13*:233 (1972).
159. P. J. Sams. E. Wyn-Jones, and J. Rassing, J. Chem. Soc. Faraday II *69*:180 (1973).
160. J. E. Rassing and E. Wyn-Jones, Chem. Phys. Lett. *21*:93 (1973).
161. J. E. Rassing, P. J. Sams, and E. Wyn-Jones, J. Chem. Soc. Faraday Trans. 2 *70*(7):1247 (1974).
162. G. T. Tiddy, M. F. Walsh, and E. Wyn-Jones, J. C. S. Comm. 252 (1979).
163. J. Gettins, P. L. Jobling, M. F. Walsh, and E. Wyn-Jones, J. C. S. Faraday II *76*:794 (1980).
164. B. Cabane, in *Surfactant Solutions* (R. Zana. ed.), Marcel Dekker. New York, 1987, p. 57.
165. A. Guinier and G. Fournet, *Small Angle Scattering of X Rays*, Wiley-Interscience, New York, 1955.
166. K. Fontell, in *Liquid Crystals and Plastic Crystals* (G. W. Gray and P. A. Winsor, eds.), Vol. 2, Horwood, Chichester, 1974, p. 80.

167. O. Glatter and O, Kratky, *Small Angle X Rays Scattering*, Academic Press, London, 1982.
168. K. Fontell, in *Surfactant Solutions* (K. L. Mittal and B. Lindman, eds.), Vol. 1, Plenum Press, New York, 1984, p. 69.
169. K. Fontell and B. Lindman, J. Phys. Chem. *87*:3289 (1983).
170. M. C. Holmes, M. S. Leaver, and A. M. Smith, Langmuir *11*:356 (1995).
171. B. Jacrot, Rep. Prog. Phys. *39*:911 (1976).
172. E. Caponetti, D. C. Martino, M. A. Floriano, and R. Triolo, Langmuir *9*:1193 (1993).
173. J. B. Hayte, in *Physics of Amphiphiles: Micelles, Vesicles, and Microemulsions* (V. Degiorgio and M. Corti, eds.), North-Holland, Amsterdam, 1985.
174. S. S. Berr and R. R. M. Jones, J. Phys. Chem. *93*:2555 (1989).
175. H. Hoffmann, J. Kalus K. Reizlein, W. Ulbricht, and K. Ibel, Colloid Polym. Sci. *260*:435 (1982).
176. H. Hoffmann, J. Kalus, and H. Thurn, Colloid Polym. Sci. *261*:1043 (1983).
177. S. J. Burkitt, R. H. Ottewill, H. B. Hayter, and B. T. Ingram, Colloid Polym. Sci. *265*:619 (1987).
178. S. J. Burkitt, R. H. Ottewill, H. B. Hayter, and B. T. Ingram, Colloid Polym. Sci. *265*:628 (1987).
179. M. Kerker, *The Scattering of Light and Other Electromagnetic Radiation*, Academic Press, New York, 1969.
180. B. J. Berne and R. Pecora, *Dynamic Light Scattering*, Wiley-Interscience, New York, 1976.
181. V. Degiorgio, M. Corti, and M. Giglio, eds., *Light Scattering in Liquids and Macromolecular Solutions*, Plenum Press, New York, 1980.
182. S. J. Candau, in *Surfactant Solutions* (R. Zana. ed.), Marcel Dekker, New York, 1987, p. 147.
183. V. Tomašić, A. Chittofrati, and N. Kallay, Colloids Surfaces *A104*:95 (1995).
184. R. Zana, in *Surfactant Solutions* (R. Zana. ed.), Marcel Dekker, New York, 1987, p. 241.
185. W. J. Moore, *Physical Chemistry*, 4th ed., Prentice-Hall, Englewood Cliffs, NJ, 1972.
186. N. J. Turro, M. W. Geiger, R. R. Hautala, and N. E. Schore, in *Micellization, Solubilization, and Microemulsions* (K. L. Mittal, ed.), Vol. 1, Plenum Press, New York, 1977, p. 75.
187. R. Zana, ed., *Surfactant Solutions*, Marcel Dekker, New York, 1987.
188. C. A. Parker, *Photoluminescence of Solutions*, Elsevier, Amsterdam, 1968.
189. S. Schulman, *Fluorescence and Phosphorescence Spectroscopy: Physicochemical Principles and Practice*, Pergamon Press, Oxford, 1976.
190. N. J. Turro, *Modern Molecular Photochemistry*, Benjamin-Cummings, Menlo-Park, CA, 1978.
191. J. Lakowicz, *Principles of Fluorescence Spectroscopy*, Plenum Press, New York, 1983.
192. K. Meguro, Y. Muto, F. Sakurai, and K. Esumi, ACS Symp. Series *311*:61 (1986).
193. Y. Muto, K. Esumi, K. Meguro, and R. Zana, J. Colloid Interface Sci. *120*:162 (1987).

194. K. Kalyanasundaram, Langmuir *4*:942 (1988).

195. Y. Muto, K. Esumi, K. Meguro, and R. Zana, Langmuir *5*:885 (1989).

196. K. Esumi, H. Otsuka, and K. Meguro, J. Colloid Interface Sci. *142*:582 (1991).

197. K. Esumi, H. Otsuka, and K. Meguro, J. Langmuir *7*:2313 (1991).

198. T. Asakawa, H. Hisamatsu, and S. Miyagishi, Langmuir *12*:1204 (1996).

199. C. R. Brundle and A. D. Baker, eds., *Electron Spectroscopy, Theory, Techniques and Applications*, Vols. 1–4, Academic Press, New York, 1977–1981.

200. D. W. Dwight, T. J. Fabiksh, and H. R. Thomas, eds., *Photon, Electron, and Ion Probes of Polymer Structure and Properties*, ACS Symposium Ser. 162, Washington, DC, 1981.

201. H. Siegbahn and L. Karlsson, *Photoelectron Spectroscopy*, Handbuch der Physik, Vol. 31, 1982.

202. K. Siegbahn, J. Electron Spectrosc. Relat. Phenom. *51*:11 (1990).

203. H. Agren, Int. J. Quantum Chem. *39*:455 (1991).

204. D. Briggs and M. P. Seah, eds., *Practical Surface Analysis*, Vol. 1, *Auger and X-ray Photoelectron Spectroscopy*, 2nd ed., J. Wiley & Sons, Chichester, 1990.

205. N. H. Turner and J. A. Schreifels, Anal. Chem. *64*:302R (1992).

206. R. W. Phillips and R. G. Dettre, J. Colloid Interface Sci. *56*:251 (1976).

207. G. N. Batts, Colloids Surfaces *22*:133 (1987).

208. L. J. Gerenser, J. M. Pochan, M. G. Mason, and J. F. Elman, Langmuir *1*:305 (1985).

209. C. S. Fadley, Prog. Solid State Chem. *2*:265 (1976).

210. M. Mitsuya, Langmuir *10*:1635 (1994).

211. J. W. McBain, *Colloid Science*, D. C. Heath and Co., Boston, 1950.

212. R. Muzzalupo, G. A. Ranieri, and C. La Mesa, Colloids Surfaces A*104*:327 (1995).

213. H.-D. Dörfler and E. Müller, Tenside Surf. Det. *26*:381 (1989).

214. H. Hoffmann, G. Platz, H. Rehage, and W. Schorr, Ber. Bunsenges. Phys. Chem. *85*:255 (1981).

215. H. Hoffmann, G. Platz, H. Rehage, and W. Schorr, Ber. Bunsenges. Phys. Chem. *85*:877 (1981).

216. H. Hoffmann, G. Platz, and W. Ulbricht, J. Phys. Chem. *85*:1418 (1981).

217. W. Schorr and H. Hoffman, J. Phys Chem. *85*:3160 (1981).

218. H. Hoffmann, H. Rehage, G. Platz, W. Schorr, H. Thurn, and W. Ulbricht, Colloid Polym. Sci. *260*:1042 (1982).

219. H. Hoffmann, G. Platz, H. Rehage, and W. Schorr, Adv. Colloid Interface Sci. *17*:275 (1982).

220. H. Hofmann, Ber. Bunsenges. Phys. Chem. *88*:1078 (1984).

221. H. Hoffmann, H. Rehage, W. Schorr, and H. Thurn, in *Surfactants in Solution*, Vol. 1 (K. Mittal and B. Lindman, eds.), Plenum Press, New York, 1984, p. 425.

222. W. Schorr and H. Hoffmann, in *Physics of Amphiphiles: Micelles, Vesicles, and Microemulsion* (V. Degiorgio and M. Corti, eds.), North-Holland, Amsterdam, 1985, p. 160.

223. M. Angel and H. Hoffmann, Z. Phys. Chem. Neue Folge *139*:153 (1984)].

224. K. Tamori, K. Esumi, K. Meguro, and H. Hoffmann, J. Colloid Interface Sci. *147*:33 (1991).

225. C. T. O'Konski, ed., *Molecular Electrooptics*, Marcel Dekker, New York, 1976.

226. B. R. Jennings, ed., *Electro-Optics and Dielectrics of Macromolecules and Colloids*, Plenum Press, New York, 1979.
227. H. Scott, J. Phys. Chem. *68*:3612 (1964).
228. I. W. Osborn-Lee, R. S. Schechter, and W. H. Wade, J. Colloid Interface Sci. *94*:179 (1983).
229. I. W. Osborn-Lee, R. S. Schechter, W. H. Wade, and Y. Barakat, J. Colloid Interface Sci. *108*:60 (1985).
230. T. Asakawa, M. Mouri, S. Myagishi, and M. Nishida, Langmuir *4*:136 (1988).
231. T. Asakawa, S. Sunazaki, and S. Miyagishi, Colloid Polym. Sci. *270*:259 (1992).
232. J. J. Bikerman, *Surface Chemistry*, 2nd ed., Academic Press, New York, 1958.
233. P. Becher, *Emulsions: Theory and Practice*, 2nd ed., Reinhold, New York, 1965, p. 381.
234. R. Heusch, Fette-Seifen-Anstrichm. *72*:969 (1970).
235. C. Weser, GIT Fachz. Lab. *24*:642–848, 734–742 (1980).
236. H. J. König and G. Hartmannn, Textiltechnik *37*:259 (1987).
237. C. C. Addison, J. Chem. Soc. 535 (1943).
238. C. C. Addison, J. Chem. Soc. 252 (1944).
239. C. C. Addison, J. Chem. Soc. 477 (1944).
240. C. C. Addison, J. Chem. Soc. 98 (1945).
241. C. C. Addison, J. Chem. Soc. 354 (1945).
242. A. F. H. Ward and L. Torday, J. Chem. Phys. *14*:453 (1946).
243. E. J. Burcik, J. Colloid Sci. *5*:421 (1950).
244. E. J. Burcik, J. Colloid Sci. *8*:520 (1953).
245. E. J. Burcik and R. Newman, J. Colloid Sci. *9*:498 (1954).
246. H. Lange, J. Colloid Sci. *20*:50 (1965).
247. X. Y. Hua and M. J. Rosen, Colloid Interface Sci. *124*:652 (1988).
248. J. van Hunsel and P. Joos, Colloid Polym, Sci. *267*:1026 (1989).
249. D. E. Hirt, R. K. Prudhomme, B. Miller, and L. Rebenfeld, Colloids and Surfaces *44*:101 (1990).
250. M. F. Cox, AOCS/CSMA Conference, Hershey, PA, Oct. 29, 1989.
251. P. Mukerjee and T. Handa, J. Phys. Chem. *85*:2298 (1981).
252. W. D. Harkins, T. F. Young, and L. H. Cheng, Science *64*:333 (1926).
253. W. D. Harkins and H. F. Jordan, J. Am. Chem. Soc. *52*:1751 (1930).
254. B. B. Freud and H. Z. Freud, J. Am. Chem. Soc. *52*:1772 (1930).
255. C. Huh and S. G. Mason, Colloid Polymer Sci. *253*:566 (1975).
256. L. Wilhelmy, Ann. Physik (Leipzig) *119*:177 (1863).
257. J. F. Padday and D. R. Russell, J. Colloid Sci. *15*:503 (1960).
258. T. W. Richards and L. B. Coombs, J. Am. Chem. Soc. *37*:1656 (1915).
259. T. W. Richards and L. B. Coombs, J. Am. Chem. Soc. *43*:827 (1921).
260. S. Sugden, J. Chem. Soc. 1483 (1921).
261. C. D. Holcomb and J. A. Zollweg, J. Colloid Interface Sci. *134*:41 (1990).
262. P. Z. Rehbinder, Z. Phys. Chem. *111*:447 (1924).
263. J. Kloubek, Tenside *5*:317 (1968).
264. J. Kloubek, J. Colloid Interface Sci. *41*:1 (1972).
265. J. Kloubek, J. Colloid Interface Sci. *41*:7 (1972).
266. J. Kloubek, J. Colloid Interface Sci. *41*:17 (1972).

267. K. Lunkenheimer, R. Miller, and J. Becht, Colloid Polym. Sci. *260*:1145 (1982).

268. K. J. Mysels, Langmuir *2*:428 (1986).

269. C. D. Holcomb and J. A. Zollweg, J. Colloid Interface Sci. *134*:41 (1990).

270. Y. Rotenberg, L. Boruvka, and A. W. Neumann, J. Colloid Interface Sci. *93*:169 (1983).

271. S. H. Anastasiadis, J. K. Chen, J. T. Koberstein, A. F. Siegel, J. E. Sohn, and J. E. Emerson, J. Colloid Interface Sci. *119*:55 (1987).

272. F. K. Hansen and G. R‰dsrud, J. Colloid Interface Sci. *141*:1 (1991).

273. J. E. Verschaffelt, Proc. Konikl. Akad. Wetenschap. Amsterdam *21*:357 (1919).

274. H. C. Parreira, J. Colloid Sci. *20*:44 (1964).

275. E. A. Hauser, H. E. Edgerton, B. M. Holt, and J. T. Cox, J. Phys. Chem. *40*:973 (1936).

276. H. Dunken, Ann. Phys. *41*:567 (1942).

277. W. D. Harkins and F. E. Brown, J. Am. Chem. Soc. *41*:499 (1919).

278. J. L. Lando and H. T. Oakley, J. Colloid Interface Sci. *25*:526 (1967).

279. R. Kumar and S. G. T. Bhat, Tenside Surf. Det. *24*:86 (1987).

280. L. Holysz and E. Chibowski, Tenside Surf. Det. *25*:377 (1988).

281. B. Vonnegut, Rev. Sci. Instrum. *13*:6 (1942).

282. H. T. Patterson. K. H. Hu, and T. H. Grindstaff, J. Polymer Sci. *34*:31 (1971).

283. J. L. Cayias, R. S. Schechter, and W. H. Wade, in *Adsorption at Interfaces* (K. L. Mittal, ed.), ACS Symposium Series 8, 1975, p. 234.

284. R. Defay and G. Pètrè, in *Surface Colloid Science* (E. Matijevic, ed.), Vol. 3, Wiley-Interscience, New York, 1971, p. 27.

285. A. P. Brady and A. G. Brown, in *Monomolecular Layers* (H. Sobotka, ed.), Symp. Amer. Assoc. Advan. Sci., Washington, 1951, p. 33.

286. R. Van den Bogaert and P. Joos, J. Phys. Chem. *83*:2244 (1979).

287. E. Rillaerts and P. Joos, J. Colloid Interface Sci. *88*:1 (1982).

288. J. Van Havenbergh and P. Joos, J. Colloid Interface Sci. *95*:172 (1983).

289. J. H. Clint, E. L. Neustadter, and T. Jones, Dev. Pet. Sci. *13*:135 (1981).

290. S. J. Sugden, J. Chem. Soc. *121*:858 (1921).

291. S. J. Sugden, J. Chem. Soc. *125*:27 (1924).

292. R. L. Bendure, J. Colloid Interface Sci. *35*:238 (1971).

293. W. D. Harkins, T. F. Young, and L. H. Cheng, Science *64*:333 (1926).

294. K. Lunkenheimer, Tenside Det. *19*:272 (1982).

295. P. J. Cram and J. M. Haynes, J. Colloid Interface Sci. *35*:706 (1971).

296. W. A. Gifford, J. Colloid Interface Sci. *64*:588 (1978).

297. K. Lukenheimer and K. D. Wantke, Colloid Polymer Sci. *259*:354 (1981).

298. K. Lukenheimer, J. Colloid Interface Sci. *131*:580 (1989).

299. E. Bartholomé and K. Schäfer, Melliand Textilber. *31*:487 (1950).

300. E. A. Boucher, T. M. Grinchuck and A. C. Zettlemoyer, J. Colloid Interface Sci. *23*:600 (1967).

301. A. M. Mankowich, J. Colloid Interface Sci. *25*:590 (1968).

302. A. C. Zettlemoyer, V. V. Subba Rao, J. Colloid Interface Sci. *29*:172 (1969).

303. P. Lenard, R. V. Dallwitz-Wegener, and E. Zachmann, Ann. Phys. [4] *74*:381 (1924).

304. N. Thoai, J. Colloid Interface Sci. *62*:222 (1977).

305. A. H. Ellison and W. A. Zisman, J. Phys. Chem. *60*:416 (1956).
306. Th. F. Tadros, J. Colloid Interface Sci. *74*:196 (1980).
307. W. Asche, Seifen-Öle-Fette-Wachse *112*:543 (1986).
308. J. M. Andreas, E. A. Hauser, and W. B. Tucker, J. Phys. Chem. *42*:1001 (1938).
309. S. Fordham, Proc. Roy. Soc. (London) *194*:1 (1948).
310. K. Prochaska, Z. Górski, and J. Szymanowski, Tenside *27*:233 (1990).
311. R. Miller and K.-H. Schano, Tenside Surf. Det. *27*:238 (1990).
312. W. D. Harkins and A. E. Alexander, in *Technique of Organic Chemistry*, 3rd ed. (A. Weissberger, ed.), Interscience, New York. 1959.
313. W. D. Harkins and F. E. Brown, J. Am. Chem. Soc. *41*:519 (1919).
314. K. Shinoda and H. Nakayama, J. Colloid Sci. *18*:705 (1963).
315. K. Shinoda, M. Hato, and T. Hayashi, J. Phys. Chem. *76*:909 (1972).
316. S. W. Morrall, A. D. Clauss, and T. J. Adams, AOCS Annual Meeting, Toronto, May 10–14, 1992.
317. M. J. Rosen and X. Y. Hua, J. Colloid Interface Sci. *139*:397 (1990).
318. X. Y. Hua and M. J. Rosen, J. Colloid Interface Sci. *141*:180 (1991).
319. P. Joos, J. P. Fang, and G. Serrien, J. Colloid Interface Sci. *151*:144 (1992).
320. A. Mehreteab, G. Broze, and J. Rouse, AOCS Annual Meeting, Toronto, May 10–14, 1992.
321. A. Müller and M. Albrecht, Tenside Surf. Det. *27*:399 (1990).
322. J. K. Borchardt and C. W. Yates, AOCS Annual Meeting, Toronto, May 10–14, 1992.
323. M. K. Bernett and W. A. Zisman, J. Phys. Chem. *63*:1911 (1959).
324. C. La Mesa and B. Sesta, J. Phys. Chem. *91*:1450 (1987).
325. V. Glöckner, K. Lunkwitz, and D. Prescher, Tenside Surf. Det. *26*:376 (1989).
326. N. Funasaki and S. Hada, J. Colloid Interface Sci. *73*:425 (1980).
327. N. Funasaki and S. Hada, Chem. Lett. 717 (1979).
328. N. Funasaki and S. Hada, Bull. Chem. Soc. Jpn. *49*:2899 (1976).
329. E. Gorodinsky and S. Efrima, Langmuir *10*:2151 (1994).
330. K. Esumi and S. Ogiri, Colloids Surfaces *A94*:107 (1995).
331. N. L. Jarvis and W. A. Zisman, J. Phys. Chem. *63*:727 (1959).
332. Matubayasi, K. Motomura, S. Kaneshina, M. Nakamura, and R. Matuura, Bull. Chem. Soc. Jpn. *50*:523 (1977).
333. K. Motomura, I. Kajwara, N. Ikeda, and M. Aratono, Colloids Surfaces *38*:61 (1989).
334. Zhao Guo-Xi and Zhu Bu-Yao, Colloid Polymer Sci. *261*:89 (1983).
335. Zhu Bu-Yao and Zhao Guo-Xi, Hua Xue Tong Bao (Chemistry) No. *6*:341 (1981).
336. H. M. Scholberg, R. A. Guenther, and R. I. Coon, J. Phys. Chem. *57*:923 (1953).
337. D. R. Taves, Nature *211*:192 (1966).
338. D. R. Taves, Nature *220*:582 (1968).
339. D. R. Taves, Nature *217*:1050 (1968).
340. D. R. Taves, Talanta *15*:1015 (1968).
341. W. S. Guy, D. R. Taves, and W. S. Brey, Am. Chem. Soc. Symp. Series *28*:117 (1976).

342. F. A. Ubel, S. D. Sorenson, and D. E. Roach, Am. Ind. Hyg. Assoc. J. *41*:584 (1980).
343. P. Venkateswarlu, L. Singer, and W. D. Armstrong, Anal. Biochem. *42*:350 (1971).
344. L. Singer and R. H. Ophaug, Anal. Chem. *49*:38 (1977).
345. P. Venkateswarlu, Anal. Biochem. 68:512 (1975).
346. E. Kissa, Tenside Surf. Det. *26*:372 (1989).
347. E. Kissa, Clin. Chem. *33*:253 (1987).
348. E. Kissa, Environ. Sci. Technol. *20*:1254 (1986).
349. D. Barceló, Anal. Chim. Acta *263*:1 (1992).

Index

ۻ

Milton Keynes UK
Ingram Content Group UK Ltd.
UKHW021634071024
449327UK00020BA/1299